T0309332

Eratosthenes and the Measurement of the Earth's Circumference (*c.* 230 BC)

Eratosthenes and the Measurement of the Earth's Circumference (c. 230 BC)

CHRISTOPHER A. MATTHEW

OXFORD
UNIVERSITY PRESS

Great Clarendon Street, Oxford, OX2 6DP,
United Kingdom

Oxford University Press is a department of the University of Oxford.
It furthers the University's objective of excellence in research, scholarship,
and education by publishing worldwide. Oxford is a registered trade mark of
Oxford University Press in the UK and in certain other countries

© Christopher A. Matthew 2023

The moral rights of the author have been asserted

Published in the United States of America by Oxford University Press
198 Madison Avenue, New York, NY 10016, United States of America

British Library Cataloguing in Publication Data
Data available

Library of Congress Control Number: 2023930924

ISBN 978-0-19-887429-4

DOI: 10.1093/oso/9780198874294.001.0001

Printed and bound by
CPI Group (UK) Ltd, Croydon, CR0 4YY

Acknowledgements

I wish to express my gratitude to Professor Miroslav Filipovic, Professor Ray Norris, and Doctor Nicholas Tothill, of Western Sydney University, for their encouragement and support during the research for this project. I also wish to thank Allison Smith for her unwavering care, support, and interest during the time of this project, and Dr Bill Franzen, lecturer of mathematics at the Australian Catholic University, for all of the assistance and interest that he has given during the research of this topic. Additionally, I wish to thank Werner Schreiner from skaphe.de for his assistance and support in the manufacture of a replica sundial from the third century BC, which helped to bring several aspects of this research to life. I would also like to thank Doctor Ain de Horta from Western Sydney University for his assistance with the re-creative experiments that formed part of this research. Furthermore, I would like to thank Michael Lewis for providing the images of his use of a replica *dioptra*, and Ken Sheedy from the Australian Centre for Ancient Numismatic Studies for the images of a coin from Cyrene, which can be found within this work. Finally, I would like to thank all of the other friends and colleagues who have given their support and feedback during the research and writing processes which have brought this project to fruition.

CM
2022

Contents

Abbreviations

Ancient Sources

Author Abbreviation	Author	Title Abbreviation	Title
Ach. Tat.	Achilles Tatius	*Arati Phaenom.*	*Introduction to Aratus' 'Phenomena'*
		Leucip.	*Leucippe and Clitophon*
Ael.	Aelian	*VH*	*Varia Historia*
		NA	*On the Nature of Animals* (De Natura Animalium)
Ael. Tact.	Aelian (Tacticus)	*Tact.*	*Tactics*
Aet.	Aetius		*Opinions of Philosophers*
Alcm.	Alcman	*Par.*	Partheneion
Amm.	Ammianus Marcellinus		*Roman History*
Anonymous Texts		*Epit. Aris.*	*The Letter of Aristeas*
		P. Hib.	*The Hibeh Papyrus*
		P. Oxy.	*The Oxyrhynchus Papyrus*
		Schol. in Ar. Ran.	*Scholia on Aristophanes 'Frogs'*
		Schol. in Ptol. Alma.	*Scholia on Ptolemy 'Almagest'*
AP			*Palatine Anthology*
Ap. Rhod.	Apollonius of Rhodes	*Argon.*	*Argonautica*
App.	Appian	*B Civ.*	*Civil Wars*
		Syr.	*Syrian War*
Apth.	Aphthonius	*Prog.*	*Progymnasmata*
Ar.	Aristophanes	*Av.*	*Birds*
		Eccl.	*The Assemblywomen*
		Ran.	*Frogs*
Archim.	Archimedes	*Meth.*	*Method of Mechanical Theorems*
		Psam.	*The Sand Reckoner*
Arist.	Aristotle	*Ath. Pol.*	*The Athenian Constitution*
		Cael.	*On the Heavens*

		De An.	On the Soul
		Eth. Nic.	Nicomachean Ethics
		Metaph.	Metaphysics
		Mete.	Meteorology
		Oec.	Economics
		Ph.	Physics
		Pol.	Politics
		Sens.	Sense and Sensibilia
Aristarch. Sam.	Aristarchus of Samos	*De Mag.*	On the Sizes and Distances of the Sun and Moon
Arr.	Arrian	*Anab.*	Anabasis of Alexander the Great
		Frag. Phy.	Fragments on Physics
		Tact.	Tactics
Asclep.	Asclepiodotus	*Tact.*	Tactics
Ath.	Athenaeus	*Diep.*	The Learned Banqueters
Autol.	Autolycus	*De Sph.*	On the Moving Sphere
		Ort. et Occ.	Risings and Settings
Caes.	Caesar	*BC*	The Civil War
Cass. Dio	Cassius Dio		Roman History
Cic.	Cicero	*Acad.*	Academic Questions
		De Or.	On Oration
		Div.	On Divination
		Nat. D.	On the Nature of the Gods
		Rep.	Republic
Clem. Al.	Clement of Alexandria	*Strom.*	Miscellanies
Cleom.	Cleomedes	*De motu*	On the Circular Motions of Celestial Bodies
CPI			Corpus of Ptolemaic Inscriptions
Curt.	Quintus Curtius		The Campaign of Alexander the Great
Dem.	Demosthenes		Orations
Did.	Didymos	*in D.*	On Demosthenes
Dio Chrys.	Dio Chrysostom	*Or.*	Orations
Diod. Sic.	Diodorus of Sicily		Library of History
Diog. Laert.	Diogenes Laertius		Lives of Eminent Philosophers
Dion. Cyz.	Dionysius of Cyzicus	*On E.*	On Eratosthenes
Epiph.	Epiphanius	*Mens.*	On Weights and Measures
Euc.	Euclid	*El.*	Elementa
		Phae.	Phaenomena

Continued

Continued

Author Abbreviation	Author	Title Abbreviation	Title
Eus.	Eusebius	*Chron.*	*Chronicle*
		DE	*Demonstration of the Gospel*
		PE	*Preparation of the Gospel*
		PG	*Patrologia Graeca*
Eutoc.	Eutocius	*In Arch. circ. dim.*	*Commentary on the Works of Archimedes and Apollonius*
Flor.	Florus	*Epit.*	*Epitome*
Gal.	Galen	*Hipp. De Nat. Hom.*	*On Hippocrates 'Nature of Man'*
		Hipp. Epid.	*On Hippocrates 'Epidemics'*
		Inst. log.	*Introduction to Logic*
		Phil. Hist.	*History of Philosophy*
Gell.	Aulus Gellius	*NA*	*Attic Nights*
Gem.	Geminus	*Isagoge*	*Introduction to Phaenomena*
Hdn.	Herodian	*Gr.*	*Technical Remains*
Hdt.	Herodotus		*The Histories*
Hero.	Heron of Alexandria	*Dioptr.*	*Dioptra*
		Met.	*Metrological Tables*
Herod.	Herodas		*Mimes*
Hes.	Hesiod	*Op.*	*Works and Days*
		Theog.	*Theogony*
Hipparch.	Hipparchus	*In Ar. et Eud.*	*Commentary on Aratus and Eudoxus*
Hippol.	Hippolytus	*Refut.*	*Refutation of All Heresies*
Hom.	Homer	*Il.*	*Iliad*
		Od.	*Odyssey*
Hor.	Horace	*Carm.*	*Secular Songs*
Hsch.	Hesychius	*Lex.*	*Lexicon*
Hyg.	Hyginus	*Ast. Po.*	*Poetic Astronomy*
Hypsicl.	Hypsicles	*Anaph.*	*On Ascensions*
Jer.	Jerome	*Chron.*	*Chronicle*
Joseph.	Josephus	*AJ*	*Jewish Antiquities*
		Ap.	*Against Apion*
		BJ	*The Jewish War*
Just.	Justin	*Epit.*	*Epitome of the Philippic History of Pompeius Trogus*

Longinus		*Subl.*	*On the Sublime*
Luc.	Lucan		*Pharsalia*
Luc.	Lucian	*Pro Lapsu*	*A Slip of the Tongue in Greeting*
Lucr.	Lucretius		*On the Nature of Things*
Lyd.	John Lydus	*Mens.*	*On the Months*
Macr.	Macrobius	*In Somn.*	*Dream of Scipio*
		Sat.	*Saturnalia*
Marc.	Marcian of Heraclea	*Perip.*	*Periplus of the External Sea*
Mart. Cap.	Martianus Capella	*Phil.*	*The Marriage of Philology and Mercury*
Nicom.	Nicomachus	*Arith.*	*Introduction to Arithmetic*
		Harm.	*Harmonics*
Pallad.	Palladius	*Agric.*	*Agricultural Works*
Papp.	Pappus	*in Ptol. Alma.*	*Commentary on Ptolemy 'Almagest'*
		Syn.	*The Collection*
Paus.	Pausanius		*Description of Greece*
Philo	Philo of Alexandria	*in Mos.*	*On Moses*
Phlp.	Philoponus	*In Mete.*	*Commentary on Aristotle's Meteorology*
Pl.	Plato	*Ap.*	*Apology*
		Cra.	*Cratylus*
		Cri.	*Crito*
		Epin.	*Epinomis*
		Grg.	*Gorgias*
		L.	*Letters*
		Lach.	*Laches*
		Leg.	*Laws*
		Men.	*Meno*
		Phd.	*Phaedo*
		Resp.	*Republic*
		Tht.	*Theaetetus*
		Ti.	*Timaeus*
Plin. (E)	Pliny the Elder	*HN*	*Natural History*
Plut.	Plutarch	*Alex.*	*Life of Alexander*
		Caes.	*Life of Caesar*
		De E.	*On the 'E' at Delphi*
		De Facie	*On the Face of the Moon*
		De Is. et Os.	*On Isis and Osiris*
		De Pyth. Or.	*On the Pythia's Prophecies*

Continued

Continued

Author Abbreviation	Author	Title Abbreviation	Title
		Marc.	Life of Marcellus
		Mor.	Moralia
		Nic.	Life of Nicias
		Per.	Life of Pericles
		Plac. Philos.	Doctrines of the Philosophers
		Quaest. Plat.	Platonic Questions
		Sept. Sap. Conv.	Symposium of the Seven Sages
		Stromat.	Miscellaneous Fragments
		Sull.	Life of Sulla
		Them.	Life of Themistocles
Plut. *Dion.*		Ath. *Diep.*	
Polyaenus	Polyaenus	Strat.	Strategems of War
Polyb.	Polybius		Histories
Procl.	Proclus	Hyp.	Hypotyposis Astronomicarum Positionum
		In Euc.	Commentary on Euclid 'Elements'
Ps-Call.	Pseudo-Callisthenes		The Romance of Alexander
Ps. Dem.	Pseudo-Democritus	Alc.	Natural and Secret Questions
Ps. Plut.	Pseudo-Plutarch	Epit.	Letters
Ptol.	Ptolemy	Alm.	Almagest
		Anal.	The Analemma
		Geog.	Geography
		Harm.	Harmonics
Quint.	Quintilian	Inst.	Institutio Oratoria
Sen.	Seneca the Younger	QNat.	Natural Questions
		Tranq.	Tranquillity of the Mind
Serv.	Servius	A.	Commentary on Virgil's Aeneid
Sext. Emp.	Sextus Empiricus	Math.	Against the Mathematicians
SIG			Sylloge Inscriptionum Graecarum
Simpl.	Simplicius	in Cael.	Commentary of Aristotle 'On the Heavens'
		In Phys.	On Physics
Steph. Byz.	Stephanus of Byzantium		Lexicon
Stob.	Stobaeus	Ecl.	Eclogues

Str.	Strabo	*Geog.*	*Geography*
Suda			*The Suda*
Sync.	Syncellus	*Chron.*	*Chronographia*
Thdt.	Theodoret	*CAG*	*Cure for Pagan Maladies*
Theoc.	Theocritus	*Id.*	*Idylls*
Theon.	Theon of Smyrna	*Expos.*	*Aspects of Mathematics Useful for the Reading of Plato*
Theoph.	Theophrastus	*Caus Pl.*	*Enquiry into Plants*
Thuc.	Thucydides		*History of the Peloponnesian War*
Tz.	John Tzetzes	*Prol. Com.*	*Commentary on Comedy*
Val. Max.	Valerius Maximus		*Memorable Deeds and Sayings*
Veg.	Vegetius	*Mil.*	*Epitome of Military Science*
Vit.	Vitruvius	*De Arch.*	*On Architecture*
Xen.	Xenophon	*Hell.*	*Hellenica*
		Mem.	*Memorabilia*

Journals

Abbreviation	Title
A&AT	*Astronomy & Astrophysics Transactions*
AAATec	*Archaeoastronomy and Ancient Technologies*
Adv. Space Res.	*Advances in Space Research*
AHB	*Ancient History Bulletin*
Aion	*Aion*
AJA	*American Journal of Archaeology*
Am. Journ. Phil.	*American Journal of Philology*
Am. J. Phys. Anthropol.	*American Journal of Physical Anthropology*
AncSoc	*Ancient Society*
AncW	*Ancient World*
Ann. Soc. Sci. Bruxelles, Ser. A	*Annales de la Société Scientifique de Bruxelles—Série A*
AntJ	*The Antiquaries Journal*
Arch. Hist. Exact Sci.	*Archive for History of Exact Sciences*
Ath. Mitt.	*Athenische Mittielungen*
Athenaeum	*Athenaeum*
AVN	*Allgemeine Vermessungs-Nachrichten*
BCH	*Bulletin de correspondence hellénique*
Bonner Jahrb.	*Bonner Jahrbucher*

Continued

Continued

Abbreviation	Title
Br. J. Hist. Sci.	*The British Journal for the History of Science*
BSA	*Biblioteca di Studi Antichi*
BZ	*Byzantinische Zeitschrift*
Centaurus	*Centaurus*
Chiron	*Chiron*
CJ	*Classical Journal*
Cl. Phil.	*Classical Philology*
Class. Antiq.	*Classical Antiquity*
Configurations	*Configurations*
CQ	*Classical Quarterly*
DHA	*Dialogues d'histoire ancienne*
G&R	*Greece & Rome*
H. M. Inst. R. Fr.	*Histoire et mémoires de l'Institut royal de France*
Hesperia	*Hesperia*
HIST MATH	*Historia Mathematica*
Historia	*Historia: Zeitschrift für Alte Geschichte*
Isis	*Isis*
J. Astron. Hist. Herit.	*Journal of Astronomical History and Heritage*
J. Forensic Leg. Med.	*Journal of Forensic and Legal Medicine*
J. Hist. Astron.	*Journal for the History of Astronomy*
J. Inst. Navig.	*Journal of the Institute for Navigation*
J. R. Asiat. Soc.	*The Journal of the Royal Asiatic Society of Great Britain and Ireland*
Jahrb. Berlin Museen	*Jahrbuch der Berliner Museen*
Janus	*Janus*
JHS	*Journal of Hellenic Studies*
JRS	*Journal of Roman Studies*
Klio	*Klio*
Math. Mag.	*Mathematics Magazine*
MDAIK	*Mitteilungen des Deutschen Archäologischen Instituts, Abteilung Kairo*
MedAnt	*Mediterraneo Antico*
Mém. Acad. Inscript. et belles-lettres	*Mémoires de l'Académie de Inscriptions et belles-lettres*
Mem. Acad. Sci. Inst. Fr.	*Mémoires de l'Académie des sciences de l'Institut de France*
Mil. Psychol.	*Military Psychology*
MMAB	*Metropolitan Museum of Art Bulletin*
Nikephoros	*Nikephoros*
Orb. Terr.	*Orbis Terrarum: Journal of Historical Geography of the Ancient World*
PAPS	*Publication of the American Philosophical Society*

Philologus Suppl.	*Philologus Supplement*
Philos. Antiq.	*Philosophia Antiqua*
Philos. Trans. R. Soc.	*Philosophical Transactions of the Royal Society of London*
Phronesis	*Phronesis*
PNAS	*Proceedings of the American Academy of Arts and Sciences*
Publ. Astron. Soc. Pac.	*Publications of the Astronomical Society of the Pacific*
Q. Jl R. Astr. Soc.	*Quarterly Journal of the Royal Astronomical Society*
QAL	*Quaderni di archeologia Libia*
Renaiss. Stud.	*Renaissance Studies*
SAK	*Studien zur Altägyptischen Kultur*
SAOC	*Studies in Ancient Oriental Civilization*
SciAm	*Scientific American*
TAPA	*Transactions of the American Philological Association*
TPR	*The Town Planning Review*
Trans. R. Soc. Edinburgh	*Transactions of the Royal Society of Edinburgh*
VDI	*Vertuic drevnej istorii*
Vist. Astron.	*Vistas in Astronomy*
ZÄS	*Zeitschrift für Ägyptische Sprache und Altertumskunde*
Zeitschr. d. Gesell. f. Erdkunde zu Berlin	*Zeitschrift der Gesellschaft für Erdkunde zu Berlin*

Illustrations

Tables

Preface

In the late third century BC, Eratosthenes, then chief librarian of the Great Library of Alexandria in Egypt, calculated the circumference of the Earth based upon the difference between the shadow cast by the pointer (*gnomon*) of a sundial in Alexandria at midday on the Summer Solstice, and compared the angle created by the shadow to the knowledge that at the city of Syene to the south no shadow was cast by a *gnomon* at the same time on the same day. Accounts of the attempt by Eratosthenes to accurately calculate the circumference of the Earth have taken on a near-legendary status—appearing in academic texts and journals, online videos, popular television programmes, and children's books.[1] However, the accounts of Eratosthenes' work are, unfortunately, lacking several specific details which would allow for the accuracy of his results to be initially determined. This is particularly the case in trying to convert the results, which were given in an ancient unit of measure (the *stade*), into a modern equivalent. This is because there were several different systems of measurement in use across the ancient Greek world that incorporated a unit called a *stade*, but these units were all of different lengths. This has led to considerable scholarly debate, almost since the time that the calculations were made, over the size of the unit that Eratosthenes was using, and subsequently the accuracy of his results.

The size of Eratosthenes' *stade* has been the topic of considerable scholarly debate—in part because scholars writing from an astronomical perspective rarely deal with all of the historical, philological, archaeological, and linguistic evidence, and scholars writing from a historical perspective rarely engage fully with the astronomical evidence. Consequently, all prior examinations on this topic are incomplete, inconclusive or incorrect. However, a critical re-examination of Eratosthenes' experiment will allow for the accuracy of his result to be determined.

---

[1] Citations of the many academic works on Eratosthenes can be found throughout this work. Placing the search term 'Eratosthenes measurement of the Earth' into YouTube or Google results in numerous hits. An excellent example of a discussion of Eratosthenes in a documentary can be found in *Episode 1: The Shores of the Cosmic Ocean* of Carl Sagan's award-winning television series *Cosmos* (00:30:05–00:36:43). For examples of works on Eratosthenes that are pitched at younger readers, see: K. Lasky, *The Librarian who Measured the Earth* (New York, Little Brown Books, 1994) and M. Gow, *Measuring the Earth: Eratosthenes and His Celestial Geometry* (Berkeley Heights, Enslow, 2010).

The aim of this project is to re-evaluate the results of Eratosthenes' calculations through a re-examination of all of the available evidence—astronomical, philological, historical, and archaeological. This, in turn, will allow for engagement with the long-running scholarly debate over this topic to be undertaken in order to identify any errors or omissions in these prior theories, and account for these errors. This project will examine the available data to compile the first multi-disciplinary re-evaluation of Eratosthenes' work in order to address two key research objectives:

1. Determine the size of the *stade* that Eratosthenes used in his calculations.
2. Determine the accuracy of Eratosthenes' calculation of the circumference of the Earth.

As a result, this project will provide the first multi-disciplinary analysis of this important event in the history of astronomy and the ancient attempts to understand our place in a wider universe.

Through a critical examination of the available evidence, it can be shown that Eratosthenes was using a *stade* of 180 m in length in his determination of the Earth's circumference. A unit of this size has never been applied to an analysis of Eratosthenes' work before now, and constitutes an original contribution by this new research. The use of a unit of this size is confirmed through a comparison of stated distances between locations in Iran, Afghanistan, and Egypt that are attributed to Eratosthenes with 'on the ground' measurements obtained using satellite mapping software (Google Earth). This has also never been applied to the examination of Eratosthenes, and constitutes another original contribution of this research.

Furthermore, an examination of several key ancient passages, which have been ignored by many previous scholars, demonstrates that the result of Eratosthenes' calculations was a circumference of around 224,000 *stadia*—as opposed to the 250,000 or 252,000 *stade* figures that form the basis of all prior studies into this topic—and that this was based upon a latitudinal distance, determined using sundials, between the two locations which formed a fundamental element of his determination of the size of the Earth. The results of this investigation show that Eratosthenes' calculations were highly accurate—with a margin of error of <1%.

1

Historical Background

1.1 The Sources

There are numerous sources available for the study of Eratosthenes' calculation of the circumference of the Earth. The primary ancient text is Cleomedes'
On the Circular Motions of Celestial Bodies (Κυκλικὴ Θεωρία Μετεώρων or
De motu circulari coporum cealestium) which, while written several hundred years after the time of Eratosthenes, is the earliest surviving account that
details Eratosthenes' work, the assumed knowledge he was working from, his
experiment and calculations, and his results (Fig. 1).[1]

Cleomedes recounts that Eratosthenes used observations of the shadows
cast by sundials in two locations on the same day to determine the difference in
the angle crated by those shadows, which represented a fraction of a full circle.
He then applied that fraction to the distance between those two sites to determine the circumference of the Earth. Through his calculations, Eratosthenes
concluded that the circumference of the Earth was 250,000 *stadia*.[2] Details of
Eratosthenes' work is also found in other ancient texts, mainly from the first
and second centuries AD such as *Geography* by Strabo, Ptolemy's *Almagest*
and *Geography*, and Pliny the Elder's *Naturalis Historia* (Natural History).[3]
The main biographical information on the life and career of Eratosthenes is
found in an even later ancient text—the Byzantine Era lexicon, *The Suda*, from

---

[1] The exact title of Cleomedes' work, and even if there was one, is unknown. The exact date of
the work is also unknown, but can be no earlier than the first century AD based upon things that are
referred to in the text. The Latin title given here, *De motu circulari corporum caelestium*, comes from
Renaissance-era manuscripts of the text, and the Greek title, Κυκλικὴ Θεωρία Μετεώρων, comes from
Ziegler's 1891 Edition published by B.G. Teubneri in Lipsiae. See also: A.C. Bowen and R.B. Todd
(trans.), *Cleomedes' Lectures on Astronomy* (Berkeley, University of California Press, 2004), pp.1–4. It
should also be noted that almost none of the ancient texts, and translations, are derived from original
source documents, but rather from Medieval-era copies and manuscripts of these texts at the earliest.
This poses potential problems as the texts have been translated, and copied by hand, from their original
Greek and Latin, then possibly into Arabic in the early Middle Ages, before being translated back into
Greek and Latin in the time of the Renaissance. Consequently, transcription errors cannot be fully
ruled out.

[2] Cleom. *De motu*, 1.10.

[3] Str. *Geog.* 1.2.1–2, 1.2.3–7; Ptol. *Alm.* H68; Ptol. *Geog.* 1.7.1, 7.5.12; Plin. (E) *HN* 5.132.

Eratosthenes and the Measurement of the Earth's Circumference (c.230BC). Christopher A. Matthew, Oxford University Press.
© Christopher A. Matthew (2023). DOI: 10.1093/oso/9780198874294.003.0001

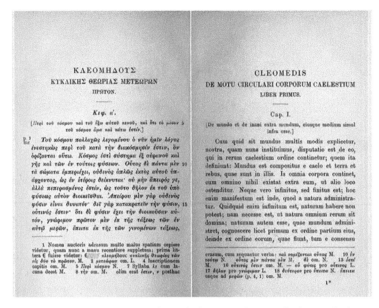

Fig. 1 The opening pages of Zeigler's 1891 edition of Cleomedes' text on astronomy.

the tenth century.[4] *The Suda* is set out in an encyclopaedic style with more than 30,000 entries—many drawing information from ancient texts that have not survived to the present day. These ancient texts have formed the basis for many of the modern examinations into Eratosthenes and the size of the *stade*.

However, these later texts have also added to the scholarly controversy over Eratosthenes due to them presenting slightly different figures to those given by Cleomedes. Many of the later ancient texts give their results in later, Roman, units of measure (for which there is also some debate over their exact size), contain possible transcription errors from when the manuscripts of these texts were copied by hand during the earlier parts of the modern world, and/or contain potential errors in translation. In some cases, these issues have compounded the problems with attempting to understand the accuracy of Eratosthenes' work. For example, several later ancient writers, mainly from the first century AD, give Eratosthenes' result as 252,000 *stadia*, rather than the 250,000 found in Cleomedes.[5] A much later writer, Marcian of Heraclea

---

[4] *Suda s.v.* Ἐρατοσθένης.
[5] For example, see: Gal. *Inst. log.* 12.2; Plin. (E), *HN* 2.112; Str. *Geog.* 2.5.7, 2.5.34; Theon. *Expos.* 3.3; Vit. *De Arch.* 1.6.9; Hero. *Dioptr.* 36.

from the fourth century AD, states that Eratosthenes' result was 259,200 *stadia*.[6] However, Gosselin dismissed this figure as mathematical convenience by demonstrating that it divides evenly into 720 *stadia*/degree, or 12 *stadia*/minute, or 1/5 of a *stade*/second—which would 'very much simplify the calculation of latitude and longitude'.[7] Letronne, on the other hand, saw this figure as a transmission error from the copying of the manuscript and suggested that the ancient Greek figures for 259,200 (Μκε´ιθσ) should actually be those for 252,200 (Μκε´ιβσ)—an unrounded version of the value given in the texts from the first century AD.[8] Regardless of the reason, the higher value of 259,200 *stadia* for the result of Eratosthenes' calculation seems unlikely to have been the result of his experiment and does not warrant any further investigation other than in an examination of the transmission of the results over the centuries (see section 5.2).

Some modern scholars have based their examinations of Eratosthenes on the acceptance of the value of 252,000 *stade* as the correct result (as opposed to the 250,000 figure found in the earliest reference in Cleomedes) without considering potential causes for the difference in the figures such as errors by the later ancient authors and/or their sources, or the possibility that these later authors were offering a value that conformed to the ideals of Platonic numerical perfection. Plato states that the figure 5,040—that is, 252,000 divided by 50—is the ideal number for use in all matters pertaining to the state, as it is divisible by all numbers from one to ten.[9] The acceptance of one value for the result over the other, with little analysis and/or justification for the acceptance of that number, has led to many modern theories on Eratosthenes' work that do not conform to all of the available sources of evidence. As will be shown, there is evidence that has not been examined by prior scholars that demonstrates that both the 250,000 figure, and the 252,000 figure, are later traditions that are attributed to Eratosthenes, but are not the true value of the results of his calculations (see section 5.2). This makes all of the previous studies into the work of Eratosthenes either incomplete or incorrect.

One of the first of these major modern examinations was published by Viedebantt in 1915.[10] Viedebantt based his analysis on works by earlier scholars, such as Letronne (1822), Hultsch (1882), and Lehmann-Haupt (1896)

[6] Marc. *Perip.* 1.4.
[7] P.F.J. Gosselin, *Recherches fur la Geographie Systematique et positive des Anciens* (Paris, National Institute of France, 1798), p.14.
[8] J.A. Letronne, 'Mémoire sur cette question: Les Anciens ont-ils exécuté une mesure de la terre postérieurement à l'établissement de l'école d'Alexandrie?' *H. M. Inst. R. Fr.* 6 (1822), p.277.
[9] Pl. *Leg.* 737e–744d.
[10] O. Viedebantt, 'Eratosthenes, Hipparchos, Poseidonios' *Klio* 14.14 (1915), pp.207–256.

who had conducted some of the earliest examinations into the size of units of measure in the ancient Greek and Near Eastern world.[11] However, Viedebantt's examination contained several errors that influenced the size of the *stade* that he attributed to Eratosthenes; such as working from the assumption that Eratosthenes' *stade* was three-quarters of the Egyptian *stade*, and basing his calculations on an incorrect distance between the cities of Alexandria and Syene—which was an integral part of Eratosthenes' calculations.[12] Thus, while detailed in its examination, Viedebantt's conclusions are flawed in that they do not correlate with other pieces of evidence, and the analysis ultimately reaches an incorrect value for the size of the units used by Eratosthenes (although Viedebantt would have been unaware of that at the time of publication).

Subsequent examinations have viewed the evidence provided in the ancient texts, and in prior scholarship such as that by Viedebantt, in various ways, and this has resulted in several schools of thought emerging within the academic community over the accuracy of Eratosthenes' calculations and the size of the *stade*.[13] Diller, for example, suggested that the *stade* used by Eratosthenes was 159.2 m in length.[14] This would give Eratosthenes' calculations a result of 39,800 km (based on Cleomedes' value of 250,000 *stadia* for the Earth's circumference)—or a margin of error of −0.5 per cent when compared to the current estimate of the polar circumference of 40,007 km.[15] Dreyer, on

---

[11] Letronne, *Mémoire*, p.315; F. Hultsch, *Griechische und Römische Metrologie* (Berlin, Weidmann, 1882), pp.30–34, 697; C.F.F. Lehmann-Haupt, *Das altbsbylonische Maß und Gewichtssystem als Grundlage der antiken Gewichts, Münz und Maßsysteme* (Leiden, Ausz. Aus den Akten des 8 internat. Orientalistenkomgresses, 1896).

[12] For his assumed size of Eratosthenes' *stade*, see: Viedebantt, 'Eratosthenes', p.209; for his distance between Alexandria and Syene, see: Viedebantt, 'Eratosthenes', p.215.

[13] For example, on the various interpretations over the size of the *stade*, see: A. Diller, 'The Ancient Measurements of the Earth' *Isis* 40.1 (1949), pp.6–9; A. Diller, 'Julian of Ascalon on Strabo and the Stade' *Cl. Phil.* 45.1 (1950), pp.22–25; I. Fischer, 'Another Look at Eratosthenes' and Posidonius' Determinations of the Earth's Circumference' *Q. Jl R. Astr. Soc.* 16 (1975), pp.152–167; D. Rawlins, 'Eratosthenes' Geodesy Unraveled: Was There a High-Accuracy Hellenistic Astronomy?' *Isis* 73.2 (1982), pp.259–265; D. Engels, 'The Length of Eratosthenes' Stade' *Am. Journ. Phil.* 106.3 (1985), pp.298–311; E. Gulbekian, 'The Origin and Value of the Stadion Unit used by Eratosthenes in the Third Century B.C.' *Arch. Hist. Exact Sci.* 37.4 (1987), pp.359–363; J. Dutka, 'Eratosthenes' Measurement of the Earth Reconsidered' *Arch. Hist. Exact Sci.* 46.1 (1993), pp.55–66; A.C. Bowen, 'Cleomedes and the Measurement of the Earth: A Question of Procedures' *Centaurus* 45 (2003), pp.59–68; N. Nicastro, *Circumference: Eratosthenes and the Ancient Quest to Measure the Globe* (New York, St. Martins Press, 2008); D. Rawlins, 'Eratosthenes' Too-Big Earth and Too-Tiny Universe' *Dio* 14 (2008), pp.3–12; C.C. Carman and J. Evans, 'The Two Earths of Eratosthenes' *Isis* 106.1 (2015), pp.1–16; D.A. Shcheglov, 'The So-Called "Itinerary Stade" and the Accuracy of Eratosthenes' Measurement of the Earth' *Klio* 100.1 (2018), pp.153–177.

[14] Diller, 'Ancient Measurements', p.8.

[15] For the current polar circumference of the Earth, see: Defense Mapping Agency, *Supplement to Department of Defense World Geodetic System 1984 Technical Report: Part I—Methods, Techniques, and Data Used in WGS 84 Development*, US Naval Observatory, Washington DC, 1987, pp.3–46.

the other hand, suggests that the *stade* used by Eratosthenes was 157.5 m in length—resulting in a circumference of 39,375 km, or a margin of error of −1.6 per cent.[16]

Other examinations have merely compounded the errors found in the earlier works or, in an attempt to rectify the errors that the authors suggest they have identified in the earlier works, have made their own attempts to determine the size of Eratosthenes' *stade*. Fischer, for example, suggested that the size of Eratosthenes' *stade* fell within a range of 148–158 m—which would encompass the values offered earlier by both Diller and Dreyer.[17] Alternatively, both Rawlins and Engels, suggest that the *stade* used by Eratosthenes was much larger at 185 m in length—which would result in a circumference of 46,250 km, or a margin of error of +13.5 per cent.[18] Gulbekian offered a *stade* of 166.7 m in size, while Dutka suggested it was 157.5 m—the same as that offered by Dreyer (for a more detailed examination of many of the modern examinations into Eratosthenes' calculation of the Earth's circumference, see the section 'Previous Estimates of the Size of Eratosthenes' *Stade*' in Chapter 4).[19]

Other, more general, works on the history of astronomy also usually contain at least a brief section on Eratosthenes' calculations and results. However, they rarely deal with the underlying debate, usually accept (or at least recount) the position of only one side, and usually do not offer any insight into providing a greater understanding of the topic.[20] Some scholars, for example, simply state that Eratosthenes' result was 40,000 km.[21] This must assume the use of a *stade* 160 m in size and the acceptance of Cleomedes' figure for the circumference of 250,000 *stadia*. O'Neil states that the result could be in any of three possible different sizes of the *stade* without even referring to what these different sizes are.[22] Yet other examinations of Eratosthenes have concentrated more on the methodology, and motivation, behind Eratosthenes' diverse range of research topics rather than his calculations.[23] While some of these works have

---

[16] J.L.E. Dreyer, *A History of Astronomy from Thales to Kepler* (London, Constable, 1953), p.176.

[17] Fischer, 'Another Look at Eratosthenes', pp.152–167.

[18] Rawlins, 'Too-Big Earth', p.10; Engels, 'Length of Eratosthenes' Stade', pp.298–311.

[19] Gulbekian, 'Stadion Unit Used by Eratosthenes', pp.359–363; Dutka, 'Eratosthenes' Measurement', pp.55–66.

[20] For example, see: Dreyer, *History of Astronomy*, p.176.

[21] C. Sagan, *Cosmos* (London, MacDonald & Co., 1980), p.14; D. Dueck, *Geography in Classical Antiquity* (Cambridge, Cambridge University Press, 2012), p.73; J. Bennett, M. Donahue, N. Schneider, and M. Voit, *The Cosmic Perspective* (San Francisco, Pearson, 2014), p.63.

[22] W.M. O'Neil, *Early Astronomy from Babylonia to Copernicus* (Sydney, Sydney University Press, 1986), pp.58–59.

[23] For example, see: A.D. Pinotsis, 'A Comparative Study of the Evolution of the Geographical Ideas and Measurements until the Time of Eratosthenes' *A&AT* 24 (2005), pp.127–138; A.D. Pinotsis, 'The Significance and Errors of Eratosthenes' Method for the Measurement of the Size and Shape of the Earth's Surface' *J. Astron. Hist. Herit.* 9 (2006), pp.57–63; L. Taub, *Science Writing in Greco-Roman*

little direct bearing on determining the accuracy of Eratosthenes' calculations, they do provide useful information for understanding Eratosthenes himself, and the times he was working in. However, regardless of the nature and extent of their examination, all of these works have provided inconclusive results in regard to the accuracy of Eratosthenes' calculations and/or the size of the unit of measure that he was working in.

This is, in part, due to the methodologies employed in many of these modern studies. For example, many of the prior examinations into the work of Eratosthenes have modelled their examination in a retrospective manner. Some examinations have taken one of the figures provided in the ancient texts (either 250,000 *stadia* or, more commonly, the later figure of 252,000 *stadia*) and then attempted to work backwards, dividing this value by, for example, a perceived value for the distance between Alexandria and Syene to try to determine the size of the unit of measure that Eratosthenes was using. Other examinations alter the details of Eratosthenes' experiment—by suggesting, for example, that some of the values credited to Eratosthenes were different to those outlined in the ancient accounts, or that the distance between Alexandria and Syene was considered to be a different value to that given in the ancient accounts. Such alterations are then used to make Eratosthenes' results fit to a scholar's preconceived idea of the size of the *stade*—rather than finding the *stade* that actually fits with the ancient accounts. As a result, many of these examinations use such approaches to make their conclusions fit to known units of measure from the ancient Greek world—or something close to them. Unfortunately, many of these units were either no longer in use by the time of Eratosthenes, or were not commonly used across the ancient Greek world in general. Another consequence of such methodologies is that the conclusions of some scholars claim that Eratosthenes used a unit of measure for which there is no evidence of its use in the ancient literary or archaeological record.

What these prior works do highlight, however, is one major underlying flaw with the majority of the re-examinations of Eratosthenes' work that have been undertaken so far: namely that the scholarly debate has not been resolved by the work of scientific writers who have not considered the nuances of the ancient Greek and Latin in which the original results were written, nor fully engaged with the historical, philological, topographical or archaeological evidence. Similarly, historical writers commenting on Eratosthenes have not fully engaged with the scientific and mathematical issues of the experiment or the calculations. This has resulted in a large corpus of work that fails

Antiquity (Cambridge, Cambridge University Press, 2017); M. Berrey, *Hellenistic Science at Court* (Berlin, De Gruyter, 2019).

to present a hypothesis for the understanding of the accuracy of Eratosthenes' calculations, and the size of the *stade* that he used, that considers all of the available evidence.

Rarely have modern scholars attempted to examine Eratosthenes' work in the other direction—i.e. looking at what units of measure were in use in Eratosthenes' own time and then applying them to the figures that are given in the ancient texts. One of the most recent advances to this method of analysis was the identification of the size of some of the units of measure in use in the later periods of Greek history by Matthew in 2012 and 2015. Matthew identified two systems of measure that were in use from the late Classical Period in the fifth century BC and across the Hellenistic Period that encompassed the time of Eratosthenes.[24] The two systems of measurement incorporate a *stade* of 180 m and 191 m in size, respectively. Importantly, neither of these measurements has been applied to an examination of the size of the *stade* used by Eratosthenes and/or the accuracy of his results. However, clearly only one, and possibly none, of the previously offered sizes for the *stade* used by Eratosthenes can be correct. What is required is engagement with all of the prior studies, determination of where their conclusions do not align with other sources of evidence, re-examination of the ancient texts in their original languages, correlation of theories with all of the available evidence, and the employment of new methodologies such as satellite mapping, to create the first comprehensive study to determine the size of the *stade* used by Eratosthenes and the accuracy of his results.

In order to engage with the scholarly debate over this topic, and to address the research objectives, a re-evaluation of Eratosthenes' calculations using a variety of methods from across several disciplines will be undertaken:

PHILOLOGICAL: The ancient texts will be reviewed in their original languages (Greek and Latin) and compared to the modern works on Eratosthenes to determine where passages have been mistranslated or misinterpreted, and where passages may have multiple meanings, and any issues that arise from these discrepancies will be applied to a new determination of Eratosthenes' results and their accuracy. An example of such potential errors can be seen in the possible transmission errors given in the ancient account of Marcian of Heraclea (see previous) which may contain errors in the figures it provides for Eratosthenes' result.

[24] See: C.A. Matthew, *A Storm of Spears: Understanding the Greek Hoplite at War* (Barnsley, Pen & Sword, 2012), pp.53–57, 179–196; C.A. Matthew, *An Invincible Beast—Understanding the Hellenistic Pike-Phalanx at War* (Barnsley, Pen & Sword, 2015), pp.35, 69–80, 148–153, 159–160.

HISTORICAL: All of the previous examinations into Eratosthenes' calculation of the circumference of the Earth offer theories and models that
favour a *stade* of a certain size over all other possibilities. Commonly,
however, these theories have not engaged with all of the available evidence, and are somewhat selective in the use of what is used to support
their claims. These models will be compared to all of the available evidence to determine where and how these prior studies are lacking in
support for their arguments. Additionally, evidence for a *stade* introduced into Egypt by the army of Alexander the Great in the fourth
century BC[25] will be examined and applied to the recalculation of Eratosthenes' circumference. Importantly, this *stade* has never been applied in
any prior examination of the work of Eratosthenes, and so presents an
entirely new analysis of this topic through the application of this unit of
measure to the recalculation of Eratosthenes' results.

CARTOGRAPHIC: In a work on geography, Eratosthenes provided the
distances between several cities and locations in northern Iran and
Afghanistan as they were recorded by the surveyors in the army of
Alexander the Great. While this text has been lost, the details have
been preserved in the works of Strabo.[26] Engels, in his work on the
logistics of the Macedonian army, used these sources to calculate the
march rate of Alexander's army.[27] However, many of the sites listed by
Strabo have not been positively identified—possibly because the measurements have been converted into modern equivalents using a *stade*
of the incorrect size. Importantly, this is one of the few surviving ancient
texts that is directly attributable to work undertaken by Eratosthenes,
and which includes precise details of distances that he recorded (or, at
least, recounted). The distances given by Strabo will be compared to
results obtained from modern satellite imagery and mapping software
to determine the location of these sites and, as a result, the size of the
unit of measure that Eratosthenes was using. Such a process has never
been attempted before—in relation to either locating the sites in Central
Asia, or determining the size of the unit of measure that Eratosthenes was
using—and so presents an entirely original approach to understanding
the accuracy of Eratosthenes' experiment and calculations.

---

[25] See: Matthew, *Invincible Beast*, pp.35, 69–80, 148–153, 159–160.
[26] Str. *Geog.* 1.2.1–2, 1.2.3–7.
[27] D.W. Engels, *Alexander the Great and the Logistics of the Macedonian Army* (Berkeley, University
of California Press, 1978), pp.157–158.

SCIENTIFIC: Some of the modern analyses of Eratosthenes engage with potential flaws in his methodology. Some of these include the 'assumed knowledge' that Eratosthenes was basing his work upon that is not actually correct—such as assuming that the cities of Alexandria and Syene sit on the same longitudinal meridian, that Syene sits on the Tropic of Cancer, that rays of light from the Sun are parallel when they reach the Earth, and that the Earth is a perfect sphere. These errors in the fundamentals behind Eratosthenes' calculations will be examined and their impact on any potential recalculation of the accuracy of the results will be considered.

STATISTICAL: Throughout the re-examination of Eratosthenes' work and the review of previous scholarship on the topic, when any recalculations are made, or any of the results from prior examinations are analysed, margins of error for all calculations (or recalculations) will be recorded and explored, and their impact (if any) on the results will be explained.

1.2 Eratosthenes, The Man

There are few details of Eratosthenes' life and work that come from any sources that are contemporaneous with his own time. Much of what we do know of him comes from the tenth century Byzantine lexicon: the *Suda*. The biographical entry for Eratosthenes in the *Suda* states that he was born in the year of the 126th Olympiad (276 BC) in Cyrene (near present-day Shahhat, Libya) and died at the age of eighty.[28] However, Nicastro doubts that Eratosthenes was born in 276 BC as per the *Suda*.[29] Nicastro claims that the *Suda* entry states that Eratosthenes studied under Zeno of Citium in Athens as a youth, and as Zeno died in 262 BC, if Eratosthenes had been born in 276 BC, he could have only been fourteen years of age at the oldest when he was studying under Zeno.[30] Nicastro offers an alternative date for the birth of Eratosthenes in the mid-280s BC. There are a number of things wrong with such a claim. Firstly, the *Suda* entry for Eratosthenes makes no mention of Zeno. Rather it states that Eratosthenes studied under Ariston of Chios—who was a younger colleague of Zeno; having been born in the same year that Zeno began teaching in Athens (*ca.* 300 BC). Thus, there is no restriction on how old Eratosthenes may have been when he was studying in Athens. Additionally, even if it is assumed that

---

[28] *Suda s.v.* Ἐρατοσθένης.
[29] Nicastro, *Circumference*, p.138.
[30] For the career of Zeno, see: Diog. Laert. 7.1–160.

Eratosthenes studied under Zeno as well, there is no reason why this could not have occurred while Eratosthenes was in his early teens. Many young aristocrats in the ancient Greek world began their studies at an early age. Alexander the Great, and many other members of the Macedonian nobility, for example, studied under Aristotle when they were teenagers.[31] Consequently, there is little reason to doubt the date of Eratosthenes' birth that is given in the *Suda*.

Eratosthenes' father went by the name Aglaus, and the family seems to have been part of the Cyrenean upper class.[32] Eratosthenes' name translates as 'lovely strength', and such a name also suggests a noble upbringing. Cyrene had been established as a colony by people from the island of Thera (modern Santorini) in the seventh century BC and had thrived as an exporter of Silphion (Silphium)—a now extinct plant which was said to have possessed healing properties and was used in medicines, aphrodisiacs, contraceptives and perfumes.[33] The plant was rare—growing only in the region of Cyrene— and the resin of the Silphium plant was said to be worth its weight in gold. Silphium was so essential to the Cyrenean economy, and was so identifiable with the city, that the city's coinage bore an image of the plant (Fig. 2).

(a)   (b)

Fig. 2 A silver *didrachm* from Cyrene *ca.* 305–300 BC with a depiction of the Silphium plant on the reverse.[a]

[a] KYRENAICA. Kyrene. Second Revolt of the Kyrenaikans. Circa 305–300 BC. Didrachm (silver, 20 mm, 7.60 g, 6 h). Head of Karneios to l., with a ram's horn over his ear. Rev. KY-[PA] Silphium plant with fruits; to upper l. and r., star. Images courtesy of the Australian Centre for Ancient Numismatic Studies, Macquarie University—inv# 01M189.

[31] Plut. *Alex.* 7–8.
[32] *Suda s.v.* Ἐρατοσθένης.
[33] For the founding of Cyrene, see: Hdt. 4.150–159; for Silphion, see: Theoph. *Caus Pl.* 3.2.1, 6.3.3–5; Plin. *HN* 22.49.

Cyrene was one of the largest cities in the ancient Greek world. Its monopoly on a rare and valuable commodity, and its location along the coastal trade route from Egypt to the west, made it prosperous and powerful. Possibly due to the wealth of the city (and its Silphium) Ptolemy I annexed Cyrene into the Egyptian fold *ca.* 320 BC.[34] Even following annexation the city remained large, cosmopolitan, and mercantile, with an elaborate system of government. An inscription outlining the constitution of Cyrene following annexation by Ptolemy states that its political institutions included a Citizens' Assembly (*Ecclesia*) of 10,000 members (all of whom had to be over thirty years of age and in possession of 20 *minas* (*ca.* 10 kg of silver) worth of property), a Council of 101 Elders (*Gerousia*), and a Council of 500 (*Boule*) chosen at random for a two-year term. There was also a Board of Generals (*Strategoi*), nine Guardians of the Law, and five Overseers.[35] The city also seems to have been ruled by a king or an appointed governor (depending upon the time in the city's history being examined) in conjunction with these institutions and the last ruler, Magas, died of over-eating in 250 BC.[36] The city had a strong intellectual and athletic tradition (which young Eratosthenes would have been exposed to) and many poets, writers, philosophers and Olympic victors came from Cyrene over the years. Cyrene also possessed a strong military, and its power is evidenced by the fact that the city was able to rebel from Egyptian control in 313 BC, and was able to fight a war against Carthage in 308 BC, and then rebel from Egypt again in 301 BC.[37] When Ptolemy I died, the city declared its independence, and attacked Egypt yet again, as an ally of Antiochus I, around the time Eratosthenes was born (274 BC).[38]

Young Eratosthenes was tutored by the grammarian Lysanias of Cyrene and Callimachus the poet.[39] Such an education at a young age further suggests that Eratosthenes' family may have been part of the Cyrenean aristocracy. At some stage later in his early life, Eratosthenes moved to Athens where he studied philosophy, grammar, and poetry.[40] In Athens, Eratosthenes was taught by Ariston that the study of nature and logic were a waste of time, as

---

[34] Diod. Sic. 18.21.6–9.
[35] *SEG* 9.1 in P. Harding (ed.), *From the End of the Peloponnesian War to the Battle of Ipsus* (Cambridge, Cambridge University Press, 2012). See also: M. Cary, 'A Constitutional Inscription from Cyrene' *JHS* 48.2 (1928), pp.222–238; J.A.O. Larsen, 'Notes on the Constitutional Inscription from Cyrene' *CP* 24.4 (1929), pp.351–368.
[36] Ath. *Diep.* 12.550b–c.
[37] For the rebellion of 313 BC, see: Diod. Sic. 19.79.1–4; for the war against Carthage in 308 BC, see: Diod. Sic. 20.40.1–20.42.3; for the rebellion of 301 BC, see: Paus. 1.6.8.
[38] For the revolt of 275/274 BC, see: Paus. 1.7.2.
[39] *Suda s.v.* Ἐρατοσθένης.
[40] *Suda s.v.* Ἐρατοσθένης.

such truths were beyond the realm of human understanding. Instead, Ariston offered, everyone should study ethics.[41] Strabo states that Eratosthenes
regarded Ariston as one of the most important philosophers of his age.[42] However, despite this, and as Nicastro points out, Ariston's teachings do not seem
to have fully rubbed off on Eratosthenes, as many of the areas of study pur-
sued by Eratosthenes across his career are scientific in nature and therefore
very 'anti-Ariston'.[43]

It was while he was in Athens that Eratosthenes produced his first major
work, an enquiry into the mathematical basis of Platonic philosophy, called
the *Platonikos*.[44] This was followed by poetic works (*Hermes* and *Erigone*),[45]
a compilation of important dates from the time of the Trojan War (*Chrono-
graphies*), and a list of winners at the Olympic festival (*Olympic Victors*).[46] His
works brought Eratosthenes to the attention of Ptolemy III (284–222 BC) in
Egypt, who arranged for him to take up a position as Chief Librarian at the
Great Library of Alexandria in 245 BC.[47] Eratosthenes' areas of later study were
diverse, releasing further works on geography,[48] instrumentation,[49] a means
of determining prime numbers (the so-called 'sieve'),[50] a work on means,[51]
the calculation of harmonics,[52] a treatise on philosophy (*On Good and Bad*),
another on rhetoric (*On Declamation*),[53] a literary critique of the works of the
poet Homer,[54] an extensive discussion of the nature of old comedy,[55] a work
on astronomical constellations (*Catasterisms*),[56] a correction of the calendar

[41] Diog. Laert. 7.161.
[42] Str. *Geog.* 1.2.2.
[43] Nicastro, *Circumference*, p.64.
[44] Theon. *Expos.* 81.17.
[45] For a commentary on *Erigone*, see: Longinus *Subl.* 33.5.
[46] Idaeus of Cyrene had won the blue ribbon 'stadion race' at the Olympics in the year that Eratosthenes was born (Eus. *Chron.*—*The Greek Olympiads* 224). The exact dates for the publication for these,
and many other of Eratosthenes' works, are unknown.
[47] *Suda s.v.* Ἐρατοσθένης; The papyrus manuscript *P. Oxy. 1241* 2.1–8 lists Eratosthenes as being
the director of the library in Alexandria following Apollonius of Rhodes and before Aristophanes
of Byzantium (see: Hunt, S. *The Oxyrhynchus Papyri Part.X* (London, Egypt Exploration Society,
1914)). MacLeod ('Introduction: Alexandria in History and Myth' in R. MacLeod (ed.), *The Library
of Alexandria: Centre of Learning in the Ancient World* (London, IB Tauris, 2004), p.6) suggests that
Eratosthenes' appointment occurred in 235 BC.
[48] See: D.W. Roller (ed.), *Eratosthenes' Geography* (Princeton, Princeton University Press, 2010).
[49] Eutoc. *In Arch. circ. dim.* 2.88.3–96.27; Vit. *De Arch.* 9.13–14; Plut. *Marc.* 14.
[50] Nicom. *Arith.* 1.13.2.
[51] Papp. *Syn.* 7.3, 7.22, 7.29.
[52] Nicom. *Harm.* 11.6; Ptol. *Harm.* 2.14.
[53] Str. *Geog.* 1.2.1–2.
[54] Str. *Geog.* 1.2.3–7.
[55] Only fragments of this work remain. For the most recently published edition of the fragments,
see: A. Bagordo (ed.), *Fragmenta Comica: Telekleides* (Heidelberg, Verlag-Antike, 2013).
[56] *Suda s.v.* Ἐρατοσθένης; Dimitrijević and Bajić ('Mythological Origin of Constellations and their
Description: Aratus, Pseudo-Eratosthenes, Hyginus' in L.Č. Popović, V.A. Srećković, M.S. Dimitrijević,
and A. Kovačrvić (eds.), *Proceedings of the XII Serbian-Bulgarian Astronomical Conference* (Belgrade,

(*On the 8-Year Cycle*),[57] an examination of planetary orbits,[58] an examination of the winds,[59] philosophical analyses (*On the Philosophical Sects* and *On Freedom from Pain*), dialogues and grammatical works,[60] a discussion of wealth and poverty,[61] and possibly a history of the campaigns of Alexander the Great.[62]

This penchant for diversity, but apparently not expertise, gained him some criticisms. Strabo, for example, described Eratosthenes as a man who:

> . . . is constantly fluctuating between his desire to be a philosopher and his unwillingness to devote himself to this profession completely, and who therefore succeeds in advancing far enough to only have an appearance of being a philosopher, or of a man who has provided himself with this as a distraction from his regular work—either as his hobby or even entertainment . . .[63]

Conversely, Archimedes dedicated his work *The Method* to Eratosthenes—calling him 'a man of considerable eminence in philosophy and an admirer [of mathematical enquiry]'.[64] Similarly, Heron of Alexandria stated that Eratosthenes 'worked rather more accurately than others'.[65] Indeed, in a letter to an Egyptian ruler, possibly Ptolemy III, Eratosthenes presented his solution to the so-called 'Delian Problem' for how to double the size of a cube.[66]

Astronomy Society 'Rudjer Bošković', 2020), pp.129, 135) outline how the extant text may be an epitome of Eratosthenes' work written by an otherwise unknown Pseudo-Eratosthenes, but that it still constitutes 'the first preserved star catalogue of ancient Greece'.

[57] Gem. *Isagoge* 8.24; Geminus says that Eratosthenes wrote *On the 8-Day Cycle* to reconcile the Egyptian calendar of 365 days with the actual solar year of 365.25 days. He did this because, by his own time, the old calendar was so out of alignment with reality that the Spring harvest festival only occurred in Spring every 1,460 years. Under the old system, intercalatory months were added every two years except in the eighth year. Ptolemy, possibly on the advice of Eratosthenes, changed it in 239 BC to three years of 365 days, and then one year of 366 days, in a four-year cycle. Nicastro (*Circumference*, p.15) says that Eratosthenes' involvement in the restructuring of the Egyptian calendar is uncertain, but it seems more than likely due to the fact that he wrote on the subject.

[58] Gal. *Phil. Hist.* 72; Stob. *Ecl.* 1.26.5; Eus. *PE* 15.53.3.

[59] Ach. Tat. *Arati Phaenom.* 33.6; Vit. *De Arch.* 1.6.9–11.

[60] *Suda s.v.* Ἐρατοσθένης.

[61] Plut. *Them.* 27.

[62] There are several passages, which are attributed to Eratosthenes, found in the extant histories of Alexander by Plutarch (*Alex.* 3, 31) and Arrian (*Anab.* 5.5.3–5), but these may have been drawn from Eratosthenes' work on geography, rather than an actual history.

[63] Str. *Geog.* 1.2.2: δηλοῖ δὲ καὶ ἡ περὶ τῶν ἀγαθῶν ἐκδοθεῖσα ὑπ᾽ αὐτοῦ πραγματεία καὶ μελέται καὶ εἴ τι ἄλλο τοιοῦτο τὴν ἀγωγὴν αὐτοῦ, διότι μέσος ἦν τοῦ τε βουλομένου φιλοσοφεῖν καὶ τοῦ μὴ θαρροῦντος ἐγχειρίζειν ἑαυτὸν εἰς τὴν ὑπόσχεσιν ταύτην, ἀλλὰ μόνον μέχρι τοῦ δοκεῖν προϊόντος, ἢ καὶ παράβασίν τινα ταύτην ἀπὸ τῶν ἄλλων τῶν ἐγκυκλίων πεπορισμένου πρὸς διαγωγὴν ἢ καὶ παιδιάν.

[64] Archim. *Meth.* prae.2.

[65] Hero. *Dioptr.* 35.

[66] The text of the letter is preserved in a work by Eutocius of Ascalon (AD 480–540) (*In Arch. circ. dim.* 2.88.3–96.27). The solution is not as simple as it may first appear. The sides of the cube cannot simply be doubled in order for the overall size of the cube to be doubled. For example, a cube with sides

The letter also states that the details and solution to this problem were inscribed on a monument—presumably erected in the grounds of the Library.[67] Taub suggests that this method for the presentation of results—in a written document sent to the current ruler, and displayed publicly on a monument—were an important aspect of the duties of the head librarian in Alexandria.[68] According to the *Suda*, Eratosthenes' wide field of interests earned him the nickname *beta* (the second letter of the Greek alphabet), because he was second best at everything.[69] Shipley suggests that this epithet was somewhat ironic and that Eratosthenes was actually regarded as the greatest scientist of his age.[70] Regardless of how his contemporaries may, or may not, have viewed him, it was in the experiment that he undertook in Alexandria to measure the Earth's circumference that Eratosthenes' brilliance for experimentation and calculation really shone.

1.3 Alexandria

The city of Alexandria, on Egypt's Mediterranean coast, was one of the most glorious cities of antiquity. Nearly eight hundred years after its founding, Alexandria was still regarded as 'the crown of all cities'.[71] Within its environs,

1m in length has a volume of 1m^3 (1 x 1×1 = 1). However, a cube with sides 2 m in length has a volume of 8m^3 (2 x 2×2 = 8). As such, Eratosthenes needed to work out the size of a cube which had sides of a length equal to the cube root of two. This would not have been easy in a time before the common use of squares, cubes, and roots in mathematics, and Eratosthenes seems to have tackled the problem from an engineering perspective rather than a mathematical one. Vitruvius (*De Arch.* 9.13–14), for example, states that Eratosthenes solved the Delian Problem using a 'proportional instrument', which may have been some kind of pantograph. Taub (*Science Writing*, pp.56, 61–64) states that some scholars consider the 'letter' to be a fake, and Russo (*The Forgotten Revolution: How Science was Born in 300 BC and Why It Had to Be Reborn* (trans. S. Levy) (Berlin, Springer, 2000), p.201) suggests that, despite the statements for Eratosthenes solving this problem, it remained unsolved until the modern age. However, Vitruvius (*De Arch.* 9.14) additionally states that the Problem was also solved by Archytas (435–360 BC) using a mathematical method. Berrey (*Hellenistic Science*, pp.5, 130–138, 146–147. 163–179) suggests that the scientific discoveries in the Ptolemaic court had two main objectives: (a) to gain patronage, and (b) to spark the interest of the lay person (possibly by using the discoveries as a form of entertainment). Berrey also suggests (p.26) that the aim of the court scientist was to gain fame (τιμή) and renown (δόξα). Indeed, following his discussion of the Delian Problem (*De Arch.* 9.15–16), Vitruvius discusses how the works of thinkers like Eratosthenes and Archytas are of perpetual advantage to future generations, and so the writers 'derive immortality from their works'. However, it seems unlikely that the court in charge of the most advanced institution for learning in the known world at that time would see any form of discovery as 'entertainment'. It is more likely that those in court would be more interested in the scholarly value of the discoveries, even if they did not fully understand them, as a means of showcasing and promoting the intellectual superiority of Alexandria.

[67] Eutoc. *In Arch. circ. dim.* 2.88.3–96.27.
[68] Taub, *Science Writing*, p.57.
[69] *Suda s.v.* Ἐρατοσθένης; the text of the *Suda* actually reads βήματα ('platforms'), but this is most likely a typographical error for βῆτα. See: R. Pfeiffer, *History of Classical Scholarship from the Beginnings to the End of the Hellenistic Age* (Clarendon Press, Oxford, 1968), p.170, n.3.
[70] G. Shipley, *The Greek World after Alexander 323–30 BC* (London, Routledge, 2000), p.327.
[71] Amm. 22.16.7.

the city contained one of the Seven Wonders of the ancient world (the Pharos lighthouse), the resting place of the body of Alexander the Great, an institution for learning and research (the *Museion*), and one of the greatest libraries ever to have existed. For centuries, the city was one of the main centres of learning in the ancient Greek world. When Eratosthenes was appointed to the position of Chief Librarian by the Egyptian ruler, Ptolemy III, in 245 BC, he inherited a position that had only been held by two previous men. As such, Eratosthenes arrived in Alexandria at a time when the city's 'intellectual star' was beginning to rise.

Alexandria was founded by Alexander the Great in 332 BC after Egypt peacefully surrendered to the invading army of the young Macedonian king.[72] The natural harbour in the region had already been in regular use by sailors for centuries, and references to this area being used as a port go back as far back as the time of Homer.[73] This site was already occupied with the pastoral village of Rhakotis, but Alexander is said to have recognized the advantages of the site, with a deep natural harbour which could cater to the largest of ships, and a marshy lake to the south, and decided to fortify the position.[74] The local inhabitants were then relocated in order for the new city to be built.[75] Running out into the bay, from what would become the eastern end of the city shoreline, was the promontory of Loachis.[76] Offshore, and running parallel to the coastline, sat the elongated Pharos Island.[77] According to Curtius, Pharos was the initial site chosen by Alexander upon which to build his new city, but the location was abandoned as it was not large enough for what Alexander had in mind for the city that would bear his name.[78] The island was eventually connected to the mainland with the construction of a causeway, the *heptastadion*, which then created two large harbours for the city—the Great Harbour to the east of the causeway and bordered by the Loachis peninsula on the other side, and the harbour of Eunostus to the west of the causeway.[79] Passage between the two harbours was possible via two canals through the *heptastadion*, which had bridges thrown over them.[80]

---

[72] For the entry of Alexander into Egypt, see: Diod. Sic. 17.49.1; Curt. 4.7.1; Plut. *Alex.* 25–26.
[73] Hom. *Od.* 4.354–359.
[74] Str. *Geog.* 17.1.6; Diod. Sic. 17.52.1–3.
[75] Ps-Call. 1.81; Arist. *Oec.* 1352a–b; Rhakotis then became a suburb in the south-west of the city.
[76] Str. *Geog.* 17.1.6.
[77] Str. *Geog.* 17.1.6.
[78] Curt. 4.8.1–2.
[79] Str. *Geog.* 17.1.6, 17.1.10.
[80] Str. *Geog.* 17.1.6.

The ancient sources describe how it was Alexander himself who designed the layout of the new city, and decided on the locations where temples, defensive works, and the main marketplace were to be built.[81] Diodorus also states that Alexander chose the location for the construction of a palace 'notable for its size'.[82] The city was laid out in the shape of a semicircle, described as being similar to a military cloak, with the base, thirty *stadia* (approximately 5.4 km) across, along the Mediterranean coast.[83] Despite the initial design of the city being attributed to Alexander, the overall planners of the city were Deinocrates of Rhodes, in charge of architecture, and Crates of Olynthus, who designed the water supply system.[84] The ancient sources also mention Heron and Hyponomus from Libya, who constructed canals.[85] Pseudo-Callisthenes states that Alexander deferred to these architects, who then built much of the city based upon their desired measurements.[86]

The city itself was set out on a grid system, which possibly followed examples that Aristotle attributes to Hippodamus of Miletus around 500 BC.[87] The main east–west oriented road was the Canopic Way—which ran from the Gate of the Sun in the East to the Gate of the Moon in the West.[88] The major north–south thoroughfare ran south from the site for the Royal Palace complex in the north, near the coastline.[89] The first governor of the city, appointed by Alexander himself, was Cleomenes of Naukratis, who oversaw the beginning of the construction of the city following Alexander's command.[90] According to Pseudo-Callisthenes, both Cleomenes and Deinocrates advised Alexander not to build the city, as they did not have enough people to occupy the site.[91] This may explain why people were subsequently ordered to move to Alexandria from neighbouring cities to boost the size of the population.[92]

---

[81] Arr. *Anab.* 3.2; Plut. *Alex.* 26; Just. *Epit.* 11.11.3; Str. *Geog.* 17.1.6; Ps-Call. 1.80; Diod. Sic. 17.52.1.

[82] Diod. Sic. 17.52.4.

[83] Plut. *Alex.* 26; Diod. Sic. 17.52.3; Str. *Geog.* 17.1.6; Pseudo-Callisthenes (1.81) says that Alexandria was 16 *stadia* and 395 feet (approximately 3 km) across. This may be a reference to the radius of the semicircle rather than its diameter. Plans of the city, based upon archaeological surveys conducted in the nineteenth century, give the city an east–west breadth of around 5 km.

[84] For Deinocrates of Rhodes, see: Val. Max. 1.4; Plin. *HN* 5.62. For Crates of Olynthus, see: Ps-Call. 1.82, who calls him Craterus of Olynthus.

[85] Ps-Call. 1.82.

[86] Ps-Call. 1.82.

[87] Diod. Sic. 17.52.2–3; Arist. *Pol.* 1267b, 1330b; for a discussion of Hippodamus' designs (and if they were even his), see: A. Burns, 'Hippodamus and the Planned City' *Historia* 25 (1976), pp.414–428; L. Mazza, 'Plan and Constitution: Aristotle's Hippodamus: Towards an "Ostensive" Definition of Spatial Planning' *TPR* 80 (2009), pp.113–141.

[88] Nicastro, *Circumference*, p.68.

[89] For descriptions of Alexandria's gates and streets, see: Str. *Geog.* 17.1.8; Ach. Tat. *Arati Phaenom.* 5.1.

[90] See: Just. *Epit.* 13.4.11; Arist. *Oec.* 1352a; Paus. 1.6.3.

[91] Ps-Call. 1.80.

[92] Curt. 4.8.5.

The city flourished. Situated in a location where the onshore breeze made the hot Egyptian summers pleasant, and with air that is described by Diodorus and Strabo as salubrious, Alexandria turned into a thriving metropolis.[93] Later writers would comment on the crowds of people that were encountered in the city's streets. For example, Theocritus, a Sicilian contemporary of Eratosthenes, talks of streets filled with crowds of people scurrying around like ants, and streets congested with chariots.[94] Achilles Tatius, a native Alexandrian of the second century AD, wondered that any city could be found to accommodate all of the inhabitants.[95] Diodorus stated that the number of Alexandria's inhabitants surpassed that of all other cities, and cites a census figure of 300,000 free residents.[96] The population, however, was not without its restrictions and regulations. Theocritus, for example, refers to a distinct lack of crime in the city ruled by the Ptolemies, which would suggest fairly stringent enforcement of the ruler's laws.[97] Strabo refers to the presence of a large law court in the city.[98] According to a letter from Ptolemy II, the ruler who may have hired Eratosthenes, to Aristeas, an assistant to Demetrius of Phaleron, native Egyptians from the rural districts were not allowed to spend more than twenty days in Alexandria, and judges who sat in the law court were instructed that, if any case involved someone from the country, it had to be settled within five days.[99] Nicastro suggests that this was to help the city retain its Greek feel.[100] However, the letter states that this was done so that the rural population could devote themselves to agriculture, while those in the city have 'a natural tendency towards the pursuit of pleasure'.[101] It was this relatively safe lifestyle, with time for leisurely pursuits, that resulted in the construction and patronage of many of the attractions that were to be found in the city.

According to Strabo, the city was 'full of public and sacred spaces'.[102] Achilles Tatius describes walking the streets, and taking in so many sites of the 'splendid beauty of the city', that he was forced to exclaim in weariness, 'Ah! My eyes!

---

[93] Diod. Sic. 17.52.2; Str. *Geog.* 17.1.7.

[94] Theoc. *Id.* 15.4–6, 15.44–45.

[95] Ach. Tat. *Leucip.* 5.1.

[96] Diod. Sic. 17.52.6.

[97] Theoc. *Id.* 15.46–47

[98] Str. *Geog.* 17.1.10.

[99] *Epit. Aris.*110; see also: Joseph. *AJ* 12.2.12; the *Letter of Aristeas* has been dated to the second century BC, so it is not contemporary with the time it is reporting (see: O. Murray, 'The Letter of Aristeas' *BSA* 54 (1987), pp.15–29; S. Honigman, *The Septuagint and Homeric Scholarship: A Study in the Narrative of the Letter of Aristeas* (New York, Routledge, 2003), pp.3–9, 129–130. Johnstone ('A New History of Libraries and Books in the Hellenistic Period' Class. Antiq. 33.2 (2014) pp.357, 358–362) calls the *Letter* 'fiction' and 'illusory' and dismisses it as a reliable source.

[100] Nicastro, *Circumference*, p.53.

[101] *Epit. Aris.* 106–108.

[102] Str. *Geog.* 17.1.10.

We are beaten!'[103] The city contained many 'beautiful public precincts' with groves and gardens.[104] The main palace complex was located in the north of the city, close to the shoreline. The complex was so large, having been extended by successive rulers, that by Strabo's time in the first century BC, it covered one-quarter to one-third of the entire city.[105] The buildings of the palace were all connected to each other and to the harbour.[106] There was also a separate palace on the Loachis Peninsula, near a Royal Harbour, with groves and buildings of different colours, and a third palace and harbour on the small island of Antirrhodos in the middle of the Great Harbour.[107] In the south of the palace complex was the *Sema*—a necropolis which contained the resting places of kings and Alexander the Great.[108] There was also a zoological park containing the royal collection of exotic animals, which 'had never been seen before and were the objects of wonder'.[109] These animals were also shown off to the public by being paraded through the streets.[110]

Across the Great Harbour from the palace stood the Pharos Lighthouse. It was made from white marble and was several stories high.[111] Achilles Tatius described it as being 'like a mountain, almost reaching to the clouds, in the middle of the sea'.[112] The lighthouse was designed and built by Sostratus of Cnidus as a guide for ships.[113] The *Suda* states that it was built in the time of Ptolemy I in 297 BC, while Jerome's *Chronicon*, a fourth-century translation of Eusebius' *Chronicle*, dates its construction to 284 BC under Ptolemy II.[114] It is most likely that construction began under Ptolemy I and was completed during the reign of Ptolemy II. The overall cost of erecting the lighthouse was said to have been 800 *talents*, or around 24 tons, of silver.[115] The lighthouse was surmounted with a huge fire (Achilles Tatius describes it as a 'second Sun') that could be seen at a distance of 300 *stadia* (approximately 54 km) which Josephus says equalled about one day's sailing from Alexandria.[116]

---

[103] Ach. Tat. *Leucip.* 5.1.
[104] Str. *Geog.* 17.1.8, 17.1.10.
[105] Str. *Geog.* 17.1.8.
[106] Str. *Geog.* 17.1.8.
[107] Str. *Geog.* 17.1.9.
[108] Str. *Geog.* 17.1.8.
[109] Diod. Sic. 3.36.3.
[110] Ath. *Diep.* 5.201c.
[111] Str. *Geog.* 17.1.6.
[112] Ach. Tat. *Leucip.* 5.6.
[113] Str. *Geog.* 17.1.6; *Suda s.v.* Φάρος; Plin. *HN* 36.83; for the dedication of Posidippus, see: *Greek Anthology* 13.10.
[114] *Suda s.v.* Φάρος; Jer. *Chron.* 124.1.
[115] Plin. *HN* 36.83.
[116] Posidippus 13.10; Ach. Tat. *Leucip.* 5.6; Joseph. *BJ* 4.10.5.

The protected harbours, and the guiding light of the Pharos lighthouse, also meant that Alexandria was a major commercial centre. Herodas states that 'everything that exists in the world, and is produced, is in Egypt'.[117] The mercantile nature of Alexandria's prosperity was also reflected in its architecture. Strabo describes how the shoreline between the main palace complex and the *heptastadion* causeway was lined with docks, boatsheds, the Emprorion, and warehouses.[118] To cater to the needs and wants of the inhabitants of this bustling city, and the many merchants and traders that arrived on its shores, Alexandria contained numerous venues for entertainment, relaxation, and religious practice. There was a large, open-air, theatre in the north of the city with views out over the Great Harbour,[119] wrestling schools,[120] and 'a most beautiful gymnasium with porticos more than 1 *stade* (*ca.* 180 m) in length'.[121] In the west of the city was a large public necropolis with gardens.[122] In the south-west was the *Paneium*—a high point from which all of the city could be seen.[123] Also in the south-west was a large hippodrome where spectacles were held.[124] Just to the north of the hippodrome was the *Serapeum*—a religious precinct dedicated to the god, Serapis.[125] The city also contained a temple to Zeus, where evening processions were held by torchlight,[126] a temple to Isis,[127] and a temple to Poseidon.[128] There was also the tomb of the hero Proteus,[129] and two shrines honouring Alexander's close companion Hephaestion.[130]

One of the other main attractions and religious sites in the city was the temple to the Muses—the *Museion*, or Museum. The Museum was also a part of the palace complex.[131] The Museum contained a public walkway, an area of outdoor seating, and a communal dining hall for the staff who worked there.[132] The Museum was run by a priest of the Muses who was appointed by the king.[133] Many of the sites that are described in the ancient texts, and many later,

[117] Herod. 1.26–27.
[118] Str. *Geog.* 17.1.9.
[119] Str. *Geog.* 17.1.9.
[120] Herod. 1.27.
[121] Str. *Geog.* 17.1.10.
[122] Str. *Geog.* 17.1.10.
[123] Str. *Geog.* 17.1.10.
[124] Str. *Geog.* 17.1.10; Herod. 1.28.
[125] Str. *Geog.* 17.1.10; Theoc. *Id.* 1.33.
[126] Ach. Tat. *Leucip.* 5.2.
[127] Ach. Tat. *Leucip.* 5.14.
[128] Str. *Geog.* 17.1.9.
[129] Ps-Call. 1.84.
[130] Arr. *Anab.* 7.23.
[131] Str. *Geog.* 17.1.8, Herod. 1.31.
[132] Str. *Geog.* 17.1.8.
[133] Str. *Geog.* 17.1.8.

Roman-period, additions to the city, have been either located or inferred via archaeological investigations that were undertaken in the nineteenth century (Fig. 3).

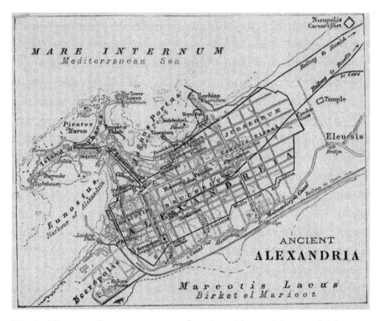

Fig. 3 Nineteenth-century map of ancient Alexandria, overlaid upon a plan of the modern city, showing confirmed and supposed locations within the ancient site.[a]

[a] Image taken from: J. Murray, *A Handbook for Travellers in Lower and Upper Egypt; Including Descriptions of the Course of the Nile through Egypt and Nubia, Alexandria, Cairo, The Pyramids, Thebes, The Suez Canal, The Peninsula of Mount Sinai, the Oases, the Fayoom, Part 1.* Seventh edition (London, John Murray, 1888), p.121.

Despite the array of sights and diversions that the city had to offer, for Eratosthenes, arriving in 245 BC, the location that he would have been most interested in was one of the places that Alexandria was famous for: his new home—the Great Library.

The Great Library of Alexandria has taken on a mantle of legend. Yet despite how widely known this institution was, even in the ancient world, few details of its origins have survived. The earliest reference to the Library is found in the *Letter of Aristeas* from the second century BC. In the *Letter*, it is stated that Demetrius of Phaleron received a vast sum of money from an unnamed king for the purpose of 'collecting together, as far as he possibly could, all of the books in the world'.[134] As Demetrius was closely connected to Ptolemy I, this would place the construction of the library at sometime between 305 BC and 282 BC. MacLeod, who doubts the authenticity of the *Letter of Aristeas*, still suggests that work on the Library began under Ptolemy I.[135]

Additionally, Demetrius had been a student of Aristotle and his successor in the peripatetic school, Theophrastus, and Strabo states that Aristotle had been the first person to collect books, and had taught the Egyptian kings how to arrange a library.[136] Barnes suggests that Demetrius took the idea of a combined school and research institute that had been set up in Aristotle's Lyceum in Athens and applied it to the Museum and Library in Alexandria.[137] Conversely, Epiphanius directly attributes the creation of the library, and the appointment of Demetrius, to Ptolemy II.[138] Nicastro suggests that the construction of the library and *Museion* were well underway by the time of Alexander's death in 323 BC.[139] However, the job of creating the library and *Museion* seems to have been first offered to Theophrastus, who declined the offer, and the job went to Demetrius around 297 BC.[140] This again would suggest a date for the construction of the Library during the reign of Ptolemy I. Erskine suggests that construction of the Library began under Ptolemy I, and was then further developed under Ptolemy II.[141]

Alternatively, Vitruvius states that:

> The Attalid kings [in Asia Minor], stimulated by their great love of philology, established an excellent library at Pergamon, and Ptolemy, motivated by passion and a great desire for the promotion of learning, collected with no less care a similar one, for the same purpose, in Alexandria about the same time.[142]

[134] *Epit. Aris.* 9; see also: Tz. *Prol. Com.* 20.
[135] MacLeod, 'Introduction', p.2.
[136] Str. *Geog.* 9.1.20, 13.1.54; Diog. Laert. 5.36. Demetrius was later arrested by Ptolemy II because he had sided with Eurydice (Ptolemy I's wife) in an issue over the succession (Diog. Laert. 5.78).
[137] R. Barnes, 'Cloistered Bookworms in the Chicken-Coop of the Muses: The Ancient Library of Alexandria' in R. MacLeod (ed.), *The Library of Alexandria: Centre of Learning in the Ancient World* (London, IB Tauris, 2004), p.62.
[138] Epiph. *Mens.* 9: 'After the first Ptolemy, the second who reigned over Alexandria, the Ptolemy called Philadelphus, as has already been said was a lover of the beautiful and a lover of learning. He established a library in the same city of Alexander, in the (part) called the Bruchion; this is a quarter of the city today lying waste. And he put in charge of the library a certain Demetrius, from Phaleron, commanding him to collect the books that were in every part of the world.'
[139] Nicastro, *Circumference*, p.50.
[140] For Ptolemy making overtures to Theophrastus, see: Diog. Laert. 5.2.37.
[141] A. Erskine, 'Culture and Power in Ptolemaic Egypt: The Museum and Library of Alexandria' *G&R* 42.1 (1995), p.38.
[142] Vit. *De Arch.* 7.pref 4.

The Pergamon Library, which contained around 200,000 works in its collection, was established by Eumenes II between 220 BC and 159 BC.[143] Johnstone argues that the entire concept of a library as an institution for research and for the collection of books did not develop until the second century BC, and that the library in Alexandria was just one of many across the Greek world, and later came to be regarded as the greatest.[144] Johnstone's argument is, however, limited through their dismissal of any piece of evidence referring to the library in Alexandria, which is not contemporary with the time to which it is referring. Thus, for example, a papyrus listing most of the Chief Librarians in Alexandria in chronological order is regarded as 'late and apocryphal' due to the date of its composition and not used in their examination.[145] Historical narratives and other sources from later in antiquity that refer to the Library in the past tense are similarly dismissed due to their date of composition.

However, while Johnstone's conclusion would correlate with the statement made by Vitruvius, a Pergamene influence on the construction of the Alexandria Library seems unlikely. If Vitruvius' statement is to be accepted, along with Johnstone's conclusions, and the Alexandrian library was built to rival that of Pergamon, then it was apparently built long after many of its known directors, including Eratosthenes, had held the office of Chief Librarian. This makes a date of construction during the reign of Ptolemy I, with possible completion under the reign of Ptolemy II—which would then account for the later attribution of its construction by Epiphanius—more than likely. This would additionally follow the timeline for the construction of other large public works in Alexandria such as the Pharos lighthouse. It is also more likely that the Pergamon library was built to rival that of Alexandria, and not the other way around (see following).

No physical remains of the Great Library or the Museum in Alexandria have ever been found, and it is uncertain whether the Library was connected to the *Museion* or not. However, the library in Pergamon, upon which the Alexandrian institution may have been based according to Vitruvius (or vice versa), was connected to a Museum.[146] If the Library was connected to the Museum, then it would have been located somewhere near the Royal Palace. This would correlate with the text of Epiphanius, which locates the Library in the district of Bruchion in the centre-north part of the city. Strabo states that the Museum was part of the palace-complex, and if the Library was attached to the

[143] Vit. *De Arch.* 7.pref 4; Str. *Geog.* 13.609.
[144] Johnstone, 'A New History of Libraries', p.349.
[145] Johnstone, 'A New History of Libraries', p.367.
[146] Vit. *De Arch.* 7.pref 4.

Museum, then the Library must have been part of the palace as well. However, considering that Strabo describes the Palace complex as taking up close to a third of the city, even if both the Library and Museum were within the bounds of the Royal precinct, they may still have not been physically connected.

Regardless of where the Library was located, its collection of works was vast. When ships entered the harbour, port officials inspected the vessels in search of books. Any that were found were confiscated at the docks, copied, and the copies were then returned to the owners so that the Library could keep the originals in the so-called 'ships' collection'.[147] Other books were purchased in the markets of Athens and Rhodes.[148] Ptolemy II may have also purchased the works once owned by Aristotle for the Library's collection.[149] Epiphanius describes how Ptolemy II similarly sent messages to 'every king and prince on Earth' asking them to send works of poetry and prose, on oratory and philosophy, medical texts and histories, and works on any other topic, to the Library.[150] Some texts were obtained using quite unscrupulous methods. Galen, for example, reports how, in the time of Ptolemy III, the king sent a request to the government of Athens asking to borrow the original works of famous playwrights such as Aeschylus, Sophocles, and Euripides. As a surety against potential loss, the Athenians required a security-deposit of 15 *talents* (45 kg of silver). When the documents arrived, Ptolemy had them copied, wrote off the funds, sent the copies back to Athens, and kept the originals in the Library.[151]

---

[147] Gal. *Hipp. Epid.* 9.239: 'Some people say that Ptolemy, the king of Egypt, brought them from Pamphylia, and such was his ambition for books, that he ordered that they should confiscate the books of all names, and they should be copied onto new papyrus, and the copied books should be given to their owners, that the books from the ships which had arrived should be brought to him, and they should be stored on the shelves for the books transported on ships, and on them should be put the title "books from the boat"'.[ἔνιοι δὲ καὶ αὐτὸν ἐκ Παμφυλίας κεκομικέναι καὶ φιλότιμον περὶ βιβλία τόν τε βασιλέα τῆς Αἰγύπτου Πτολεμαῖον οὕτω γενέσθαι φασὶν, ὡς καὶ τῶν καταπλεόντων ἁπάντων τὰ βιβλία κελεῦσαι πρὸς αὐτὸν κομίζεσθαι καὶ ταῦτα εἰς καινοὺς χάρτας γράφοντα διδόναι μὲν τὰ γραφέντα τοῖς δεσπόταις, ὧν καταπλευσάνιον ἐκομίσθησαν αἱ βίβλοι πρὸς αὐτὸν, εἰς δὲ τὰς βιβλιοθήκας ἀποτίθεσθαι τὰ κομισθέντα καὶ εἶναι τὰς ἐπιγραφὰς αὐτοῖς τῶν ἐκ πλοίων.]

[148] Ath. *Diep.* 1.3b.

[149] Ath. *Diep.* 1.3b; Barnes ('Cloistered Bookworms', p.66) suggests that the acquisition of the works from Aristotle's Lyceum occurred during the reign of Ptolemy III. On the other hand, Strabo (*Geog.* 13.1.54) and Plutarch (*Sull.* 26) say that the collection ended up in Rome in the first century BC. For a discussion of the fate of Aristotle's library, see: R.G. Tanner, 'Aristotle's Works: The Possible Origins of the Alexandria Collection' in R. MacLeod (ed.), *The Library of Alexandria: Centre of Learning in the Ancient World* (London, IB Tauris, 2004), pp.79–91.

[150] Epiph. *Mens.* 9: 'And he wrote letters and made request of every king and prince on earth to take the trouble to send those that were in his kingdom or principality I mean, those by poets and prose writers and orators and philosophers and physicians and professors of medicine and historians and books by any others.'

[151] Gal. *Hipp. Epid.* 9.239: 'How far the famous Ptolemy is interested in purchasing old books is no small thing compared to what they say he did to the Athenians. He received the works of Sophocles, Euripides and Aeschylus to simply copy them and to send them back, intact, immediately afterwards,

The *Letter of Aristeas* also outlines how the Library acquired works of Jewish scripture and then had them translated from Hebrew into Greek so that 'these books may be added to other royal books in [the] Library'.[152] This event was celebrated annually at a festival held in Alexandria by its Jewish inhabitants.[153] Similarly, Syncellus remarks how Ptolemy II collected works from the Greeks, Chaldeans, Egyptians, and Romans, and had all of the foreign-language texts translated into Greek.[154] This would suggest that, in keeping with the Library's goal of obtaining all the books in the world, the literature of cultures other than just the Greeks was also procured by the Library, and that a team of multi-lingual scribes were on staff to translate any work that came into the Library's possession.

When the Libraries of both Alexandria and Pergamon were at their height, there was stiff competition between the two to acquire texts, and manuscripts seem to have been thoroughly examined and critiqued. This led to instances where the officials from one institution openly questioned the authenticity of a text in their collection.[155] Argument also ensued over which institution rightfully maintained the peripatetic legacy of Aristotle's Lyceum, and even the very nature of the research being undertaken at both libraries. In the second century BC, for example, Crates of Mallus, working at Pergamon, described the Pergamene scholars as *kritikoi* for their critical mastery of logic, while

---

against surety of fifteen talents (approximately 390 kg) of silver. However, once he had expertly copied them on the best papyrus, he kept those which he had received from the Athenians, and he sent to them those that he had copied, begging them to keep the fifteen talents as they had received new books instead of the old ones that they had delivered. And so, as he had sent to the Athenians the new books, but had retained the old, there was nothing they could do, as they had received the money on the condition that they would keep the money if he kept books, and so they received the new books and kept the money.' [ὅτι δ' οὕτως ἐσπούδαζε περὶ τῶν παλαιῶν βιβλίων κτῆσιν ὁ Πτολεμαῖος ἐκεῖνος οἱ μικρὸν εἶναι μαρτύριόν φασιν ὃ πρὸς Ἀθηναίους ἔπραξε. δοὺς γὰρ αὐτοῖς ἐνέχυρα πεντεκαίδεκα τάλαντα ἀργυρίου καὶ λαβὼν τὰ Σοφοκλέους καὶ Εὐριπίδου καὶ Αἰσχύλου βιβλία χάριν τοῦ γράψαι μόνον ἐξ αὐτῶν, εἶτ' εὐθέως ἀποδοῦναι σῶα, κατασκευάσας πολυτελῶς ἐν χάρταις καλλίστοις, ἃ μὲν ἔλαβε παρὰ Ἀθηναίων κατέσχεν, ἃ δ' αὐτὸς κατεσκεύασεν ἔπεμψεν αὐτοῖς παρακαλῶν ἔχειν τε τὰ πεντεκαίδεκα τάλαντα καὶ λαβεῖν ἀνθ' ὧν ἔδοσαν βιβλίων παλαιῶν τὰ καινά. τοῖς μὲν οὖν Ἀθηναίοις, εἰ καὶ μὴ καινὰς ἐπεμόφει βίβλους, ἀλλὰ κατεσχήκει τὰς παλαιάς, οὐδὲν ἦν ἄλλο ποιεῖν, εἰληφόσι γε τὸ ἀργύριον ἐπὶ συνθήκαις τοιαύταις, ὡς αὐτοὺς κατασχεῖν, εἰ κἀκεῖνος κατάσχοι τὰ βιβλία, καὶ διὰ τοῦτ' ἔλαβόν τε τὰ καινὰ καὶ κατέσχον καὶ τὸ ἀργύριον.]

[152] *Epit. Aris.* 39; see also: Epiph. *Mens.* 9.
[153] Philo *in Mos.* 2.41–42.
[154] Sync. *Chron.* 1.516.3–10.
[155] For example, Didymos (*in D.* 11), who worked in Alexandria during the first century AD, outlines a debate over the authenticity of a *Philippic* from the Athenian orator Demosthenes—with the librarian in Alexandria showing that it was on the Library's shelves, but as part of the seventh book of a work attributed to Anaximenes of Lampsacus. Similarly, Didymos, a prolific writer, objected to a story found in someone's work as being absurd. To his embarrassment, a book of his own that contained the same tale was produced—presumably from the Library's collection (Quint. *Inst.* 1.8.20). Additionally, Galen (*Hipp. Epid.* 9.254–255) describes a debate that took place over the authenticity of a copy of Book III of the *Epidemics* by Hippocrates based upon notations found in the margins of one edition which were not found in another.

those at Alexandria were referred to as *grammatikoi* who only focused on vocabulary and patterns of rhythm.[156] At another time, Egypt halted the export of papyrus to Pergamon, believing that if the actual material for the copying of manuscripts were hard to come by, the collection in the Pergamon library would not be able to grow. To get around this trade embargo, the people of Pergamon invented parchment.[157] Nothing was allowed to jeopardize an institution's position—especially the possible defection of a chief librarian. When Aristophanes of Byzantium—the man who succeeded Eratosthenes as the fourth head of the Library in Alexandria—attempted to move to the court of Eumenes in Asia Minor, and most likely to a position at the rival Library in Pergamon, the ruling Ptolemy had him imprisoned.[158]

Another effect of this race to acquire books was that the vendors of texts often merged works together so that a larger work may command the payment of a higher price from one of the libraries.[159] Other works that flooded into the market were simply forgeries.[160] In Pergamon, local residences were searched for books that could be used to expand the Library's collection, and it can be assumed that a similar practice occurred in Alexandria as well.[161] Through such methods, the Library obtained a great quantity of works on a variety of topics. Strabo refers to how, when he visited the Library in the first century BC, he compared two different works on the same subject to try to determine who had plagiarized whom.[162] This indicates that the Library's collection contained multiple texts on the same subject matter. It is believed that this vast collection was sorted and stored by discipline and by author. Callimachus, a contemporary and sometime rival of Alexandria's second Chief Librarian, Apollonius of Rhodes, is said to have composed the *Pinakes*—a lost bibliographical work which catalogued the Library's vast collection across 120 volumes.[163]

---

[156] Sext. Emp. *Math.* 1.79, 1.248; see also: Dio Chrys. *Or.* 53.1.

[157] Pl. *HN* 13.70: 'due to the rivalry about libraries between king Ptolemy and king Eumenes, Ptolemy stopped the export of papyrus . . . and so the Pergamenes invented parchment.'

[158] *Suda s.v.* Ἀριστώνυμος.

[159] Gal. *Hipp. De Nat. Hom.* 5.18: 'At the time when the kings Attalus and Ptolemy vied with each other to buy books, those who obtained money from the kings started to forge titles and merge books of the writings of the most celebrated authors offered by them. Rather than two books, one on the nature of man, and the other on the condition of the healthy food, which were small, thinking someone would value them based on their size, they merged the two into one.' [ἐν γὰρ τῷ κατὰ τοὺς Ἀτταλικούς τε καὶ Πτολεμαϊοὺς βασιλέας χρόνῳ ττρὸς ἀλλήλους ἀντιφιλοτιμουμένους περὶ κτήσεως βιβλίων ἤ περὶ τὰς ἐπιγραφάς τε καὶ διασκευὰς αὐτῶν ἤρξατο γίγνεσθαι ῥᾳδιουργία τοῖς ἔνεκα τοῦ λαβεῖν ἀργύριον ἀναφέρουσιν ὡς τοὺς βασιλεῖς ἀνδρῶν ἐνδόξῳ συγγράμματα.]

[160] Gal. *Hipp. De Nat. Hom.*, 5.9.

[161] Str. *Geog.* 13.1.54.

[162] Str. *Geog.* 17.1.5.

[163] The *Suda* (*s.v.* Καλλίμαχος) calls this work 'Tables of Men Distinguished in Every Branch of Learning, and their Works, in 120 Books' (Πίνακες τῶν ἐν πάσῃ παιδείᾳ διαλαμψάντων, καὶ ὧν συνέγραψαν, ἐν βιβλίοις κ καὶ ρ); see also: Tz. *Prol. Com.* 20.

The exact number of works kept in the Library is far from certain, and the ancient source material is no real help on this matter. The *Letter of Aristeas* cites 200,000 works as the size of the collection.[164] Aulus Gellius, who accompanied Strabo to Alexandria and later became Roman Prefect of Egypt, states that 'nearly 700000 volumes were either acquired, or written in Egypt, under the kings known as the Ptolemies'.[165] The problem with attempting to determine the validity of such claims is that, at some time in the Library's history, an annex, or 'daughter library', was set up in the *Serapeum*, which had its own collection.[166] In effect, there were two libraries in Alexandria— the Great Library and the Annex. Tzetzes writes how Ptolemy, with the help of Demetrius, used royal funds to buy books from all over the world and gathered the collection into two libraries with the *Serapeum* Annex containing 42,800 works, while the Great Library contained 532,800.[167] Ammianus, who seems to be conflating the collections of the Library and the Annex, or paraphrasing Gellius, states that 700,000 books had been brought together by the unwavering energy of the Ptolemaic kings.[168] Epiphanius states that the whole collection was 54,800 works, which may be either a 10 per cent corruption of the size of the Great Library's collection, or another reference to the number of works in the Annex.[169] Bagnall questions the size of the Library's collection by pointing out that only around 450 Greek authors are known to have lived in the fourth century BC, and only about 175 are known from the third century BC. Bagnall argues that, even if each of these writers produced fifty works each, this would still come up to only 31,250 works.[170] However, there are a number of issues with Bagnall's conclusions. Firstly, it fails to consider that the rival library in Pergamon is reported to have contained 200,000 scrolls. Secondly, it fails to consider passages from sources like the *Letter of Aristeas*, which describe how works from cultures other than Greece were translated for the Library's collection. Barnes, on the other hand, makes an unjustified claim that the number of translated works in the Library's collection was small—suggesting that the total number of titles (i.e. works across multiple scrolls) in the Library was somewhere between 70,000 and 100,000.[171]

---

[164] *Epit. Aris.* 10; this figure is also cited by Josephus (*AJ* 12.2.1) who paraphrases the *Letter of Aristeas*.

[165] Gell. *NA* 7.17.3.

[166] Epiph. *Mens.* 11.

[167] Tz. *Prol. Com.* 20.

[168] Amm. 22.16.12–13.

[169] Epiph. *Mens.* 9.

[170] R.S. Bagnall, 'Alexandria: Library of Dreams' *Proceedings of the American Philosophical Society* 146 (2002), pp.348–362.

[171] Barnes, 'Cloistered Bookworms', pp.65, 67; for a discussion of the size of the Library's collection, see: Barnes, 'Cloistered Bookworms', pp.64–67.

Thirdly, Bagnall's conclusion fails to consider that a single work, by a single author, may comprise multiple volumes—such as Callimachus' massive work on just cataloguing the collection.

The Roman writer Seneca describes how 'Alexandria contains countless books and libraries whose titles their owner can hardly read through in a lifetime'.[172] Seneca uses the plural term 'libraries' (*bybliothecas*) which is most likely a reference to the two main repositories. MacLeod suggests that, from the time of Ptolemy III, there were also numerous small libraries scattered throughout Alexandria along with the Great Library and the *Serapeum* Annex.[173] Furthermore, Galen outlines how books that had recently arrived in Alexandria were not immediately placed in the collection of one of the libraries, but were kept in warehouses.[174] This was presumably done so that the new arrivals could be checked to see if they were works that the libraries already had, to check the validity and condition of the manuscripts, and to determine whether they needed to be translated or could be directly placed into one of the collections. Not only does the requirement of a warehouse indicate the sheer number of books that were coming into Alexandria as part of the procurement policy of the kings (contra to the conclusion of Bagnall), but it also means that there were, in fact, at least three large repositories of books in Alexandria.

This adds to the problems of attempting to determine the size of the Great Library's collection (and even that may be a matter of semantics when it is really 'Alexandria's collection'). Other sources outline how, when Julius Caesar invaded Alexandria in 48 BC, he attempted to destroy the naval power of his rivals by setting fire to their ships which were moored in the harbour. Several ancient sources state how this fire spread to the surrounding dockyards and also destroyed the Library. Cassius Dio, for example, states that the fire burned down the docks and the warehouses 'and also the library whose books, it is said, were the greatest in number and excellence'.[175] Plutarch, in his biography of Caesar, wrote that 'the fire spread from the dockyards and destroyed the Great Library'.[176] Conversely, there are also numerous other sources of the event which do not mention any damage being sustained by the Library.[177] Seneca, on the other hand, provides an exact figure for the number of lost

---

[172] Sen. *Tranq.* 9.5: *Quo innumerabiles libros et bybliothecas, quarum dominus vix tota vita indices perlegit.*

[173] MacLeod, 'Introduction', p.5.

[174] Gal. *Hipp. Epid.* 9.239.

[175] Cass. Dio 42.38.2.

[176] Plut. *Caes.* 49.

[177] For example, see: Luc. 10.486; Flor. *Epit.* 2.13.39; Caes. *BC* 3.111–112; App. *B Civ.* 2.90.

works when he states that '40,000 books were burned in Alexandria'.[178] The figure given by Seneca is interesting in that it is far too low to be the collection housed in the Great Library, and yet the Annex, which did contain around 40,000 works, was too far from the shore to be engulfed by a fire in the docks. It is possible that what burned in the fire was one of the storage warehouses belonging to the Great Library and not the Great Library itself.

Regardless of any damage that the Library, or its collection, may have suffered in later time-periods, when Eratosthenes arrived in Alexandria in 245 BC, he would have taken up a position at one of the leading research institutes in the world containing a vast array of works on all manner of topics. But why was such an institution established in Alexandria to begin with, and why did Alexandria become the main cultural and intellectual hub of the time? The answer to such questions is that both Eratosthenes and the Library of Alexandria were built on a tradition of Greek scientific enquiry, particularly in relation to understanding the nature of the universe and humankind's place in it, that stretched back for centuries.

1.4 Greek Astronomical Thought Before The Time Of Eratosthenes

Eratosthenes was not the first, nor the last, person in the ancient Greek world to attempt to understand the nature of the Earth and its place is a wider universe. Rather, the attempts to understand the mechanics and meanings behind the motions of the sky had a tradition that went back to some of the earliest extant Greek literature. From these beginnings, Greek religious, philosophical, and scientific thought progressed through a number of evolutionary steps towards the time of Eratosthenes in the third century BC. It was the results of these steps, and the socio-political events of the Greek world across these centuries, that created the environment where not only places like Alexandria and its Great Library would intellectually thrive, but also created a setting where the promotion of culture through scientific investigation would be seen as the pinnacle of accomplishment. It was in this atmosphere of investigative and observational enquiry that people like Eratosthenes and his predecessors formulated some of the first known attempts to understand the workings of the Heavens.

The ancient Greeks, like so many other people from so many other cultures, always sought to understand the meanings behind the movements of

---

[178] Sen. *Tranq.* 9.5.

the lights in the night sky. Whether this was from a religious perspective—assigning the position of deities to some of the celestial bodies—or from an agricultural perspective—understanding how the rising and setting of certain stars or constellations heralded a change of the seasons—is of no great concern for understanding the origins of ancient scientific ideas. Rather, what is important, is understanding that these uses of astronomical observations, regardless of any distinct lack of an attempt to determine a model for understanding why the motions of the stars occurred as they did in the first place, set the foundation for all of the more scientific investigations that were to follow.

For the ancient Greeks, some of their earliest literature contained attempts at understanding where the universe had come from, our place within that universe, and how what was seen in the sky could be understood. Some of the first of these texts were written by Hesiod (*ca.* 750–650 BC). Hesiod was born in Cyme in Asia Minor (modern Nimrut Limani, Turkey) but later settled in the city of Thespiae in the Greek city-state of Boeotia.[179] His first work, the epic poem *Theogony*, tells the story of the creation of the universe. Early in the work, in a description of the birth of the universe, Hesiod states that at first there was Chaos, but next came the Earth, the Night, and the Sky.[180] Chaos was the primordial void from which all creation sprung. Then under the influence of Eros ('attraction'), the Earth and the other celestial bodies were formed. Such notions seem to have been influenced by earlier, Near Eastern, compositions such as the Hittite *Song of Kumarbi* (*ca.* 1350 BC)—which may have still been known in Turkey (the location of the former Hittite empire and where Hesiod was from) in Hesiod's time—and the *Enuma Elis* from Babylon (pre-seventh century BC), both of which contain similar creation mythology.[181] Russo suggests that one of the important factors in the rise of Greek scientific thought 'was the new relationship established between Greek civilization and the ancient Egyptian and Mesopotamian civilizations' as 'the Greeks of the Classical Age we still behind the Egyptians and Mesopotamians from a technological perspective'.[182] Alternatively, Cooper outlines how many of these earlier, Near East works were astrological in nature, and were more concerned with connections and interactions between the heavenly and the earthly. However, demand for better predictions led to a shift away from purely mythological interpretations of the sky, to the construction of models that

---

[179] Hes. *Op.* 640.
[180] Hes. *Theog.* 116–138.
[181] On the *Song of Kumarbi*, see: H.G. Güterbock, 'The Hittite Version of the Hurrian Kumarbi Myth: Oriental Forerunners of Hesiod' *AJA* 52.1 (1948), pp.123–134; for the *Enuma Elis*, see: W.G. Lambert and S.B. Parker, *Enûma Eliš. The Babylonian Epic of Creation* (Oxford, Oxford University Press, 1966).
[182] Russo, *Forgotten Revolution*, p.28.

could predict movements mathematically based upon centuries of recorded observational data.[183] Much of this information would be drawn upon by later, scientific, thinkers in ancient Greece. While Hesiod's *Theogony* is not a scientific text in any way, shape, or form, it is one of the first ancient Greek attempts to describe how the universe was created.[184]

Hesiod's other major work, another verse composition of 800 lines titled *Works and Days*, describes the different stages in the development of civilization, from creation to Hesiod's present, with an underlying emphasis on agrarian toil. Within the work are found several passages which refer to such things as using the rising of certain stars to ascertain the best times for harvests, and how the phases of the moon, or the rising of the star Sirius, could be seen are harbingers of either good or ill.[185] Again, while this is not a scientific treatise per se, Irby-Massie and Keyser see this work as having the potential to be used as a kind of 'Farmer's Almanac' to aid them with their work and to help them understand, and identify, the changing of the seasons.[186] Certainly the principle of how certain stars become visible only at certain times of the year had to have been understood for such passages to have been included in the text. Such ideals held sway for centuries to come—although, by the first century AD, writers such as Geminus advised that what Hesiod said about the effects of stars and other celestial bodies should not be taken literally.[187] Despite what criticisms, such as that of Geminus, suggest, what can be taken from such a critique is that Hesiod's work was still being consulted nearly seven centuries after it was written.

Another pair of works, written around the same time as those of Hesiod, were the epic poems of Homer. While the tales contained in these epics were much older, the stories were not written down into a 'standard' format until the late seventh century BC.[188] Similar to the works of Hesiod, the two works

---

[183] G.M. Cooper, 'Astrology: The Science of Signs in the Heavens' in P.T. Keyser and J. Scarborough (eds.), *Oxford Handbook of Science and Medicine in the Classical World* (Oxford, Oxford University Press, 2018), p.386; for ancient comments on the Near Eastern origins of astrology, see: Plin. *HN* 7.193; Vit. *De Arch.* 9.6.2; Cic. *Div.* 1.81; Hdt. 2.82; Diod. Sic. 1.81.

[184] It is also interesting to note how closely these texts parallel many aspects of the current model of the Big Bang where the universe was a void and then, in a gigantic eruption of matter and energy, everything in the universe was instantaneously brought forth which then coalesced under the influence of gravity (i.e. 'attraction').

[185] For example, see: Hes. *Op.* 571–572, 598, 615, 618–619, 621, 770–828.

[186] G.L. Irby-Massie and P.T. Keyser, *Greek Science in the Hellenistic Era* (London, Routledge, 2002), p.82.

[187] Gem. *Isagoge* 17.13–14.

[188] Cic. *De Or.* 3.34.137; Ael. *VH* 13.14; Paus. 7.26.13; for a discussion of whether the Athenians of the seventh century created the standard versions of Homer, see: S.H. Newhall, 'Peisistratus and his Edition of Homer' *PNAS* 43.19 (1908), pp.491–510; J.A. Davidson, 'Peisistratus and Homer' *TAPA* 86 (1955), pp.1–21.

attributed to Homer—the *Iliad* and the *Odyssey*—were not scientific works. However, within the lines of verse are found some of the first descriptions of how the ancient Greeks viewed the shape and structure of the Earth, a variant on the creation myth, and numerous references to celestial objects and events. For Homer, the inhabited world was surrounded by a great river, Ocean, which was the 'beginning of everything', and from which all of Creation, even the gods, had been born.[189] The Earth itself seems to be a flat disk, although this is not directly stated and is only implied by the surrounding waters of Ocean. Below the Earth were the dark realms of Tartarus and Hades.[190] Above the Earth sat a layer of air, then a layer of 'higher air', or *aether*, and then came the hemispherical vault of the Heavens, fashioned like an upturned bowl made of bronze or iron, which was highly decorated and covered with stars.[191] The dome was said to cover the Earth and was held aloft by the Titan, Atlas.[192]

Within the decoration of the dome, Homer refers to several constellations, stars, planets and stellar clusters such as Orion, Ursa Major, The Pleiades, The Hyades and Sirius.[193] Similar to the writing of Hesiod, for Homer, the rising of Sirius heralded the coming of a period of extremely hot weather in July and August (the so-called 'Dog Days').[194] The *Odyssey* also contains a passage that may be the first recorded account of a total solar eclipse.[195] In regards to planetary references, Venus, in its guises of the morning-star and evening-star, were seen to be two separate objects by Homer.[196]

Despite some of the obvious flaws (from a modern understanding) that can be found in Homer's works, there is one important conclusion that can be made. That Homer described the sky as an upturned bowl, as opposed to another flat layer above a flat earth, shows that the Greeks had already considered, by the seventh century BC, if not earlier, that the stars in the sky followed a circular path across the Heavens. As Russo points out, as the stars seem to move all together, what this suggested to the ancient Greeks was that what was moving, making a full turn each day, was the whole sky which seemed to be

[189] Hom. *Il.* 14.200, 14.245–246.
[190] For Tartarus, see: Hom. *Il.* 8.13–16; for Hades, see: Hom. *Il.* 8.480.
[191] For the air and aether, see: Hom. *Il.* 8.554–559, 14.288; for the vault of the Heavens, see: Hom. *Il.* 2.458, 5.504, 16.364, 17.424–425, 19.351; Hom. *Od.* 1.53–54, 3.2, 9.534, 11.17, 15.329, 17.565; for the stars decorating the dome, see: Hom. *Il.* 6.108; Hom. *Od.* 9.535.
[192] Hom. *Od.* 1.53–54; Hesiod (*Theog.* 517) states that the god Zeus had given the task of holding up the Heavens to Atlas.
[193] Hom. *Il.* 18.478–488, 22.25–31; Hom. *Od.* 5.272–277; for a discussion of the astronomical events in the *Iliad* and *Odyssey*, see: E. Theodossiou, V.N. Manimanis, P. Mantarakis, and M.S. Dimitrijrvic, 'Astronomy and Constellations in the *Iliad* and *Odyssey*' *J. Astron. Hist. Herit.* 14 (2011), pp.22–30.
[194] Hom. *Il.* 22.25–31.
[195] Hom. *Od.* 20.356–357.
[196] Hom. *Il.* 22.317, 23.226; Hom. *Od.* 13.98–100.

the limit of the universe.[197] Thus, the works of both Homer and Hesiod, as well as being epic tales based on myth, legend and folklore, are also the first extant texts which demonstrate that the ancient Greeks had already begun to look at the sky and try and work out why the objects in the Heavens behaved in the way that they did. This was an aspect of these works that was not lost on later writers of antiquity. For example, several centuries later, Plato would criticize Hesiod by stating that he was more concerned with detailing the risings and settings of the objects in the sky, rather than actually understanding the revolutions of the heavenly bodies.[198] Later still, how the works of Hesiod and Homer were viewed seems to have undergone a paradigm shift. Aristotle believed that the question of initial causes in the workings of nature began with Hesiod's account of the creation of the universe that is found in *Theogony*.[199] This shows that a major consequence of the astronomical passages found in the four epic poems by Hesiod and Homer is that, as how they were viewed changed, they set the foundation for all of the scientific enquiry which was to follow.

The Greeks had to wait about a generation after the time of Hesiod and Homer before the first real scientific work on astronomy was released (or, at least, that is the impression that we get based upon the surviving literary sources). This would come in the form of the work of Thales of Miletus (624–524 BC). Again, it is from the region of Asia Minor, the same region as Hesiod, from which these works come. Nicastro suggests that, like Hesiod, Thales may have had access to eastern astronomical ideas and data via contact with the Persian Empire (which finally annexed the Greek states of Asia Minor in 546 BC).[200] According to the ancient sources, Thales wrote about astronomical phenomena in hexameter verse, understood the nature of eclipses and equinoxes, predicted a solar eclipse that was seen in Asia Minor in 585 BC, calculated the size of the Sun and Moon, calculated the height of an Egyptian pyramid, wrote a text called *On the Equinox* and many others.[201] Thales also apparently 'discovered' the constellation of Ursa Minor, considered the Moon to be Earth-like and illuminated by the light of the Sun, and divided the year into 365 days.[202] Any one of these would constitute a quantum leap in scientific

[197] Russo, *Forgotten Revolution*, p.860.
[198] Pl. *Epin.* 990a.
[199] Arist. *Metaph.* 983b–987a.
[200] Nicastro, *Circumference*, p.31.
[201] *Suda s.v.* Θαλῆς; see also: Hdt. 1.74; Plin. *HN* 100.12.53; Diog. Laert. 1.23–24; Cic. *Rep.* 1.16.25; for a discussion of the date of the solar eclipse predicted by Thales, see: Panchenko, 'Thales and the Origin', pp.11–16. Shipley (*Greek World*, p.350) doubts that anyone, including Thales, could have accurately predicted an eclipse until Meton of Athens had worked out the nineteen-year cycle for the motions of the Sun and Moon in 432 BC (see: *Suda s.v.* Μέτων).
[202] Diog. Laert. 1.23–24, 1.27; Plut. *Plac. Philos.* 2.24–28; Stob. *Ecl.* 1.1.26, 1.26.2.

thinking beyond the attribution of earthly events to the rising and setting of stars that are found in Hesiod and Homer. With Thales, the Greek literary tradition began to move into the realms of a more critical and analytical investigation of the Heavens in order to understand the mechanics behind them to a greater extent.

However, this is not to say that some of the ideals outlined in the earlier works no longer held sway. In Thales, for example, the Earth is still considered to be a flat disk surrounded by water.[203] This echoes the descriptions of Ocean found in Homer. Many other aspects of Thales' writing also seem to border on the more religious and astrological, rather than the scientific and astronomical. This may be because, as Irby-Massie and Keyser suggest, that before the advent of the Hellenistic Age *ca.* 400 BC, it was very difficult, if not impossible, for the ancient Greeks to distinguish between such things as science, philosophy, magic, astrology, and religion.[204] Bowen similarly suggests that science in ancient Greece served as a model for philosophy due to the way it focused on explaining what knowledge was, how such knowledge was obtained, and how knowledge was communicated.[205] However, where Thales expanded on earlier works by the likes of Hesiod and Homer is by stating that the flat Earth was also suspended in the universe by resting on the water of the great Ocean as well, and that it was from this water that all life sprung and eventually returned.[206] Thus, while Thales' work does still contain many elements that have carried forward from the earlier texts, his writings expand on those elements to try to understand the 'bigger picture' of the structure of the universe.

The principles behind a more critical and analytical examination of the universe, and the Earth's place within it, were taken further by Thales' student Anaximander of Miletus (610–546 BC). Anaximander is said to have authored a number of works that examined the mechanics of the natural world such as: *On Nature, On the Circuit of the Earth, On the Fixed Bodies,* and *The Globe.*[207] He is also credited with introducing a type of sundial to the Greeks that was designed for the purpose of determining solstices, times, seasons, and

---

[203] Arist. *Metaph.* 983b20–22.

[204] Irby-Massie and Keyser, *Greek Science*, p.1.

[205] A.C. Bowen, 'The Exact Sciences in Hellenistic Times: Texts and Issues' in D. Furley (ed.), *Routledge History of Philosophy Vol. II: From Aristotle to Augustine* (London, Routledge, 1999), p.287; for a discussion of the connections between the work of Thales and the origins of Greek scientific thought, see: D. Panchenko, 'Thales and the Origin of Theoretical Reasoning' *Configurations* 1.3 (1993), pp.387–414.

[206] Arist. *Metaph.* 983b20–22; Diog. Laert. 1.27; similar to the parallels between Hesiod's creation mythology and the modern Big Bang theory, it is interesting to note the parallels between Thales' suggestion that all life came from the water, and modern evolutionary theory which suggests that life began in the nutrient-rich waters of the primordial Earth.

[207] *Suda s.v.* Ἀναξίμανδρος.

equinoxes.[208] Anaximander is said to have discovered an equinox and solstices (which suggests that they were not fully discovered by his mentor, Thales) and came up with the segregation of the day into hours.[209] Such things may have been determined using the sundial that he designed. However, Herodotus states that the Greeks got the sundial and the twelve hours of the day from the Babylonians.[210] This does not discount Anaximander from using Near Eastern data, as per Hesiod and Thales, and either refining the Babylonian sundial or designing an entirely new type of instrument based upon his needs, wants, and any flaws he may have seen in its earlier form. Regardless of where Anaximander may, or may not, have obtained the ideas for his sundial from, if this attribution is correct, this would be the first instance of the creation of an instrument for the purpose of undertaking observations of astronomical phenomena in the ancient Greek world.

Anaximander also used his observations to further the understanding of the world and the wider universe. He is said, for example, to have been the first person to make a drawing of the inhabited part of the world—most likely in his work *The Globe*.[211] For Anaximander, the Earth was at rest in the centre of the universe due to 'indifference'—meaning that it could not move in two opposite directions at the same time—and was shaped like a flat cylinder with a diameter three times that of its height.[212] Hahn suggests that the shape of Anaximander's world was a by-product of him being part of the generation that was witness to the construction of the first large-scale stone temples—with cylindrical drums as part of their columns—in Greek Asia Minor.[213] Regardless of the influence, the inhabited part of the world was located on one end of this cylindrical Earth and, unlike Thales who had suggested that the Earth floated in water, Anaximander suggested that it was supported by nothing at all.[214] There was still water surrounding the land on Anaximander's cylindrical Earth, and he took the evolutionary concepts of Thales even further and

---

[208] *Suda s.v.* Ἀναξίμανδρος, Γνώμων, Ἡλιοτρόπιον; Eus. *PE* 10.14.11.

[209] *Suda s.v.* Ἀναξίμανδρος.

[210] Hdt. 2.109.

[211] Diog. Laert. 2.2.

[212] *Suda s.v.* Ἀναξίμανδρος; Hippol. *Refut.* 1.6; Plut. *Stromat.* Fr.2; Arist. *Cael.* 295a13; for a discussion of the concept of 'centricity' and the possible reasons for why the early Greeks believed that the Earth was at the centre of the universe, see: M. Munn, 'From Science to Sophistry: The Path of the Sun, the Shape of the World, and the Place of Athens in the Cosmos' in A. Pierris (ed.), *Physis and Nomos: Power, Justice, and the Agonistical Ideal of Life in High Classicum—Proceedings of the Symposium Philosophiae Antiquae Quartum Atheniense*, 4–12 July 2004 (Patras, David Brown Book Co., 2007), pp.111–133.

[213] R. Hahn, *Anaximander and the Architects: The Contributions of Egyptian and Greek Architectural Technologies to the Origins of Greek Philosophy* (Albany, State University of New York Press, 2001), pp.200–210.

[214] Hippol. *Refut.* 1.6; Plut. *Stromat.* Fr.2.

suggested that all living things had come from the water, and that man had evolved from fish.[215] The base substance behind the creation of everything was the 'Infinite'—a substance that could not be determined.[216]

Anaximander also attempted to determine the sizes, motions, and characteristics of the objects in the sky. For him, the stars were circles of fire.[217] He suggested that the Sun was twenty-seven times larger than the Earth, and the Moon is nineteen times larger, and that the phases of the moon were caused by our view off it being slowly blocked—like looking down a pipe as the end of it is slowly closed.[218] While many of these suggestions are clearly incorrect—the diameter of the Sun, for example, is actually about 109 times that of the Earth—the important thing about the work of Anaximander is that it continued to expand, rather than just recount, the work that had come before. Not only did he disagree with the teachings of his mentor over the shape and nature of the world, but through the use of instrumentation, he was able to begin to undertake observations that refined the understanding of the workings of the Heavens.

The torch of astronomical investigation was then taken up by Anaximander's own pupil, Anaximenes of Miletus (586–526 BC).[219] For Anaximenes, the base substance of the universe was also infinite. However, unlike Anaximander, Anaximenes stated that this infinite substance was determinable and that it was air.[220] It was this air that held the Earth aloft in the universe.[221] Anaximenes held that the Earth was flat—like a table—and had the other heavenly bodies revolving around it.[222] The stars, Sun, and Moon were all said to be composed of fire.[223] In a further refinement of the critical analysis of the sky, the work of Anaximenes additionally contains the first attempt to understand, and explain, the retrograde motion of the planets. For Anaximenes the planets occasionally reversed their course due to the resistance of the compressed air that he saw as the foundational material of the universe.[224] While this, and much else of Anaximenes' work, is incorrect, the examination of retrograde motion demonstrates a further refinement in the processes that were beginning to be undertaken to understand the workings of the sky.

---

[215] Hippol. *Refut.* 1.6; Aet. 5.19, Plut. *Sept. Sap. Conv.* 8.8.4; Plut. *Stromat.* Fr.2; this is even closer to modern evolutionary theory than the earlier work of Thales.
[216] Plut. *Stromat.* Fr.2.
[217] Hippol. *Refut.* 1.4–6; Aet. 2.13.
[218] Hippol. *Refut.* 1.5–6.
[219] *Suda s.v.* Ἀναξιμένης.
[220] Hippol. *Refut.* 1.7.
[221] Plut. *Stromat.* Fr.3.
[222] Hippol. *Refut.* 1.7; Aet. 3.10.
[223] Hippol. *Refut.* 1.7; Aet. 2.20, 2.22, 2.25.
[224] Aet. 2.23.

Pythagoras of Samos (570–495 BC) may have also been a pupil of Thales. According to the *Suda*, Pythagoras went to Egypt to study—which would then correlate with many of his predecessors drawing on Near Eastern information.[225] He later settled in Croton in southern Italy where his school of philosophical enquiry thrived.[226] Pythagoras posited that the universe could only be understood through the mathematics of whole numbers. For him the beginning of all reality is the so-called *Monad* which begat the Infinite *Dyad*. The *Dyad*, in turn, begat numbers, which led to mathematical signs, then to planes, solids, and finally animate bodies.[227] The animated world was composed of four primal elements—earth, air, fire and water—which, through numerical perfection, created everything.[228] Diogenes Laertius states that Pythagoras was the first person to suggest that the world was spherical— although according to Theophrastus this was Parmenides (another pupil of Thales), while the Stoic philosopher Zeno said that it was Hesiod.[229] On this spherical Earth, Pythagoras is said to have hypothesized about the existence of the Antipodes—an inhabited land on the opposite side of the planet where the people were upside down (from Pythagoras' perspective).[230] Theon of Smyrna states that Pythagoras was the first person to notice retrograde motion in the planets—although this would conflict with the attribution of this observation to Anaximenes unless it is assumed that both Pythagoras and Anaximenes were studying under Anaximander at the same time and that they worked on this idea together.[231] The five observable planets (Mercury, Venus, Mars, Jupiter, and Saturn) were given the same status as the Sun and Moon, and all of them were said to orbit around the centre at proportionally scaled distances.[232] For Pythagoras, the Earth itself either occupied this central position, or was itself orbiting a great central fire.[233] Regardless, and in contrast to earlier models, the Moon was said not to be made of fire (as per Anaximenes), but was correctly described as receiving its light from the Sun. What does not seem to have occurred, however, was following this principle to its furthest conclusion in regard to the light from the planets, or working out how the Sun could shine on the Moon if the Earth occupied the central position in the universe. Despite this seeming lack of follow-through on the part of some of the conclusions that were made, Pythagoras, and the school of thought which carried on

---

[225] *Suda s.v.* Αἴγυπτος.
[226] Cic. *Rep.* 2.25.
[227] *Suda s.v.* Πυθαγόρας.
[228] *Suda s.v.* Πυθαγόρας.
[229] Diog. Laert. 8.48, 9.21.
[230] *Suda s.v.* Πυθαγόρας.
[231] Theon. *Expos.* 12–18.
[232] Stob. *Ecl.* 1.22.1.
[233] Stob. *Ecl.* 1.22.1.

his tradition, along with its mathematical mysticism, was the first to postulate a non-geocentric model of the solar system.

The works of Pythagoras, and that of Thales, were referred to in the writings of Xenophanes of Colophon (570–478 BC).[234] Despite growing up in the same region of western Turkey where so much intellectual thought had flourished before, Xenophanes fled his homeland, either as a fugitive or as an exile, and settled in the cities of Zancle and Catana in Sicily, which had been established as colonies by the city-states of Asia Minor.[235] It is most likely that it was here that he founded his own school of philosophical enquiry.[236] Xenophanes argued that the underlying substance of the universe was the Divine.[237] For Xenophanes, the universe was both spherical and finite, and was governed by two extreme states of being: wet, symbolized by water (ὕδωρ), and dry, symbolized by earth (γῆ).[238] All matter, including life, was said to alternate between these two extremes as it died and was reborn.[239] The Sun and stars, for example, were said to be made of burning clouds which were extinguished when they set, and were reignited into life again when they rose the next day.[240] McKirahan sees this as a rebuke of the theory of Anaximenes that the universe was infinite and composed of air.[241] Xenophanes also argued against long held religious beliefs, stating that the Divine are beyond human morality, are not anthropomorphic, cannot die or be born, have no divine hierarchy, and do not intervene in human affairs.[242] McKirahan argues that, while Xenophanes seems to be rejecting long held religious traditions, he is not actually questioning the presence of a divine entity. Rather the base substance of the universe is the Divine, and Xenophanes' philosophy is a commentary on earlier writers and their ideas on divinity.[243] Xenophanes also suggested the possible existence of an infinite number of worlds in the Heavens—preceding the discovery of the first exoplanet by more than 2,000 years—and for the first time offering that there was more than just the Earth as one of the planets in the universe.[244] As such, the writings of Xenophanes hold a special place among Greek astronomical texts in that, rather than trying to attribute the motions and influences

---

[234] Diog. Laert. 1.23, 1.111, 8.36, 9.18, 9.20.
[235] Diog. Laert. 9.18.
[236] Eus. PE 15.2, 15.17.
[237] Arist. Metaph. 986b21–24.
[238] Thdt. CAG 4.5; Hippol. Refut. 1.12.
[239] Hippol. Refut. 1.12.
[240] Aet. 2.13, 2.20.
[241] R.D. McKirahan, Xenophanes of Colophon: Philosophy Before Socrates (Indianapolis, Hackett Publishing Company, 1994), p.65.
[242] Clem. Al. Strom. 5.110, 7.22. See also: C. Osborne, Presocratic Philosophy: A Very Short Introduction (Oxford, Oxford University Press, 2004), p.62.
[243] McKirahan, Xenophanes, p.61.
[244] Diog. Laert. 9.19.

of the Heavens to the Divine, Xenophanes actually alters the basis of the Divine to suit what could be observed in the sky.

A possible student of Xenophanes was Heracleitus of Ephesus (535–475 BC)—although some accounts describe him as a self-educated man.[245] He was from the ruling aristocratic family, but had abdicated from the position of monarch in favour of his brother.[246] In his writings, Heracleitus was most critical of the works of Hesiod, Pythagoras, and Xenophanes, stating that 'the knowledge of the most famous people . . . is but opinion.'[247] Despite his criticisms, Heracleitus seems to have supported many of the ideas proposed by Xenophanes. While Heracleitus stated that the base element of the universe was fire,[248] he believed in the idea of a 'unity of opposites' and that a characteristic feature of the universe was that of change, or flux, similar to that of Xenophanes.[249] This state of change was kept in harmony in accordance with the *logos*—the reason for everything being the way it was.[250] Heracleitus argued that the heavenly bodies were flames set into hemispherical 'bowls' that were more distant than the Sun and had their concave sides facing the Earth.[251] Based upon this model, Heracleitus suggested that the phases of the moon were the result of the gradual turning of the Moon's bowl, and the lunar eclipses were caused by the Moon's bowl occasionally turning away from the Earth.[252] It is in one of Heracleitus' passages discussing the fiery nature of the universe that the word *kosmos* (κόσμος) is first used in an extant piece of Greek literature to refer to the universe as a whole and ordered structure.[253] Thus, with his notions of the *logos* and the *kosmos*, Heracleitus set down the ideas that there was one underlying principle governing the motions of all things. This idea would cause ripples, rifts, and controversy in ancient Greek political, religious, and intellectual circles for years to come.

Not long after Heracleitus had died, Empedocles of Akragas (494–434 BC) wrote two books on nature, with about 2,000 lines of epic verse, which seemed to have combined elements of different theories that had come before, dismissed others, and offered some original ideas about the structure of the

---

[245] *Suda s.v.* Ἡράκλειτος.
[246] Diog. Laert. 9.6.
[247] Diog. Laert. 9.1; see also: Clem. Al. *Strom.* 5.9.3.
[248] Diog. Laert. 9.8; see also: Clem. Al. *Strom.* 5.103.3; Plut. *De E.* 338d–e.
[249] Arist. *Eth. Nic.* 8.2 1155b4.
[250] Sext. Emp. *Math.* 7.133; Hippol. *Refut.* 9.9.1, 9.9.5.
[251] Diog. Laert. 9.10.
[252] Diog. Laert. 9.10.
[253] The passage, recounted in the later writings of Clement of Alexandria (*Strom.* 5.103.3), states: 'This kosmos, which is the same for all, no one of the gods or men has made. But it always was, and will be, an ever-living fire.'

universe.[254] For Empedocles, the Heavens were made of air and the Sun was made of fire.[255] The universe was shaped like an egg, and the Heavens were a revolving crystalline sphere to which the stars were fixed.[256] The Earth was surrounded by a sphere divided in half—with a fiery side (day) and an air side (night).[257] In this model the Moon was judged to be twice as far from the Sun as it was from the Earth.[258] Empedocles theorized that the Earth was kept in its place within this structure by the revolution of the Heavens.[259] It was also suggested that the Sun itself was not made of fire, but reflected the fire of a second Sun, like light off water.[260] In another example of a prediction that was to be later proved scientifically, Empedocles said that light had a finite speed and reached the space between the Sun and Earth before it reached the Earth itself.[261] With Empedocles, while the geocentric model of the universe was still accepted, the examination of the nature of the Heavens moved to incorporate the interactions of the celestial bodies in regard to how they may (or may not) have been able to illuminate each other, and how light could travel within the ordered structure of the cosmos.

A similar sentiment is found in the works of Anaxagoras of Clazomenae (500–428 BC). Like others that had gone before, Anaxagoras drew on eastern knowledge and, according to the *Suda*, went to Egypt to study before moving to Athens.[262] For Anaxagoras, the Earth was flat and was supported by air (*aether*)—echoing the earlier model of Anaximander.[263] He claimed that due to the rotation of the *aether*, stones were torn from the Earth and were then kindled into the stars, the Sun (which was said to be a red-hot fiery stone) and the Moon (which was said to be an Earth-like illuminated solid, bigger than the Peloponnese, covered with plains and ravines).[264] The rotation of the heavenly sphere also caused shooting stars—which were thought to be sparks caused by rotational friction.[265] Following Empedocles, Anaxagoras stated that the light that came from the Moon was actually a reflection of the Sun.[266] Additionally, Anaxagoras, in agreement with other theories that

---

[254] *Suda s.v.* Ἐμπεδοκλῆς.
[255] Aet. 2.6, 2.11.
[256] Diog. Laert. 8.77; Aet. 2.13, 2.31.
[257] Plut. *Stromat.* Fr.10; Aet. 2.11.
[258] Aet. 2.31.
[259] Arist. *Cael.* 295a13.
[260] Plut. *Stromat.* Fr.10; Aet. 2.20, 2.21; Plut. *De Pyth. Or.* 400b.
[261] Arist. *Sens.* 446a26–446b2; Arist. *De An.* 2.4.418b21–23.
[262] *Suda s.v.* Αἴγυπτος.
[263] Hippol. *Refut.* 1.8.
[264] Aet. 2.13, 2.25, 2.30; Hippol. *Refut.* 1.8; Pl. *Cra.* 409a; Plut. *De Facie* 929b; *Suda s.v.* Ἀναξαγόρας.
[265] Hippol. *Refut.* 1.8.
[266] Hippol. *Refut.* 1.8; Pl. *Cra.* 409a; Plut. *De Facie* 929b; Aet. 2.25, 2.30.

had been proposed by mathematicians, said that the phases of the Moon were caused by the Moon following the path of the Sun, which illuminates it, and that eclipses were caused by the Moon falling within a region of shadow when the Earth was between the Sun and the Moon. Solar eclipses were said to be caused by the Moon being between the Sun and Earth.[267] Anaxagoras also believed that there were heavenly bodies closer to the Earth than the Moon and that these also caused eclipses.[268] However, while a few of these ideas are somewhat correct in principle, the most original, and controversial, aspect of Anaxagoras' theories was that he moved away from accepting the idea of a divine influence on the workings of the universe, moved into the realm of an entirely scientific approach, and stated that the mind and matter were the guardians of all things.[269] Such views got Anaxagoras into considerable trouble. The *Suda* states that he was exiled from Athens 'for introducing a novel belief in the divine', while Plutarch says that he was imprisoned, and another philosopher, Protagoras, was exiled, 'because in those days people refused to accept the findings of philosophers and astronomers who would remove the divine from the workings of the heavens and, instead, ascribe it to causes, forces and properties'.[270]

Despite the controversial nature of much of what Anaxagoras had said, his ideas do seem to have become dispersed among the intellectual communities of the ancient Greek world, and some aspects of his model of the universe were taken up by the school of thinkers who flourished in southern Italy who carried on the Pythagorean tradition. This was particularly so for Anaxagoras' model of how eclipses worked.[271] For the Pythagoreans, the whole system of the Heavens was in harmony physically, geometrically and numerically and so they went to great lengths to construct models for the workings of the

---

[267] Aet. 2.29; Arist. *Cael.* 293b29.

[268] Aet. 2.29.

[269] *Suda s.v.* Ἀναξαγόρας.

[270] *Suda s.v.* Ἀναξαγόρας; Plut. *Nic.* 23; Protagoras was a sophist philosopher who was a proponent of 'relativism'—stating that in all things there are two opposing arguments or positions. One of his most famous, and controversial, sayings was that 'Man is the measure of all things: of the things that are, that they are, of the things that are not, that they are not' (πάντων χρημάτων μέτρον ἐστὶν ἄνθρωπος, τῶν μὲν ὄντων ὡς ἔστιν, τῶν δὲ οὐκ ὄντων ὡς οὐκ ἔστιν) (Sext. Emp. *Math.* 7.60; see also: Pl. *Tht.* 152a). In another agnostic work, titled *On the Gods*, Protagoras wrote: 'Concerning the gods, I have no means of knowing whether they exist or not, nor of what sort they may be, because of the obscurity of the subject, and the brevity of human life' (περὶ μὲν θεῶν οὐκ ἔχω εἰδέναι, οὔθ᾽ ὡς εἰσὶν οὔθ᾽ ὡς οὐκ εἰσὶν οὔθ᾽ ὁποῖοί τινες ἰδέαν· πολλὰ γὰρ τὰ κωλύοντά με εἰδέναι, ἥ τε ἀδηλότης καὶ βραχὺς ὤν ὁ βίος ἀνθρώπου). The agnostic position of Protagoras was what aroused ire in the Athenians, and Protagoras' works were collected and burned in the marketplace when he was exiled (Cic. *Nat. D.* 1.23.6). The so-called 'Decree of Diopeithes' of the 430s BC, which made provisions for the prosecution of those who did not give due regard to the Divine and/or taught about the motions of the heavens, was instituted by the political opponents of the Athenian politician, Pericles, possibly due to his connections with Anaxagoras (Plut. *Per.* 32; Pl. *Ap.* 26d–27a; see also: R.W. Wallace, 'Private Lives and Public Enemies: Freedom of Thought in Classical Athens' in A.L. Boegehold and A.C. Scafuro (eds.), *Athenian Identity and Civic Ideology* (Baltimore, Johns Hopkins University Press, 1994), pp.127–155).

[271] Arist. *Cael.* 293b29.

universe which made theories fit with the ideals of geometric and numerical perfection.[272] Geminus goes as far as to suggest that it was the Pythagoreans who first came up with the idea of studying the Heavens through the concept of circular and uniform motions.[273] This then influenced the way in which the Pythagoreans saw the universe. The Sun, for example, was seen as a perfect sphere.[274] The universe as a whole was also made of ten perfect concentric spheres. There was a great fire positioned in the centre of the Pythagorean universe, as it was believed that the centre should be occupied by the worthiest thing, and fire was seen as more worthy than the Earth.[275] The Earth was just one of the bodies revolving around this central fire, and this was what produced night and day.[276] Philolaus (470–385) expanded on this and said that there was a fire in the centre of the universe, and another sphere of fire at the furthest extent which enclosed everything.[277] Between these two fires were the Earth, a 'Counter-Earth' on the direct opposite side of the central fire, the Moon, Sun, the five planets, and the stars.[278] All of these bodies sat on their own rotating spheres. Goldstein and Bowen argue that the attempts by the Pythagoreans to construct a 'perfect' model should not be seen as any attempt to precisely measure the celestial motions, but rather an attempt to impose an 'ethical and aesthetic order' to physical phenomena.[279] However, not all Pythagoreans accepted this model, and Hicetas of Syracuse (400–335 BC) argued against the dynamic models that had come before and offered that it was only the Earth that was rotating, and that this caused the apparent motion of the stars and planets.[280] This shows that even within a single school of thought, models were not unanimously accepted and that ideas could be questioned and refined in order to reach a complete 'truth'.

For Leucippus of Abdera (480–420 BC), the universe was still geocentric, and the Earth was flat—shaped like a tambourine.[281] The universe was infinite, and the Earth was positioned in the centre of a series of concentric hemi-spherical 'bowls' which represented the orbits of the heavenly bodies.[282]

---

[272] Arist. *Metaph.* 986a1; Hippol. *Refut.* 1.2.2.
[273] Gem. *Isagoge* 10.2–21; Russo (*Forgotten Revolution*, p.86) suggests that the idea of a heavenly sphere is Pythagorean. However, as has been shown, there are earlier references to this idea.
[274] Aet. 2.22.
[275] Arist. *Cael.* 293a15; see also: Aet. 2.7.
[276] Arist. *Cael.* 293a15; see also: Aet. 2.7.
[277] Aet. 2.7, 2.20, 3.11, 3.13.
[278] Aet. 2.7.
[279] B.R. Goldstein and A.C. Bowen, 'A New View of Early Greek Astronomy' *Isis* 74.3 (1983), p.333.
[280] Cic. *Acad.* 2.39; Diog. Laert. 8.85.
[281] Diog. Laert. 9.30.

The Sun was positioned on the outermost orbit from the centre, the Moon was closest to the Earth, and the other celestial bodies were in between.[283] Despite the errors in Leucippus' model, he also came up with a few novel, and somewhat accurate, ideas. For example, Leucippus attempted to account for the obliquity of the zodiac by suggesting that it was at an even angle across the sky because the Earth was inclined towards the south.[284] Cosmologically, Leucippus proposed that the Earth and the planets had formed from small particles in the void, whirling around in a circular motion, colliding, and sticking together.[285] This is essentially a correct (albeit basic) summary of how planets form out of the spinning proto-planetary discs that surround young stars.

Many of these ideas, and those of previous thinkers, are also found in the writings of Democritus of Abdera (460–370 BC), who wrote works titled *The Great Diacosmos* and *On the Nature of the Cosmos*.[286] Democritus may have also composed a separate work on the constellations—a text later cited by Vitruvius.[287] Democritus may have been a student of both Anaxagoras and Leucippus, but was also schooled in Eastern traditions via being educated by Persian, Egyptian, and Indian philosophers.[288] For Democritus the Earth was flat (as per Leucippus) but was hollow in the middle.[289] Democritus also followed Leucippus in stating that the Earth was inclined towards the south— which was why the southern parts of the known world (eg. Egypt and North Africa) were hotter.[290] The Sun and stars were fiery stones and the Moon was Earth-like as it seemed to have valleys.[291] Democritus departed from Leucippus in regards to the order of the celestial bodies—giving an order, from the centre, of Earth, Moon, Sun, other planets, and finally the fixed stars.[292] Democritus attempted to account for the apparent differences in motions of the heavenly bodies by suggesting that the reason why the Sun and Moon pass through different signs of the zodiac is because the speed of rotation for the different spheres that designate their orbits was slower closer to the Earth.[293] Similar to Leucippus and his ideas of planetary formation, Democritus also put forward ideas that are conspicuously modern. Democritus theorized that

---

[282] Diog. Laert. 9.31.
[283] Diog. Laert. 9.31.
[284] Diog. Laert. 9.31.
[285] Diog. Laert. 9.30–31.
[286] *Suda s.v. Δημόκριτος.*
[287] Vit. *De Arch.* 9.5.4.
[288] *Suda s.v. Δημόκριτος.*
[289] Aet. 3.10.
[290] Aet. 3.12.
[291] Aet. 2.13, 2.20, 2.25.
[292] Hippol. *Refut.* 1.13.
[293] Lucr. 5.621–636.

there were an infinite number of worlds of all different sizes—some inhabited, some not, some with single suns, some with two.[294] This idea predicts many things that have since been proven, such as exoplanets and binary star systems, and also predicts the existence of life elsewhere in the cosmos (which has yet to be proven).

The study of empirical astronomy, and the attempts to construct models that explained the workings of the Heavens, found its most vocal detractor in Plato of Athens (428–348 BC). Plato had travelled to Egypt to learn from the Egyptians and Hebrews 'about all things that exist'.[295] Plato suggested that 'the true astronomer must be the wisest of men . . . the man who investigates the seven revolutions',[296] but seems to have found issue with both the methodologies and motivations which had formed all previous thought on the topic. For Plato, the study of astronomy should not rest on the attempt to understand the motions of the celestial objects and the workings of the heavenly spheres. Rather, astronomy should strive to attain something much higher and considerably more esoteric. Plato argued that to study things empirically, based on observation and calculated models as prior astronomers and philosophers had done, did not enlighten. In his *Republic*, Plato stated:

> 'I, for my part, cannot think of any other study to be the one that makes the soul look upwards except that which is concerned with the real and the invisible, and if anyone attempts to learn anything that is perceivable . . . he will never, as I hold, learn, because no object of sense admits of knowledge . . .'[297]

Plato additionally stated that the objects in the Heavens were considered to be more perfect and beautiful than anything else that was visible, but that they were, in fact, far inferior to those things that were 'true', far inferior to their movements with speed and slowness, in true number and in true form. He called such things 'true objects' which could be understood through reason and intelligence, not just by observation.[298] Indeed Plato, in something of an existential circular argument, argued that you cannot understand science without first knowing what it is, nor can you claim to understand it.[299] Similarly, in his work *Phaedro*, Plato derided the work of Anaxagoras for looking at things empirically and not with the soul.[300] In *Timaeus*, Plato stated that

---

[294] Hippol. *Refut.* 1.13.
[295] *Suda s.v.* Αἴγυπτος; see also: Plut. *De Is. et Os.* 10.
[296] Pl. *Epin.* 990a.
[297] Pl. *Resp.* 7.529a.
[298] Pl. *Resp.* 7.530a.
[299] Pl. *Men.* 80d–e.
[300] Pl. *Phd.* 97b–99b.

those who call for proof 'would forget the difference of human and divine nature. For only the divine has the knowledge and the power [to understand such things].'[301] While in *Republic*, Plato argued that geometrists confuse the requirements of geometry, which leads to astronomy as a study of solids in motion, with the requirements of daily life, and consider that they are doing something practical, whereas knowledge is the real object of the discipline.[302] Rather than an empirical approach, Plato advocated the study of the reasoning behind the motions of the heavenly objects by suggesting that the patterns in the Heavens should be used as illustrations to facilitate the study of the 'higher objects' in the same way that plans and diagrams aid with the understanding of geometry.[303] In *Timaeus*, Plato argued that the purpose of scientific investigation was not to understand the world that could be experienced through the senses, but to discover the underlying plans of the Divine which had imposed order and beauty onto what would otherwise be a chaotic and, to Plato, evil universe. Plato stated that 'to attempt to tell of [the workings of the cosmos] without a visible representation of the heavenly system would be a labour in vain.'[304] To that end, Plato recommended that the study of astronomy should be based on problems, in a similar manner to the geometric propositions of Euclid, rather than observations, as this would dispense with the starry heavens, and would 'convert the natural intelligence of the soul from a useless to a useful possession.'[305]

Hetherington suggests that Plato's astronomical views were a product of the socio-political environment in which he lived.[306] Certainly, the condemnation of individuals like Anaxagoras for presenting 'heretical thought' would have had a strong influence on Plato. Additionally, Athens was undergoing a time of great upheaval and political change in Plato's day. Following the surrender of Athens to the Spartans at the end of the Peloponnesian War in 404 BC, the Spartans had installed a board of 'Thirty Tyrants', members of the Athenian aristocracy with oligarchic tendencies, to replace the former democratic system of the city.[307] Plato had initially considered joining the side of the Thirty Tyrants, but then decided against such an affiliation, as he did not like the way the Tyrants were running the state.[308] Plato then decided to embark of a life of

---

[301] Pl. *Ti.* 68d.
[302] Pl. *Resp.* 7.527a.
[303] Pl. *Resp.* 7.529d–530c.
[304] Pl. *Ti.* 40d.
[305] Pl. *Resp.* 7.530b.
[306] N.S. Hetherington, 'Plato's Place in the History of Greek Astronomy: Restoring *both* History and Science to the History of Science' *J. Astron. Hist. Herit.* 2 (1999), p.93.
[307] Xen. *Hell.* 2.3.15–16; Arist. *Ath. Pol.* 35.1.
[308] Pl. *L.* 7.324c–d, 7.325a–c.

philosophy rather than focus on a political career.[309] The Thirty Tyrants only remained in power for less than a year before being expelled by the Athenian populace, and democracy was restored, but Plato had already set himself on his philosophical path. During this time (*ca.* 399 BC), Socrates was also placed on trial on charges of both impiety against the gods (ala Anaxagoras) and for corrupting the youth of the *polis*.[310]

During his trial, Socrates avoided talking about astronomy altogether. Xenophon says:

> he did not even discuss that topic so favoured by other speakers—the nature of the universe . . . [and he] avoided speculation on the so-called 'cosmos' of the professors, how it works, and the laws that govern the phenomena in the heavens. Indeed, he would argue that to trouble one's mind with such issues is utterly foolish.[311]

This suggests that Socrates was very aware of, and warry of, the charges that had previously been levelled against Anaxagoras, and may be reflective of much of the Athenian mindset at the time. Some of the Thirty Tyrants had also been students of Socrates, and his trial may have been as much a political statement against the actions of the Tyrants as it was against Socrates' religious and scientific views. Indeed, Socrates seems to have been quite critical of the democratic form of government—especially according to Plato.[312] The trial, and subsequent death, of Socrates would have a strong influence on the young Plato who was then embarking on his philosophical career. The transient nature of government and the running of the city-state, and the seemingly fickle nature of the Athenian populace, also seem to have had a strong influence on how Plato viewed the workings of both the world and the Heavens.

As Munn points out, there was a growing tension in Athens at the time between new scientific models and popular ideology over the understanding of the *physis* (natural growth) that governed the world according to nature, and the *nomoi* (laws/customs) upon with the city-state was based.[313] The trial of Socrates suggests that, at this time, the study of astronomy was seen as a useless pursuit—which was why Socrates attempted to convince the court that what he had done in the past had nothing to do with this line of thinking. For thinkers like Plato, in the aftermath of the trial and the turmoil of the Thirty Tyrants,

---

[309] Pl. *L.* 7.325b–326b.
[310] Xen. *Mem.* 1.2.29–38.
[311] Xen. *Mem.* 1.1.11.
[312] Xen. *Mem.* 1.2.9; Pl. *Cri.* 47c–d; Pl. *Lach.* 184e; Pl. *Grg.* 503c–d, 515d–517c; Pl. *Ap.* 25a–b.
[313] Munn, 'Science to Sophistry', p.111.

the best type of state was one where the *physis* and the *nomoi* were in close accord and ordered according to the principles seen in nature. Consequently, any form of 'scientific' writing was never far from the popular ideals of the time, and may have even endeavoured to justify popular beliefs. Munn suggests that what the Athenian populace really wanted was clear guidance rather than the subtleties of mathematical astronomy. Scientific thought had presented a number of big questions, but then had failed to answer those questions adequately. The result of this was that some felt betrayed by philosophers and astronomers who had promised so much, but who seemed to have delivered so little.[314] This may then account for the trials of Socrates and Anaxagoras, and the viewpoints of Plato.

For Plato the Heavens were governed by harmonic, geometric, and numerical perfection, and only a person with a solid understanding of such concepts had the slightest chance of understanding their workings. In *Epinomis*, Plato stated that:

> it is not easy to understand the courses of those other stars [i.e. planets] ... but we should, for this purpose provide persons of such ability as we can find, considering that much initial learning and practice is required, followed by continuous work during childhood and adolescence. It follows that mathematics is essential.[315]

Within Plato's flawless mathematical construct, much like others had previously argued, the universe was spherical—a shape 'most perfect'—and the whole universe was made up of 'perfect bodies'.[316] The Earth was a sphere (contra the flat Earth of Leucippus and Democritus) in the centre of the Heavens, held aloft by the uniformity of the substance of Heaven (as opposed to air or *aether* as had been suggested by others).[317] Each of the heavenly bodies sat on one of a set of concentric revolving spheres—described as being like spindle whorls of ever decreasing size set one inside of the others.[318] Each of the spheres contained a spiral-like twist to account for the retrograde motion of

---

[314] Munn, 'Science to Sophistry', p.133.

[315] Pl. *Epin.* 990c; along a similar vein, Russo (*Forgotten Revolution*, pp.199–200) states that 'if the person who obtains a mathematical result also knows its only possible application, it doesn't matter if the result is exact: a reasonable approximation is enough, as was generally the case in the mathematics of Pharaonic Egypt or Old Babylonia, which did not distinguish, for example, between exact and approximate area formulas. But when the result obtained is consciously kept internal to the theory, that is, when it must be applied, often indirectly, to a variety of problems not known a priori, a mathematician's rigor becomes essential.'

[316] Pl. *Ti.* 33b, 34a; Pl. *L.* 7.334a–342a.

[317] Pl. *Phd.* 109a.

[318] Pl. *Resp.* 10.616b–617d; Pl. *Ti.* 34a.

the planets.[319] In the sixth century AD, Simplicius composed an astronomical treatise, based up Eudemus' lost *History of Astronomy* from the fourth century BC, in which he claimed that it was Plato who first set astronomers the task of attempting to understand the working of the Heavens through circular motion.[320] This seems unlikely due to how many models prior to the time of Plato had been based upon celestial spheres, and many scholars doubt the validity of the claims made by Simplicius.[321]

However, Simplicius does bring to light an apparent contradiction in the Platonic model—how can a universe based upon perfect spheres also contain twists in those spheres to account for retrograde motion? In answer to this Simplicius outlines how some astronomers, adhering to the Platonic idea of heavenly perfection, held that what was observed in the sky was not reality, but only an 'appearance'.[322] This follows the ideas that Plato presented in the *Republic* that reality was not really visible, but only existed in a universe of 'ideas'.[323] Hetherington suggests that Plato 'had a good sense of the qualitative problem posed by planetary motions, but begged off presenting anything approaching a detailed, quantitative solution', had 'intellectual infatuation with circular motion', and calls his astronomy 'highly abstract, with an emphatic contrast between the domain of the visible heavens and that of intelligible truth and reality'.[324] A result of this universe of 'ideas' was that the Platonists could continue to try to understand the nature of things, based upon a model of numerical and harmonic perfection, and anything that deviated from this ideal was simply considered not to be real.

Plato states that some of what the Greeks understood of the 'spheres' of the orbits of planets had come from eastern observers in Egypt and Syria.[325] By knowing the patterns that were visible in the sky, it was clear to Plato that the Heavens were in motion.[326] The order of the heavenly spheres for Plato, from the centre, was the Earth, Moon, Sun, Venus, Mercury, other planets, and then the stars.[327] Plato suggested that the stars—which were massive in size, bigger

---

[319] Pl. *Ti.* 39b.

[320] Simpl. *in Cael.* 488.18–24, 492.28–493.5.

[321] See: O. Neugebauer, *The Exact Sciences in Antiquity* (New York, Dover Publications, 1969), p.152; Hetherington ('Plato's Place', pp.89–90) suggests that the trail of transmission runs from Eudoxus' work on Plato (now lost), which was then summarized by Eudemus (lost), which was, in turn, commented on by Sosigenes in the second century AD (lost), which was then used by Simplicius in the sixth century AD.

[322] Simpl. *in Cael.* 488.18–24.

[323] Pl. *Resp.* 7.514a–517a.

[324] Hetherington, 'Plato's Place', pp.89, 92.

[325] Pl. *Epin.* 987a–b, 988a.

[326] Pl. *Resp.* 7.530b.

[327] Pl. *Ti.* 37e, 40a–b; Pl. *Epin.* 990b–c.

than the sun, but very distant—were fixed onto their heavenly sphere as they did not seem to move like the planets did.[328] The rotations of the heavenly spheres was governed by the *Morae* (Fates), who dictated the course of everything in the cosmos.[329] On each of the spheres stood a Siren who sang a single note, and each of the Sirens' notes were different, so that when they were combined they formed the Harmony of the Heavens.[330] Thus, for Plato, the Divine was an essential element to the workings of the universe, and it was only the divinities who had the ability to really understand it all. Whether this was influenced by the punishments imposed against thinkers such as Protagoras and Anaxagoras (whom Plato is specifically critical of) is unclear. Plato argued that mortal men could not know the true nature of the Heavens as they are unable to rise high enough to see things that are above the air—in the same way that a creature that lives deep in the sea cannot know about the world unless it comes to the surface—whereas the Divine can rise high enough and see the 'big picture'.[331] Such a sentiment can also be seen in Eratosthenes' early poem, *Hermes*, where the eponymous god was able to travel far above the Earth and then look back upon it.[332]

For Plato, time itself did not exist before the creation of the universe and was brought into existence along with the Sun, Moon, and planets.[333] As part of the perfect construction of the cosmos, the orbits of the heavenly bodies were designed as a way of defining and preserving the numerical properties of time.[334] Perfection also influenced the motions of the heavenly bodies themselves as their movements were the way that they were because this was the correct and proper way for them to behave.[335] With Plato, although much of this writings contain elements similar to those found in many prior works, the underlying sentiment is almost a regressive motion back towards the earliest statements of Hesiod and Homer where the reasoning behind the motions of the universe are attributed to the Divine in conjunction with the ideals of perfection. Whether this was the prevailing mindset in Athens—following the prosecutions of agnostic thinkers like Socrates and Anaxagoras—is not known.

However, with the advent of the fourth century BC, the study of empirical astronomy continued in the east. Eudoxus of Cnidus (390–337 BC), for example, had studied in southern Italy under the Pythagoreans, in Athens

---

[328] Pl. *Epin.* 982c, 983a.
[329] Pl. *Resp.* 10.617d.
[330] Pl. *Resp.* 10.617c.
[331] Pl. *Resp.* 109b–110a.
[332] Ach. Tat. *Arati Phaenom.* 153a, 153c, 157c.
[333] Pl. *Ti.* 37d.
[334] Pl. *Ti.* 37e–39d.
[335] Pl. *Phd.* 97b–99b.

under Plato, and in Egypt where he had studied in the temples.[336] Shipley suggests that Eudoxus may have been the first person to rely on Babylonian concepts in his manner of enquiry.[337] However, this seems unlikely in face of the ancient texts that refer to many early astronomers either being educated in the east, or drawing upon eastern works. Eudoxus wrote many texts on astronomy including works titled *The Cycle of Eight* and *Astronomia*, and determined the sidereal periods of the five visible planets.[338] Eudoxus may have also undertaken calculations for the determination of the so-called 'Great Year'—a 12,960-year period between the conjunctions of the five visible planets at a given point on the celestial sphere.[339] This may have led Eudoxus to create the first ever 'star globe' of the Heavens—although the evidence for this is very circumstantial and is based upon notations found on two manuscript star maps (Aberistvid NLW 735 and MS B 24/163) from the eleventh and twelfth centuries, respectively.[340] Vitruvius also credits Eudoxus with the creation of the *arachne*—the series of lines inscribed on sundials which set out the hours, solstices and equinoxes.[341] If this is the case, then Eudoxus must have engaged with fairly complex calculations and/or long-period observations in order to compose a system that would accurately, and mathematically, account for the motions of the heavenly spheres and the shapes and locations of the lines on the face of the sundial.

Indeed, the use of a sundial that was accurately inscribed with the lines marking out the Solstices, equinox and hours as designed by Eudoxus may have contributed to the shift in Greek thought away from the flat Earth model found in the earlier descriptions of the universe. As Munn points out, in a flat Earth model, events such as sunrise and sunset would correspond to different points on the eastern and western horizons depending upon an individual's location in the planar world. Similarly, the location of the Sun when it was at its apogee would also differ depending upon location. With a sundial like the one that Anaximander is said to have earlier created, the shadow cast by the *gnomon* would trace a different pattern depending upon how close the instrument was positioned to a north–south meridian line which ran through the exact centre of a flat Earth. For example, when set up close to the meridian, the shadow cast by the *gnomon* at midday would be perpendicular to

[336] Diog. Laert. 8.86–87.
[337] Shipley, *Greek World*, p.351.
[338] *Suda s.v.* Εὔδοξος; Diog. Laert. 8.86–91.
[339] See: Serv. A. 3.284.
[340] See: Dimitrijević and Bajić, 'Mythological Origin of Constellations', pp.133–134.
[341] Vit. *De Arch*. 9.8.1; Vitruvius also states that some people credit the creation of the *arachne* to Apollonius of Perga (262–190 BC).

the equinoctial line. If positioned away from the meridian, the shadow would incline towards the closest edge of the world, and Eudoxus' *arachne* would be considerably asymmetrical; with the arcs marking the solstices being splayed more widely on the side that was closest to the edge of the world, and more diminished on the opposite side.[342] However, on a spherical Earth, every point on its surface is on a meridian line and so the *arachne* will be symmetrical and the same for all locations at the same latitude. For the Greeks, this would not become evident until they had undertaken enough widespread observations using sundials to disprove the concept of a flat Earth. Munn suggests that this occurred in the time of Eudoxus.[343]

Seneca states that Eudoxus' knowledge of planetary motion was obtained while he was in Egypt, and that he introduced such knowledge to the Greeks.[344] Indeed, ancient Greek astronomy seems to have taken a shift away from the Divine ideals of Plato at this time into something even more empirical in the form of a more detailed attempt to account for the retrograde motions of the planets.[345] Vitruvius goes as far as to suggest that older, Babylonian, astronomical knowledge was also transmitted to the Greeks around this time by Berossus (350–280 BC)—who had settled on the island of Kos, then under control of the Ptolemies, and established a school of astrology.[346] This information could have then been additionally drawn upon by Eudoxus and others. Goldstein and Bowen see ancient Greek astronomical thought as being delineated into two distinct phases: the quasi-divine reasonings behind the motions of the Heavens, and an emphasis on calendrical cycles, found in texts from Hesiod to Plato, and then the more scientific and mathematical concerns, mostly focused on planetary motion, which began with Eudoxus and continued into later periods.[347] This claim has merit. While prior works like those of Democritus and Leucippus did engage with the heavenly spheres to try to account for such things as the retrograde motion of the planets, it is under Eudoxus that the examination of such phenomena becomes more mathematical. Another possible reason for demarking the transition from one phase of Greek astronomical thinking to another with the works of Eudoxus is that his works constitute the first fully extant astronomical treatises that have survived in the literary record. All prior works are known only from fragments and/or references to them in later texts.

---

[342] Munn, 'Science to Sophistry', pp.123–124.
[343] Munn, 'Science to Sophistry', p.124.
[344] Sen. *QNat.* 7.3.2.
[345] Nicastro, *Circumference*, p.33.
[346] Vit. *De Arch.* 9.6.2.
[347] Goldstein and Bowen, 'A New View', pp.330–334.

Retrograde motion was accounted for by Eudoxus by providing a total of twenty-seven different nested spheres to the structure of the heavenly bodies. The Earth was still positioned at the centre of the universe, and the Sun, Moon, and stars each had a single sphere upon which they were set, as they did not exhibit any retrograde motions. The sphere of the Moon was inclined to that of the Sun and stars to account for the different path that the Moon took across the sky, while the Sun's path passed through the signs of the zodiac.[348] Each of the five visible planets had four spheres (or circles) which governed their path. The first sphere was concentric to that of the fixed stars and determined the general rotation of the planet's path. The second sphere was inclined so that the path of the planet moved through the signs of the zodiac. The third circle was small and set perpendicular to the zodiacal path (i.e. an epicycle) and the planet's movement around this smaller circle resulted in its retrograde motion. Finally, a fourth circle inclined the epicycle so that the resultant retrograde motion would also move slightly up and down in the sky.[349]

While others had previously attempted to account for the retrograde motion of the planets, Eudoxus' model involved a significantly more complex construction to try to account for not only the back-and-forth motions of the planets, but also the inclination of some of the heavenly bodies into one comprehensive system. Nicastro goes as far as to state that Eudoxus ushered in the 'beginning of scientific astronomy'.[350] However, while this could be argued, as Nicastro also points out, there was also a problem. Both the Sun and Moon appear to change in size over the course of a year. This indicates that that they are not at a set distance from the Earth as they are in the Eudoxan model.[351] However, for Eudoxus, both bodies seemed to be of a set size, and Archimedes later recounted how Eudoxus had measured the difference in the size of the Sun and Moon and suggested that the Sun was nine times bigger.[352] Similarly, Hetherington points out that the four homocentric circles of the Eudoxan model cannot account for the length of planetary retrograde motion, nor changes in latitude, with any level of accuracy.[353] Despite any issues, support for the Eudoxan model was long-lived, and some elements of Eudoxus' works seem to have been quite accurate. Helikon of Cyzicus, whom Plato says was a pupil of Eudoxus, was awarded 60 kg of silver by the tyrant of Syracuse, Dionysius II, for correctly predicting the time of a solar eclipse using the

[348] Arist. *Metaph.* 8.1073b17–1074a15.
[349] Arist. *Metaph.* 8.1073b17–1074a15.
[350] Nicastro, *Circumference*, p.33.
[351] See: Nicastro, *Circumference*, p.33; see also: Hetherington, 'Plato's Place', p.102.
[352] Archim. *Psam.* 3.
[353] Hetherington, 'Plato's Place', p.101.

Eudoxan model.[354] However, despite these seemingly accurate characteristics of the Eudoxan model, there were continued attempts to refine it.

Eudoxus' pupil, Callippus of Cyzicus (370–300 BC), who had studied under both Eudoxus and Aristotle in Athens, refined the Eudoxan model by adding more epicycles to better account for the retrograde motion of the planets as he found that the twenty-seven spheres of the model of his former teacher were insufficient to account for their movements. According to Aristotle, Callippus added two spheres to the paths for both the Sun and Moon, and one each for the planets Mercury, Venus and Mars.[355] This brought the total number of spheres in the revised Eudoxan model to thirty-four.[356] Callippus additionally analysed the length of the seasons—finding them to be ninety-four, ninety-two, eighty-nine, and ninety days in length. Based upon his measurement of the length of the year, Callippus composed the first astronomical calendar, and devised a seventy-six-year 'Callippic Cycle' containing 27,759 days or 940 months (twenty-eight of which were intercalary) which began in 330 BC.[357] Dividing the total number of days in the cycle (27,759) by the seventy-six-year period of the cycle, provided Callippus with a more accurate figure for the length of the year of 365.25 days.[358]

Aristotle (384–322 BC) also engaged much more deeply with prior attempts to describe the workings of the universe than had come before. Hetherington suggests that Aristotle would not have been content with the purely geometrical models of the universe that had been proposed previously, and additionally wanted to understand the forces behind the motions of the celestial bodies.[359] Bechler suggests that Aristotle was attempting to apply a mathematical

[354] On Helikon, see: *Suda s.v.* Ἑλικώνιος; Pl. *L.* 13.360c; for the eclipse, see: Plut. *Dion.* 19.
[355] Arist. *Metaph.* 1073b32–38; Simpl. *in Cael.* 497.17–24.
[356] For an examination of the complexities of the models of Eudoxus and Callippus, see: H. Mendell, 'Reflections of Eudoxus, Callippus and their Curves: Hippopedes and Callippopedes' *Centaurus* 40 (1998), pp.177–275.
[357] Gem. *Isagoge* 8.57–60.
[358] The seventy-six-year Callippic Cycle was incorporated into the famous Antikythera Mechanism—an astronomical 'clock' from the second or first century BC. The instrument, discovered in 1901, has been heralded as the world's first analogue computer and is a complex device of more than thirty gears housed in a wooden case which can be used to predict the positions of astronomical objects, the time of eclipses, and the timing of the four big Pan-Hellenic festivals in Greece. Freeth ('Decoding an Ancient Computer' *SciAm* 301.6 (2009), p.83) suggests that the Device either originated in Pergamon—the location of the Library which was the main rival to the one in Alexandria—or that it may have been built by Hipparchus. One of the faces of the Mechanism contains a dial for monitoring the Callippic cycle, and Callippus' seventy-six-year cycle is specifically mentioned in the instructions for the Mechanism which are inscribed upon it. For more details of the Antikythera Mechanism, see: M.K. Papathanassiou, 'Reflections on the Antikythera Mechanism Inscriptions' *Adv. Space Res.* 46 (2010), pp.545–551; X. Moussas, 'The Antikythera Mechanism: The Oldest Mechanical Universe in its Scientific Milieu' in D. Valls-Gabaud and A. Boksenberg (eds.), *The Role of Astronomy in Society and Culture: Proceedings of IAU Symposium No.260, 2009* (Cambridge, IAU Publishers, 2011), pp.135–148.
[359] Hetherington, 'Plato's Place', p.101.

solution to the Eudoxan model so that it would fit to physical reality, rather than improve, or even reject, Eudoxus' more geometric construct.[360] Gysembergh suggests that one of Aristotle's works was an examination into the 'Great Year'—which drew on the earlier work of Eudoxus.[361]

Aristotle had been a student of Plato,[362] and while he seems to have followed the concept of a geometrical perfection to the universe based on circles and spheres, Aristotle also rejected some of the ideas that had been proposed by his former teacher. For Aristotle, circular motion was the natural way of things, and this movement resulted in what was observed in the sky and how the universe was structured.[363] This may have been because, for Aristotle, there was a geometric necessity within his model for there to be a fixed centre and an axis around which all of the heavenly spheres turned. The Earth was spherical, said Aristotle, not the flat Earth sitting in water as had been suggested by Anaximenes, Anaximander, and Democritus, as that which has weight has a natural tendency to move towards the centre.[364] The notion of the Earth sitting in water was rejected as water needs something to sit upon, and water is lighter than Earth, and not vice versa.[365] The Moon was also spherical as evidenced by its phases and eclipses.[366] Additionally, the stars were spherical in shape, 'as others maintain', but Aristotle claimed that they did not rotate—as the Moon seemed not to do.[367] Aristotle further claimed that the stars were not made of fire as some suggested, but the air underneath the star's sphere was made hot by the friction of its rotation, and this caused the stars to shine.[368] Aristotle debunked the idea that the stars made sounds and created a harmony in the universe (re Plato) because the Sun and Moon do not make sounds, and the stars should make no sound as they appeared to not actually be moving themselves, but were imbedded in a moving sphere.[369] Aristotle's understanding of the motions of the heavenly spheres upon which the stars and planets sat had, like for many others, come from the east as the Greeks had 'a number of accepted facts relating to each of the stars' which went back many years to the Egyptians and Babylonians.[370]

---

[360] Z. Bechler, 'Aristotle Corrects Eudoxus: *Met* 1073b39–1074a16' *Centaurus* 15 (1970), pp.113–123.
[361] V. Gysembergh, 'Aristotle on the "Great Year", Eudoxus, and Mesopotamian "Goal Year" Astronomy' *Aion* 35 (2013), pp.111–123.
[362] *Suda s.v.* Ἀριστοτλές.
[363] Arist. *Cael.* 1.2.3.
[364] Arist. *Cael.* 269b30, 297a1.
[365] Arist. *Cael.* 269b30.
[366] Arist. *Cael.* 291b23.
[367] Arist. *Cael.* 291b23.
[368] Arist. *Cael.* 2.289a20.
[369] Arist. *Cael.* 2.290b12.
[370] Arist. *Cael.* 292a9.

Observations of occultations had demonstrated to Aristotle the order of the heavenly spheres by showing that the planets were further out than the Moon.[371] The rotational speed of each of the spheres was proportionate to its size, and Aristotle suggested that the furthest sphere, that of the stars, moved the fastest.[372] The spheres of the planets were slower than that of the stars and had several components to them to account for retrograde motion.[373] The rotation of the spheres had an effect on the mechanics of the universe and, in a precursor to some of Newton's laws of motion, Aristotle stated that 'something that is in continuous motion will cause motion in another thing.'[374]

As with the theory of Leucippus, Aristotle saw the existence of both light and heavy particles in the early universe prior to the commencement of universal rotation.[375] The Earth was at the centre of these concentric spheres since, much like with the idea behind spherical perfection, everything fell inward—perpendicular to the surface of the Earth.[376] This followed on from an earlier statement of Xenophanes that accounted for gravity by suggesting that the inside of the Earth extended inward infinitely.[377] However, for Aristotle the Earth did not rotate as Hicetas of Syracuse had suggested. As proof of this, Aristotle stated that, if the Earth did rotate, then the stars would exhibit retrograde motion.[378] This shows that while the ancient Greeks commonly placed the stars on the furthest sphere from the Earth in attempts to construct a geometrically perfect model for the universe, the true distances to the stars was still far beyond their limit of understanding, and the workings behind what was observed in the sky for both the stars and planets was far from understood.

Autolycus of Pitane (360–290 BC) seems to have been writing in Athens between 335 BC and 300 BC and so is a contemporary of Aristotle. Autolycus is the author of some of the earliest extant astronomical texts. One, *On the Moving Sphere* (Περὶ κινουμένης σφαίρας), is considered to be the oldest surviving mathematical treatise to come from ancient Greece and contains basic theorems, based on geometrical principles, on the motions of spheres that could be used by astronomers.[379] Autolycus' other work, *On Risings and Settings*

---

[371] Arist. *Cael.* 292a9.
[372] Arist. *Cael.* 2.289b10.
[373] Arist. *Cael.* 2.289b10.
[374] Arist. *Ph.* 8.10 267a21–b9, b17–b26; see also: Arist. *Metaph.* 8.1073a23–b17.
[375] Arist. *Cael.* 295a13.
[376] Arist. *Cael.* 295a13.
[377] Arist. *Cael.* 294a20.
[378] Arist. *Cael.* 295a4.
[379] For example, see: Autol. *De Sph.* 1–7, 9, 11.

(Περὶ ἐπιτολῶν καὶ δύσεων), examined the risings and settings of the celestial bodies. The work seems to be both an expansion of his work on the motions of the sphere, and also seems to be based on the work of Eudoxus, as Autolycus advocated the Eudoxan model of homocentric spheres.[380] Autolycus seems to have recorded observations over more than a year. For example, in *Risings and Settings* he states that any star that rises and sets always does so at the same point on the horizon.[381] From such observations, Autolycus also found issue with some of the findings of the Eudoxan model, as it could not account for the apparent changes in size and brightness that could be observed with Mars and Venus.[382] Thus, Autolycus begins to build on the more mathematical works of Eudoxus, identified flaws in the earlier conclusions, and continued the more scientific examination of the Heavens rather than the more existential approach that had been promoted by Aristotle.

At around the same time that Aristotle died, Euclid of Alexandria (325–270 BC), although more well known as being the 'father of geometry', also composed a work on astronomy—the *Phaenomena*. According to an Arabic biography, Euclid may have been born in Tyre, in modern Lebanon, but moved to Egypt when he was very young, and arrived in Alexandria about a decade after the city's founding (*ca.* 322 BC). Some scholars see this later biographical text as fictitious and consider Euclid to have been a born and bred Alexandrian.[383] Regardless, Euclid would have grown up in Alexandria under the rule of Ptolemy I, lived through the reign of Ptolemy II, and would have been an older contemporary of Eratosthenes. The ancient writer Proclus states that Euclid was younger than the followers of Plato, but older than Archimedes and Eratosthenes.[384] Euclid would have also been familiar with both the Museum and the Library in Alexandria, and may have even worked in the Library. Proclus, for example, states that Euclid collected many of the ideas of his predecessors and, 'reduced to invincible demonstrations such things as had been demonstrated by others with a weaker arm'.[385] One place where Euclid would have had access to these former works that he built on would have been the Great Library.

Proclus also states that Euclid was an adherent of the Platonic school of thought and had composed many mathematically based treatises on such

---

[380] Autol. *Ort. et Occ.* 1.1–5; see also: Autol. *De Sph.* 4.
[381] Autol. *Ort. et Occ.* 1.6–7, 2.5–6.
[382] Simpl. *in Cael.* 504.22–25.
[383] See: T. Heath, *The Thirteen Books of Euclid's Elements* (New York, Dover, 1956), p.4; T. Heath, *A History of Greek Mathematics* (New York, Dover, 1981), p.355.
[384] Procl. *In Euc.* 2.4.
[385] Procl. *In Euc.* 2.4.

things as optics, harmonics, music, divination, and fallacies—as well as his work on geometry.[386] Russo, sees the time of Euclid as the transition to more 'exact science' (compared to setting Eudoxus as the transition point as others have done), and says that the study of astronomy, in close connection with mathematics, only begins in the fourth century BC.[387] This seems evident in the way in which Euclid applied aspects of geometry to the study of the Heavens, and the resultant *Phaenomena* was a work on 'spherical astronomy'. In Euclid's model, the universe was spherical and revolved uniformly about its axis. This was evidenced by the segments of the sky being equal in shape and so, Euclid argued, the shape of the universe could not be a cylinder or a cone.[388] The Earth was situated at the centre of this spherical construct.[389] The stars rotated with circular motion, and were fixed to one sphere. This was evidenced by them rising and setting at the same place at the same time from year to year.[390] A revolution of the whole universe (i.e. a year) was seen as the time between the first rising of a star until it rose again.[391]

One thing that seems to confirm Proclus' statement of how Euclid drew on prior sources is that he seems to refer to the observations of stars at many different northern latitudes. Others had raised the ideas of an inclination of the Earth's axis in the past—suggesting how this accounted for the warmer temperatures that were experienced in Africa. However, Euclid took the evidence of stellar observations and applied this as proof of an inclination in his examination of the rotating spherical universe. For example, Euclid refers to the pole-star (Polaris) and noted how stars near the Northern Celestial Pole were always visible as they moved about the Pole in smaller circles, while those closer to the Arctic circle grazed the horizon because their circle of rotation was larger.[392] Stars below the Arctic circle rose and set because part of their circle was above the Earth, and part of it was below the terrestrial horizon.[393] For stars sitting on the Equinox circle, the amounts of time that they spent above and below the horizon were equal.[394] From the application of spherical geometry to such observations, Euclid was able to outline the details of the celestial horizon, the meridian circle through the poles, the 'tropics', the zodiac circle, and the equinoctial circle.[395]

[386] Procl. *In Euc.* 2.5.
[387] Russo, *Forgotten Revolution*, pp.22, 78.
[388] Euc. *Phae.* Preface.
[389] Euc. *Phae.* 1.
[390] Euc. *Phae.* Preface.
[391] Euc. *Phae.* Preface.
[392] Euc. *Phae.* Preface.
[393] Euc. *Phae.* Preface.
[394] Euc. *Phae.* Preface.
[395] Euc. *Phae.* Preface.

Another relative contemporary of Eratosthenes was Aristarchus of Samos (310–230 BC). Aristarchus also moved to Alexandria, became an 'Alexandrian by adoption', and became a student of Strato—who had held the position of third head of Aristotle's Lyceum in Athens before moving to Alexandria to become the tutor to Ptolemy II.[396] Similarly, Aristarchus would later become tutor of the young Ptolemy III.[397] Vitruvius calls Aristarchus a man 'proficient in all branches of science' and credits him with the invention of the hemispherical style of sundial—the type that Eratosthenes would later use in his calculation of the circumference of the Earth.[398] Not much is known in regard to the breadth of Aristarchus' works, little of it survives (even in fragments), and the references to it in later ancient works are sometimes questionable. Aetius, for example, says that Aristarchus stated that colour was light impinging on a substratum.[399] How much more detail Aristarchus went into in his study of optics is not fully known. A critique of some of Aristarchus' work, written by Pappus in the fourth century AD, also outlines, and then attempts to debunk, six propositions made by Aristarchus concerning the size of the Moon, its phases, eclipses, and its position relative to the Sun and Earth.[400]

However, Aristarchus' most famous, and somewhat controversial, contribution to ancient astronomy was to come up with a new model for the workings of the universe which did away with the geocentric model which had held sway for so long. The text of this novel concept has been lost, and details are only obtained second hand through later references to them by Archimedes. In his work *Psammites* ('The Sand-Reckoner'), Archimedes describes how Aristarchus had made the Sun synonymous with the 'central fire' found in

---

[396] Aet. 1.15; Diog. Laert. 5.58.

[397] *Suda s.v.* Ἀρίσταρχος.

[398] Vit. *De Arch.* 1.1.17, 9.8.1; Vitruvius (*De Arch.* 9.6.2) also credits the invention of this type of Sundial to Berossus (350–280 BC), who had established a school of astrological enquiry on the Ptolemy-controlled island of Kos.

[399] Aet. 1.15.

[400] Papp. *Syn.* 6.37; the six propositions of Aristarchus are: (1) the Moon receives its light from the Sun. This concept is also mentioned by Vitruvius (*De Arch.* 9.2.3); (2) the Earth is in the position of the point and centre of the Moon's sphere (this is similar to what Euclid had said about the centre of a circle—Euc. *El.* 1.16—but, depending upon interpretation, could be a suggestion that the Moon orbits the Earth rather than moves along its own concentric heavenly sphere); (3) when the Moon is half full, the line between light and dark points towards the Earth; (4) when the moon is half full, its distance from the Sun is less than a full quadrant (i.e. 90°) by 1/30th of a quadrant (i.e. 90/30=3°). In other words, the Moon is situated at an angle of (90°−3°=) 87° relative to the Sun [the real value is 89° 50′]; (5) the breadth of the Earth's shadow is equal to that of two Moons; and (6) the Moon subtends an angular diameter of 1/50 a sign of the zodiac [or 1/720 of the zodiacal circle, or 1/2°] (see also: Archim. *Psam.* 3 where Archimedes adds that Aristarchus calculated the diameter of the Sun to be >18 <20 x the size of the Moon). For a detailed discussion of Aristarchus' work, especially his examination of the sizes of, and distances to, the Sun and Moon, see T. Heath, *Aristarchus of Samos: The Ancient Copernicus* (New York, Dover Publications, 2004).

some early models of the universe, suggested that the Sun and stars were fixed—with the stars being at a very great distance (an idea that Archimedes rejected), and that the Earth revolved around the Sun.[401] Russo sees these calculations not only as the first documented case of the use of trigonometric methods in the study of astronomy, but also as essentially a work on geometry rather than an attempt to understand the motion of the Heavens.[402]

Aristarchus' heliocentric model may have been quite controversial—although this is far from certain. Certainly, some elements of it were not new. Both Heraclides and Hicetas had previously suggested that the Earth was in motion, and Aristarchus may have simply expanded on this idea and placed the Sun at the centre of things.[403] It is commonly held, based upon a passage from Plutarch, that Aristarchus was charged with impiety for suggesting a Sun-centred model. However, as Russo points out, this may be a corruption of the text.[404] Plutarch's text, which is in the form of a dialogue, reads, in many modern editions:

> do not, my good fellow, enter into an action against me for impiety in the style of Cleanthes, who thought it was the duty of the Greeks to indict Aristarchus of Samos on the charge of impiety for moving the Hearth of the Universe—this being the result of his attempt to explain phenomena by suggesting that the Heavens remain at rest, and the Earth revolves in an oblique circle, while it rotates, on its own axis, at the same time.[405]

While the description of Aristarchus' heliocentric model in the later part of the passage correlates with that of Archimedes, it is the earlier parts of the passage that are more problematic. For example, who is actually indicted on a charge of impiety seems to have been altered. The text seems to have been changed for the publication of Plutarch's work by Gilles Ménage in the early seventeenth century—not long after the trials of both Galileo and Giordano Bruno for heresy. Russo suggests that the nominative and accusative forms of the names were changed—from Ἀρίσταρχος to Ἀρίσταρχον, and from Κλεάνθη to Κλεάνθης, respectively—thus altering who were the subject and object of the sentence, so that it was Aristarchus who was accused by Cleanthes, and not the

[401] Archim. *Psam.* Pref.; see also: Aet. 2.24; Sext. Emp. *Math.* 10.174; Plut. *Quaest. Plat.* 8.1(1006c).
[402] Russo, *Forgotten Revolution*, pp.66, 79.
[403] Stob. *Ecl.* 1.21.182.20–21, 1.24.204.21–25; Cic. *Acad.* 2.39; Diog. Laert. 8.85.
[404] Russo, *Forgotten Revolution*, p.82.
[405] Plut. *De Facie* 6 [ὥσπερ Ἀρίσταρχον ᾤετο δεῖν Κλεάνθης τὸν Σάμιον ἀσεβείας προσκαλεῖσθαι τοὺς Ἕλληνας, ὡς κινοῦντα τοῦ κόσμου τὴν ἑστίαν, ὅτι τὰ φαινόμενα σῴζειν ἀνὴρ ἐπειρᾶτο, μένειν τὸν οὐρανὸν ὑποτιθέμενος, ἐξελίττεσθαι δὲ κατὰ λοξοῦ κύκλου τὴν γῆν, ἅμα καὶ περὶ τὸν αὐτῆς ἄξονα δινουμένην.]

other way around, as a heliocentric model went against current Church doc-
trine in the seventeenth century.[406] If this is the case, and Aristarchus is the
real subject of the sentence, then it was really Aristarchus who filed charges
against Cleanthes—who had been a vocal critic of Aristarchus and had published a work titled *Against Aristarchus*[407]—possibly for defamation in relation
to his refutation of the heliocentric model. However, it is this notion of a prosecution of Aristarchus that still holds today, and the altered text is found in
almost all modern translations.

Despite this, Aristarchus' idea did not catch on immediately with everyone, and there were those who favoured the idea, and those who opposed
it. It certainly seems to have had its detractor in Cleanthes. Sextus Empiricus
similarly mentions how there were only 'some' who accepted the heliocentric concept.[408] In a different work, Plutarch states that Aristarchus had only
presented the Sun-centred idea as a hypothesis (ὑποτιθέμενος μόνον), while
Seleucus of Seleucia (*ca.* 150 BC) took its ideas of an infinite universe without
a sidereal sphere as definite (ἀποφαινόμενος).[409] As Nicastro points out, the
possible reason for the rejection of a heliocentric model was that many people could not accept that the Earth was moving through space, and could not
explain why things fell to Earth if it was not the centre of everything, nor why,
if the Earth was rotating, the trajectories of flying objects were not affected.[410]
Russo additionally points out how the Aristarchan model would remove the
function of the sphere of stars that surrounded the universe, and could not
account for why, if the Earth was moving, it caused no observable parallax.[411] According to Archimedes, in Aristarchus' model, the lack of parallax is
accounted for by the suggestion that the ratio (λόγος) between the distance
from the Earth's orbit to the sphere of stars, and the centre of the universe to the
sphere of stars, was almost zero—and idea that Archimedes disagreed with on
mathematical grounds.[412] It is also likely that, even though Aristarchus placed
the heavenly bodies in their correct configuration, his model of the universe
was still based on the perfect spheres of Plato and the Pythagoreans, and so
could not account for such things as retrograde motion any better than any of
the geocentric models that had come before. Thus, with Aristarchus, the study

---

[406] Russo, *Forgotten Revolution*, p.82.
[407] Diog. Laert. 7.174.
[408] Sext. Emp. *Math.* 10.174.
[409] Plut. *Quaest. Plat.* 8.1(1006c); for a discussion of Seleucus and the heliocentric model, see: Russo, *Forgotten Revolution*, pp.311–315.
[410] Nicastro, *Circumference*, p.33.
[411] Russo, *Forgotten Revolution*, pp.86–87.
[412] Archim. *Psam.* 5; Archimedes argued (*Psam.* 5) that the ratio between two lengths cannot be zero. However, as per Aristarchus, the ratio can be close to zero.

of the workings of the Heavens takes a dramatic leap in a new and novel (and somewhat correct) direction. However, the failure of that model to account for all of the observable phenomena in the sky meant that it could not be held as truth.

However, the die had been cast, and in the time of Eratosthenes, other theories for the configuration of the universe began to appear which also altered the standard geocentric model. The geometer Apollonius of Perga (262–190 BC), for example, offered a model to account for the retrograde motion of the planets by suggesting that the planets orbited the Sun, with each planetary orbit containing an epicycle, and the Sun, in turn, orbited the Earth which was placed back in the centre of the universe.[413] Furthermore, interest seems to have increased at this time in not only attempting to determine the size of the Sun and Moon, but also of the Earth as well. Russo suggests that the need to describe quantitatively the whole of the known world became an acute need following the sudden expansion of the Greek world by the conquests of Alexander the Great.[414] The work of Archimedes of Tralles (287–212 BC), an intellectual compatriot of Eratosthenes, outlines such things as the concept that the Earth is larger than the Moon, and the Sun is larger than the Earth, and a calculated size for the Sun which was 30x greater than that of the Moon.[415] As part of his attempt to understand the motions of the Heavens, Archimedes constructed an orrery that represented the motions of the celestial spheres.[416] Archimedes also detailed how there had been previous attempts to determine the circumference of the Earth, and the results of his own attempt to do so.[417] Thus, the calculation of the circumference of the Earth seems to have been something of a vogue topic at the time when Eratosthenes was coming into his intellectual prime and may account for why this was one of his chosen research areas.

Even a cursory examination of Greek astronomical and cosmological ideas prior to the time of Eratosthenes highlights some interesting characteristics. Firstly, almost all of the Greek astronomical thought that comes from before the advent of the Hellenistic Age in the fourth century BC, and the establishment of institutions like the Alexandria Library, come from thinkers and philosophers who lived or worked in city-states that either bordered on,

---

[413] Ptol. *Alm.* 12.1.
[414] Russo, *Forgotten Revolution*, p.67.
[415] Archim. *Psam.* 1–4.
[416] Cic. *Rep.* 1.14.21–22.
[417] Archim. *Psam.* 1–2.

or were at one time part of, the Persian Empire. These thinkers, as some ancient texts attest, would have drawn upon previous works that had been composed by cultures that were a part of that empire—Babylon, Egypt, and India—which, while mainly astrological and/or calendrical in nature, laid the foundations for a more scientific examination of the Heavens. Russo suggests that Greeks who later moved into Egypt and Mesopotamia at the time of Alexander's conquests found there a level of technology higher than their own.[418] This is something of an exaggeration as the Greeks had been drawing on eastern texts and information for centuries. Even philosophers from places outside of direct Persian influence—such as Plato and Aristotle in Athens—still drew on such information. However, where the difference lies is that the Greeks took this earlier information, and then engaged with it in a new way in order to 'explain' rather than to just 'understand'.

The result of centuries of engagement with these earlier Eastern texts, and the formulation of so many different models which attempted to explain the workings of the celestial bodies, was that by the time Eratosthenes took up his position in the Library of Alexandria in 245 BC, he would have been able to draw upon a large corpus of work for any topic he chose to research. Strabo, referring to a quote by Hipparchus, states that Eratosthenes had access to many historical works 'with which he was well supplied if he had a library as big as Hipparchus says it was.'[419] Just in regard to the shape of the Earth, there would have been works in the collection of the Library that suggested a flat disk (Homer, Thales, Anaximenes, Anaxagoras), a flat cylinder or tambourine shape (Anaximander, Leucippus), flat, but hollow (Democritus), or spherical (Pythagoras, Xenophanes, Plato, Aristotle, Euclid). In regard to the position of the Sun in relation to the Earth, there were many works that suggested a geocentric model of the universe—with various configurations for the order of the heavenly bodies (Anaximenes, Pythagoras, Anaxagoras, Philolaus, Leucippus, Democritus, Aristotle), and relatively new models containing the heliocentric model (Aristarchus), or one where the Sun orbited the Earth (Apollonius).

As Nicastro points out, by the time of Eratosthenes, despite how old some of these models were, none of them had been disproven.[420] Most seem to have accepted the idea of a spherical Earth and even Eratosthenes, in his early poem *Hermes*, has the god fly away from the Earth and then turn around and look at it. It is spherical with five lateral zones: cold at the poles, hot at the equator, and

---

[418] Russo, *Forgotten Revolution*, p.29.
[419] Str. *Geog.* 2.1.5.
[420] Nicastro, *Circumference*, p.36.

temperate in between.[421] Interestingly, in *Hermes* the Earth is still positioned in the centre of the universe. It is uncertain if Eratosthenes ever fully embraced the heliocentric model of Aristarchus, and his calculation of the circumference of the Earth is not reliant upon a correct understanding of the configuration and motions of the Heavens. Rather, its only real requirements were that the Sun was placed at a great distance so as to make its rays almost parallel when they reach the surface of the Earth, and that the Earth itself was fundamentally spherical in shape (see section 3.2.4).

The evolutionary train of Greek astronomical thought, and many of the socio-political events that occurred across the history of Greece from the seventh century to the fourth, also provide clues as to why it was a place like Alexandria, and not somewhere like Athens, that developed into a seat of intellectual enquiry and learning in the fourth and third centuries BC. Russo argues that the ancient Greek scientific revolution owes much to the development of deliberative and judicial rhetoric—the art of arguing and convincing in assemblies and law courts which had developed in the Greek states.[422] Russo also argues that the scientific revolution depended on the diffusion of written culture and so would not have been possible before the fourth century, and ties the developments in scientific thought only to 'certain forms of democracy'.[423] There are many issues with this assumption. Firstly, there is clear evidence of both scientific thought and a strong literary tradition in many different genres, prior to the fourth century BC. Cerri argues that a connection with rhetoric implies that the origins of debative reasoning may go back to the oral culture of pre-seventh-century BC Greece.[424] This would seem to correlate more closely with the extant literary record. Secondly, much of Greek astronomical thought, even in its earliest and most basic literary forms, comes from places that were not governed as democracies. Shipley argues that open debate between citizens was part of the culture of most city-states, whether large or small, democracy or oligarchy, and many of the cities of Asia Minor where scientific thought came from had also co-existed alongside monarchies.[425] Some cities, such as Halicarnassus, were monarchies themselves. As Barnes states:

[421] Ach. Tat. *Arati Phaenom.* 153a, 153c, 157c; see also: E. Hiller, *Eratosthenis carminum reliquiae* (Leipzig, Teubner, 1872), pp.56–68; as Nicastro points out (*Circumference*, p.86), the notion dividing the Earth into equal parts may have been influential in Eratosthenes' later work on geography in which the known world was divided by a parallel running through the Gibraltar, Athens and Rhodes (which are not on the same parallel in reality) and the regions above and below this line were separated into *sphragides* ('seals') in which all land was distributed equally.

[422] Russo, *Forgotten Revolution*, p.172; see also: Shipley, *Greek World*, p.359.

[423] Russo, *Forgotten Revolution*, pp.29, 172, 197.

[424] G. Cerri, 'Le scienze esalte nel mondo antico' *MedAnt* 1.2 (1998), pp.363–380.

[425] Shipley, *Greek World*, p.59.

'Athenian democracy (and for that matter the Roman Republic) never created a public library. It was only the cultural pretensions of tyrants, kings,
and emperors which could conceive anything like a universal collection of
books.'[426] Furthermore, if democracy was so essential for the establishment
and continuance of scientific intellectual thought, as Russo suggests, this does
not explain why Alexandria, situated in a region controlled by a monarchy,
thrived as the seat of learning and research in the ancient Greek world.

Another cause for doubting some of Russo's claims is that when Athens
was at the height of its democratic period, the *polis* seems to have actually suppressed intellectual thought that was seen as subversive. The trials of
Anaxagoras and Socrates, and the astronomical basis of the charges and tes-
timonies, bear witness to a culture in that city-state that was somewhat toxic
towards cosmological investigations and whose focus was more on the artistic. Intellectual pursuits, such as the composition of plays, would not only
showcase a place like Athens as a focal centre of the arts, but also simultaneously honour the gods, and provide the Athenian state with political avenues
for keeping their people united. As such, playwrights in Athens were well
patroned. Yet this does not seem to have been enough for many intellectuals
in Athens. The poet Pindar had moved to the court of the tyrants of Sicily,
so too had Plato, and the playwrights Euripides and Agathon had gone to
Macedon.[427] Even Aristotle, who had once had a library within his Lyceum in
Athens, moved to Macedon where he was employed as the tutor for the young
Alexander and other youths from among the nobility[428]—many of whom, like
Ptolemy, would go on to rule parts of Alexander's fragmented empire following
his death in 323 BC.

As Hetherington points out, much of Greek society supported playwrights
well, but scientific thinkers were not as generously supported until the establishment of the Museum and Library in Alexandria under the Ptolemies.[429]
Even physicians and architects in classical Athens could charge fees, but
philosophers could only really earn an income via the establishment of
schools—such as Plato's Academy and Aristotle's Lyceum—and drawing fees
from pupils.[430] This may also account for why much of the early Greek scientific thought was fostered in the regions of Asia Minor or, even if schools were
founded by philosophers in Athens itself, why these flourished elsewhere in

[426] Barnes, 'Cloistered Bookworms', p.76.
[427] Ar. *Ran.* 83; *Schol. in Ar. Ran.* 85.
[428] Plut. *Alex.* 7–8.
[429] Hetherington, 'Plato's Place', pp.98–99.
[430] Hetherington, 'Plato's Place', pp.98–99.

the Greek world. Erskine suggests that the establishment of the Library and Museum in Alexandria was following a similar principle of support to that which was given to the arts in Athens—albeit on a much grander scale—by providing not only generational patronage of the arts and the sciences, but also by creating the facilities where such work could be undertaken.[431]

Indeed, the tutelage of young Ptolemy by Aristotle may have had an impact on the foundation of the Alexandria Library. Nicastro, for example, sees a direct impact by Aristotle and suggests that the Aristotelian model of learning favoured analysis based on a comparison of different sources, occasionally written, and this became the rationale for the Alexandria library.[432] Furthermore, Nicastro suggests that Aristotle's library in the Lyceum in Athens may have been an inspiration for people like Alexander and Ptolemy—despite the fact that such a statement seems to ignore the ancient references to it being Ptolemy II and Ptolemy III who were responsible for the creation of the Library in Alexandria.[433] Bagnall sees a more indirect link between Aristotle and the Library. The connection is based on a passage of Strabo, which states that Aristotle taught how to form a library to the Egyptians.[434] Bagnall suggests that this statement can hardly be true as Aristotle died in 322 BC, only a year after Alexander the Great.[435] However, this conclusion does not seem to consider that the idea of a library, and the arranging and use of texts and information along the Aristotelian model, would have been taught to Ptolemy in Macedon as part of his education as a youth. Bagnall, alternatively, finds a link between Aristotle and the Ptolemies through Demetrius of Phaleron—who had been a student of Aristotle's, had ruled Athens for the Macedonian King, Cassander (another of Alexander's former generals), and had then moved to Alexandria where he was instrumental in the establishment of the Library.[436] Regardless of any possible connection between Aristotle, his library, and the rulers and nobility of Macedon and Egypt, it seems clear that Macedonia had adopted a culture of learning, something that had been in a state of stagnation, if not decline, in Athens, and these ideals were then reintroduced to the Greek world following the conquests of Alexander.

Following the death of Alexander the Great in 323 BC, and the fragmentation of his empire into different regions controlled by his squabbling generals, the

---

[431] Erskine, 'Culture and Power', p.40.
[432] Nicastro, *Circumference*, p.65.
[433] Nicastro, *Circumference*, p.65.
[434] Str. *Geog.* 9.1.20.
[435] Bagnall, 'Library of Dreams', pp.350–351; see also: Erskine, 'Culture and Power', pp.39–40.
[436] For the life of Demetrius, see: Diod. Sic. 18.74.1–3; Diog. Laert. 5.75; Ael. *VH* 3.17; For Demetrius and the Library, see: *Epit. Aris.* 9; Tz. *Prol. Com.* 20.

environment was ripe for the reintroduction of Greek intellectual thought—albeit mainly for political reasons. A century earlier, at the end of the fifth century BC, Athens had not only been suppressing astronomical thought, but had also been defeated by Sparta after the lengthy, and costly, Peloponnesian War (431–404 BC). The victorious Spartans where a heavily religious and austere culture, with little interest in intellectual pursuits, and were controlling Greece as puppets of the real victor of the Peloponnesian War, Persia. Persia had entered the war in the later stages of the conflict by offering financial support to the Spartans—whose culture had prohibited the use of money centuries earlier as it was seen as a corrupting luxury. The Persian policy was to keep the Greeks fighting among themselves, as the Persians feared that an eventual winner of the conflict would turn the eyes of a united Greece against Persia and seek revenge for their invasions of 490 BC and 480 BC. The result was that Persia would support the weakest side of the conflict, help them regain a position of strength, and then change which side they were supporting—with Persian aid going back-and-forth between the major players of the conflict. This continued for decades following the 'official' end of the Peloponnesian War, kept the conflict between the city-states going, and kept the Greek world fragmented.

By the mid fourth century BC, most of the city-states of Greece were bankrupt—financially, militarily, politically, and intellectually—following decades of conflict. Areas on the fringes of the Greek world, however, regions like Sicily, southern Italy, and Macedonia, had avoided much of conflict that was occurring in central Greece and had thrived culturally. For example, the tyrants of the Greek cities of Sicily had hosted playwrights like Aeschylus of Athens as early as the fifth century BC and would later be the home of intellectuals like Archimedes. In southern Italy, the Pythagorean school of thought thrived. Macedonia had also been the host of playwrights like Euripides and thinkers like Aristotle.

Following the death of Alexander, there was rivalry between the Successors—not just on the battlefield, but to assert a cultural and intellectual dominance on the Greek world as well.[437] This was because everything in the Greek world was a contest—an *agon*. In all of the Greek states, individuals would compete in the political arena, in the law courts, in dramatic festivals, and in every other aspect of their daily lives. On a larger scale, the city-states competed militarily on the battlefield, culturally in their festivals and in their

---

[437] It is difficult to establish a clear understanding of the war of the Successor period due to the constantly shifting alliances, political intrigues, and the fragmentary nature of the surviving source material. The best ancient accounts of this period can be found in books 18 and 19 of Diodorus' *Library of History*. A good, and concise, summary of the events of the time can be found in Shipley, *Greek World*, pp.40–46.

control of religious sanctuaries, and athletically in the great sporting events of the Greek world. This competitive nature correlates, in part, with the notion that scientific enquiry was born out of the political freedom and rhetoric found in some *poleis*. However, it seems that the nature of this agonistic mindset, and its influence on the evolution of scientific thinking and investigation, was extremely more multifaceted and complex. The spirit of competition permeated every aspect of life for the ancient Greeks, and intellectual enquiry was not immune to its influences.

For the early Ptolemies, it was not enough just to be dominant on the battlefield if they wanted to project an image of being the legitimate successors to Alexander and the heirs of all of Greece's cultural and intellectual legacies. It was also not enough for the Ptolemies to claim that they were the rightful protectors of Alexander's empire because they had the body of the former king interred in Alexandria,[438] nor to have established their capital in a city founded by, and named after, their former ruler. Other Successor generals had similarly powerful claims that they could make. Seleucus, for example, held Babylon, where Alexander had died, and Antipater held Greece and Macedon—the heartland of Greek culture—and had acted as regent and protector of Alexander's older half-brother, Philip Arrhidaeus, and Alexander's posthumous son, Alexander IV.[439] Many of the Successor kings were also relatively equal in military terms.

The Ptolemies also lacked any legitimate claim to the throne of Egypt other than sheer military power and so they needed to create a strong link between themselves and Alexander—who had been crowned pharaoh—in order to remain in position. Stories were circulated, for example, that Ptolemy I and Alexander had the same father.[440] Ptolemy I also published his own account of Alexander's campaign, which greatly emphasized their relationship (although not in direct familial terms).[441] As part of an extravagant religious procession through the city streets, statues of Ptolemy and Alexander stood side-by-side, surrounded by images of other figures and deities, in a further demonstration of the supposed connection between the two.[442] Thus, for the Ptolemies, the propaganda aspect of the wars of the Successors entered a new, and intellectual, theatre. Bagnall sees the establishment of the Library as part of a

---

[438] Str. *Geog.* 17.794; Diod. Sic. 18.26–28; Paus. 1.6.3.

[439] Diod. Sic. 18.2.1–4, 18.18.1–8, 18.39.1–7; Philip Arrhidaeus was executed by Alexander's mother, Olympias, in 317 BC (Diod. Sic. 19.11.4–7), and Alexander IV was executed by Cassander in 310 BC (Diod. Sic. 19.105.1–4).

[440] Paus. 1.6.2.

[441] Arr. *Anab.* 1.1.

[442] For the procession as a whole, see: Ath. *Diep.* 5.197d–203d; for the statues of Ptolemy and Alexander, see: Ath. *Diep.* 5.201d.

Ptolemaic project aimed at making Alexandria a rival to Athens.[443] However, the objective seems to have been much larger. The establishment of both the Library and the Museum in Alexandria presented Egypt as not only the legitimate continuation of Alexander's empire, but also as the curators and protectors of all of Greek culture. Additionally, by establishing a library—an institution that may have been based on the collection of Alexander's former tutor, Aristotle—the Ptolemies could further demonstrate a link between themselves and the former king, but the royal funding of the Library would additionally differentiate it from institutions such as Aristotle's Lyceum. Johnstone sees the establishment of libraries as statements of 'aristocratic and royal munificence', and objects of 'aristocratic and royal display and propaganda', which were not so much repositories of knowledge, but 'fundamentally political institutions'.[444]

There was another aspect of this game of politics and propaganda that the Ptolemies needed to consider: the cultural needs of the many Greeks who now inhabited Alexandria. During the reigns of the first two Ptolemies, many of the Greeks who resided in Alexandria were immigrants, like Eratosthenes, who had severed their ties to their home city-states. Many of these immigrants, other than the Macedonians, would have been unused to living under monarchies and the Ptolemies may have wanted to create a cultural link back to the homelands of these people in order to promote stability. Ptolemaic coins and art, for example, are distinctly Greek. The layout of Alexandria itself may have also mirrored that of Pella—the main city of Macedonia—which would have given the Macedonian inhabitants of Alexandria a link to their homeland.[445] The establishment of the Library and Museum may have also given these people a tangible connection to their former cultural traditions and even expanded on them by promoting the patronage of the sciences as well as the arts. These two institutions then became the focal point for the application of Greek culture onto the entire city. This would have had been essential for Ptolemaic control of the region in the early years after the city's foundation, and for the continued subjugation of Egypt as a whole—which Diodorus states contained seven million people under the Ptolemies.[446] To help spread Hellenism and provide security and stability among such a vast

[443] Bagnall, 'Library of Dreams', p.349.
[444] Johnstone, 'A New History of Libraries', pp.347, 349.
[445] R.A. Tomlinson, 'The Town Plan of Hellenistic Alexandria' in P. Schmidlin, G. Monaco, M. Trojani, and D. Said (eds.), *Alessandria e il Mondo Ellenistico-Romano* (Rome, L'Erma di Bretschneider, 1995), pp.236–240.
[446] Diod. Sic. 1.31.6–8.

population, the Ptolemies settled former soldiers throughout the land, 'Greek-ness' was promoted through positions and tax incentives—especially if an Egyptian could demonstrate at least one Greek or Macedonian ancestor, and intermarriage was allowed in many areas.[447] Furthermore, the Macedonians had been viewed by some in the Greek world, the Athenians in particular, as not quite Greek.[448] However, MacLeod suggests that the Library was created by Macedonian rulers who had a vested interest in acquiring Eastern knowl-edge and then translating it as part of the dissemination of a syncretized system of Hellenic imperialism across the known world.[449] Indeed, by supporting the collection of knowledge from Greece, or by translating knowledge from other areas into Greek, and then making that knowledge available to a wider world, the Ptolemies, and the inhabitants of Alexandria, were able to project an image of the Egyptian Macedonians as the true continuation of Hellenic culture—with the Library at its very centre.

However, the creation of such institutions did not guarantee the longevity or success of either Ptolemaic policy, or of the Library and Museum. What was required was continued patronage to ensure that this claim to the cul-tural legacy of the Greek world could be made and then maintained. What partially explains why the Ptolemies were able to promote such an agenda so successfully for such a long time was that Egypt was immensely rich.[450] The Ptolemies could draw upon the wealth of Egypt to encourage the emigration of philosophers and thinkers to Alexandria, to patron research in the Library and Museum, and to collect books, through the variety of means that were under-taken, to ensure that the cultural traditions of the Greeks remained under their control. As Shipley points out, the Greeks had a long tradition of constructing buildings for public use and these structures could bring material and social benefits to the state, but could also act as a demonstration of power.[451] Ptolemy I may have also announced a large series of benefactions when he first took control of Egypt.[452] This may have been Ptolemy using the wealth of Egypt to both continue pharaonic traditions and to begin the promotion of Hellenistic

[447] For the settling of ex-soldiers, see: D.J. Crawford, *Kerkeosiris: An Egyptian Village in the Ptole-maic Period* (Cambridge, Cambridge University Press, 1971), pp.53–58; for positions and incentives, see: N. Lewis, *Greeks in Ptolemaic Egypt: Case Studies in the Social History of the Hellenistic World* (Oxford, Clarendon, 1986), p.28; for intermarriage, see: W. Clarysse, 'Greeks in Ptolemaic Thebes' in S.P. Vleeming (ed.), *Hundred-Gated Thebes: Acts of a Colloquium on Thebes and the Theban Area on the Greco-Roman Period* (Leiden, Brill, 1995), pp.1–19.

[448] Dem. 9.31.

[449] MacLeod, 'Introduction', pp.2–3.

[450] Egypt had large gold deposits in the south, but according to Diodorus (3.12.1–3.14.5) they were difficult to exploit.

[451] Shipley, *Greek World*, pp.83–84.

[452] Diod. Sic. 18.14.1–2.

traditions. This placed the Ptolemies in a strong position to claim to be the protectors of Greek culture. As Nicastro points out, the wealth of Egypt was far greater than any of the Greek city-states, which would be in no position to compete in this intellectual and cultural arena, and this may have partially been why knowledge and enquiry thrived in Alexandria.[453]

The ancient writer Strabo states that the art of persuasion was a characteristic of orators and not kings, and that royal persuasion came in the form of some sort of benefaction.[454] This sentiment is at the heart of why it was that Alexandria, and not a place like Athens, became the focus of intellectual research in the ancient world. Clearly, the art of rhetorical persuasion would have been useful in the political and judicial debates found in a place like Athens, and may have even contributed to the rise of scientific enquiry as some have suggested. However, while such a culture of verbal point and counterpoint would have created an environment in which the arts were sponsored and thrived, it would not have been enough for science. In the fickle and ever-changing world of Athenian politics, for example, the patronage of a playwright for the course of a year while they composed a play for performance at one of the city festivals would have been relatively straightforward as the government of the day, and therefore the level of support, would not change too much over the course of that year. Philosophers, reliant upon the fees charged by their respective schools, would have similarly not been at the mercy of changing political conditions in the state unless they presented something, as Anaxagoras had done, which went against the collective ideal. Thus, in Athens, patronage was a very short-term phenomenon, more suitable to the production of plays and other artistic outcomes. Under a monarchy such as was found in Egypt, on the other hand, the long-term nature of the government meant that both time and funds could be committed to research and projects into areas that would take longer to reach a result, if at all. An Athenian benefactor would want a result to be reached. As such, things like plays were sponsored. Philosophical ideas and ideals produced no real tangible outcome, and astronomical science had no guarantees other than potentially making yet another contribution to a debate over the nature of the Heavens that had been going for centuries. Therefore, the philosophers were left to their own devices, and the scientific thinkers were hardly supported at all. Yet under the Ptolemies, funds were distributed, and places of learning established, where long-term research, with no guarantee of success and/or outcome, could be undertaken. This explains

---

[453] Nicastro, *Circumference*, p.65.
[454] Str. *Geog.* 9.2.40.

why, when people such as Eratosthenes reached a result, that information was erected on a monument, and sent to the ruler in the form of correspondence: it was a means of both advertising and recording the results of research that had been funded by the state which, in turn, aided the projection of Alexandrian intellectual supremacy to the world.

Russo suggests that research such as Eratosthenes' attempt to calculate the circumference of the Earth may also have had a more practical application— that it may have been sponsored by the state, as it was necessary for the creation of accurate maps.[455] MacLeod similarly suggests that the people working in the Library, like Eratosthenes, turned to the study of mathematics and geography as it would have been useful for both the annual land surveys that were undertaken in Egypt and also for naval navigation in the Mediterranean.[456] However, while it is clear that the Egyptians had been using land surveys to determine annual tax levels for centuries, it is uncertain what benefit knowing the circumference of the Earth could bring. Alexander's empire had been surveyed extensively and accurately, and the distances between locations measured, during the campaign. Many of these locations were now being fought over by the Successors and there would have been little need to update any information on their position, or the best way to get to them, for strategic purposes. Rather, it seems more likely that research such as calculating the distance around the Earth was done more out of a desire to have the intellectuals in Alexandria engage with a topic that was currently in vogue and, by coming up with a new solution to what had been an ongoing topic of scholarly debate, assert Alexandria's cultural dominance. Some thinkers and philosophers in the Greek world seem to have actually been against the use of their ideas and research for any sort of applied purpose. Archimedes, for example, seems to have been particularly averse to the practical application of much of his work, and preferred to have much of what he had done remain entirely theoretical.[457] This suggests that one of the key elements of research in the ancient Greek world was to engage in a game of intellectual one-upmanship. It is therefore unsurprising that Eratosthenes was the first person from Alexandria to offer an answer to a question that had prompted responses and theories from thinkers elsewhere in the Greek world.

The desire to engage with topics of current scientific debate in order to establish the intellectual dominance of Alexandria in the Greek world may also explain why Eratosthenes had been appointed to the position of Chief

[455] Russo, *Forgotten Revolution*, p.276.
[456] MacLeod, 'Introduction', p.6.
[457] Plut. *Marc.* 17.

Librarian in the first place. A papyrus fragment (*P. Oxy.* 1241) that lists the directors of the Library in Alexandria places Eratosthenes as the second person to hold the position. However, there is a lacuna at the start of the text and the *Suda* details Zenodotus of Ephesus as the first Chief Librarian.[458] Both the *Suda* and *P. Oxy.* 1241 then list Apollonius of Rhodes as the second Chief Librarian (which then makes Eratosthenes the third).[459] Zenodotus was an epic poet and grammarian, and an editor of the works of Homer.[460] Apollonius was also a writer and poet and had been an associate of Callimachus—the teacher of the young Eratosthenes.[461] As such, neither of the two previous Chief Librarians had engaged with science. Eratosthenes' first major work, on the other hand, had engaged with the ideas of Platonic mathematics, which were the basis for much of the astronomical understanding of the time. The associated Museum was also a temple to the nine Muses—of which only one was for a science: Urania, the Muse of astronomy—and this provides another link between Eratosthenes' early work and the institutions in Alexandria.[462] Whether the Ptolemies intended to have different Librarians devote their works to different Muses (or several of them) is unknown. Under the first director, for example, Zenodotus worked on epic poetry, while other staff worked on comedy and tragedy.[463] However, the shift from poets and literary critics to a scientific thinker like Eratosthenes suggests that the focus of the research that was to be undertaken in Alexandria was changing by the time of the Library's third director, and this may have been part of the need to exert a level of intellectual dominance on the ancient Greek world by the Ptolemies through their support of an environment conducive to free and open scientific investigation.

The safe and somewhat leisurely lifestyle experienced by the Greeks residing in Alexandria would have also facilitated the emigration of thinkers, and the promotion of research, under the patronage of the king. This air of stability seems to have been brought about, in part, by Egypt's military strength. Polybius, for example, states that earlier Egyptian rulers had controlled territory form Egypt to the Hellespont, and this kept the other rulers of the time in check and kept Egypt stable.[464] This programme for promoting research

---

[458] See: *P. Oxy.* 1241; *Suda s.v.* Ζηνόδοτος.
[459] See: *P. Oxy.* 1241; *Suda s.v.* Ἀπολλώνιος.
[460] *Suda s.v.* Ζηνόδοτος; Tz. *Prol. Com.* 21–22.
[461] *P. Oxy. 1241*; *Suda s.v.* Ἀπολλώνιος.
[462] The nine Muses were: Calliope (epic poetry), Clio (history), Erato (love poetry), Euterpe (music), Melpomene (tragedy), Polyhymnia (sacred poetry), Terpsichore (dance), Thalia (comedy), and Urania (astronomy). See: Hes. *Theog.* 36–79.
[463] Tz. *Prol. Com.* 21–22.
[464] Polyb. 5.34.

within this steady environment of Egypt seems to have been very successful. According to Strabo, by the time Eratosthenes arrived in the city, places like the Library and Museum were crowded with people. Eratosthenes is said to have stated that 'never ... at one time, and in only one city, were there so many philosophers flourishing together as there was in my time'.[465] Similarly, Timon of Phlius called Alexandria the 'Birdcage of the Muses' and described all of the people who he said were endlessly working there.[466]

This strong tradition of research and collecting/preserving the cultural heritage of the Greek world explains why, when texts from foreign cultures were copied for the Alexandrian collection, they were translated into Greek. It seems that it may not have been enough to collect everything that had come before. Rather, the Ptolemies wanted to be seen as adding to the Greek collective knowledge base. This would also aid any claim to being the legitimate heir of the Greek world. Furthermore, acquiring all of this knowledge was one thing, but to add strength to any claim to cultural curatorship, that knowledge also had to be shared. This also partially explains why foreign texts were translated into Greek in Alexandria. Athenaeus states that the Alexandrians became the teachers to all Greeks and barbarians.[467] Furthermore, the collection in the *Serapeum* Annex was open to the public, and the inhabitants of Alexandria were encouraged to study.[468] This would suggest that many people had access to the texts that were being kept in the Library and, as Nicastro states, a public-access section to the Library would also suggest a level of literacy among some of the inhabitants of Alexandria (most likely the Greeks and higher-class Egyptians).[469] It was this educated strata of Alexandrian society, and the educated classes of the Greek world in general, who would have been the target audience for the results that came from the research undertaken in in the Library and Museum, as they would have had the leisure time to both engage with the results and any subsequent debate on the topic. This would have then aided the dissemination of the ideas behind the research as part of establishing Alexandria's position in the struggle for cultural supremacy.

The use of the Library, its collection, and the research undertaken there as a tool in a propaganda campaign for dominance in the Greek world also explains why places like the Library in Pergamon were established. The Attalids, who

[465] Str. *Geog.* 1.2.2.
[466] Ath. *Diep.* 1.22d.
[467] Ath. *Diep.* 4.184b–c.
[468] Apth. *Prog.* 12.107.
[469] Nicastro, *Circumference*, p.72.

ruled Pergamon, were somewhat latecomers to the Wars of the Successors—having broken away from the control of Lysimachus (another of Alexander's successors) in 283 BC—and while they were not always the equals of the other Successor kingdoms militarily, by also engaging on the intellectual battlefield, a ruler like Eumenes I (ruled 263–241 BC) or Attalos I (ruled 241–197 BC) could quickly rise above many of the others.[470] This would have been especially important for the Attalids as they did not possess other things—such as Alexander's body, his son, a city named after him, or where he had died—to help support any of their claims to supremacy. However, the Attalids were still able to lay claim to being protectors of Hellenism in other ways. Attalos I commemorated his victory over the Gauls (who were seen as barbarians by the Greeks) through great works of art with a distinctly Hellenic flavour.[471] Following victories in another series of conflicts, this time against the Seleucids who were in control of the city, Pergamon was able to assert its independence in 218 BC,[472] and by 200 BC, Pergamon was fighting as an ally of Rome against Macedon.[473] Later Pergamene rulers such as Eumenes II (ca. 220–158 BC) continued the tradition of upholding Greek cultural norms by sending lavish gifts to the cult centre of the god Apollo at Delphi.[474] In the war for cultural supremacy in the Greek world, the Ptolemies and the Attalids engaged in a librarial contest in an attempt to outdo each other. It is, then, hardly surprising that both lines of rulers established libraries and museums, and supported extensive research programmes. It was as a part of the Ptolemaic-sponsored programme that people like Eratosthenes were brought to Alexandria from Athens to run the Library and to conduct research into some of the leading intellectual problems of the day such as the 'Delian Problem'—which we know Eratosthenes solved while in Alexandria and had the results erected on a monument—and the determination of the size of the Earth.

---

[470] A chronological summary of the Attalid kings of Pergamon can be found in Strabo's *Geography* (13.4.1–2).

[471] For Attalos' victory over the Gauls, see: Polyb. 18.41.7–8. One of the most famous of the works of art commemorating this victory is the statue of the 'Dying Gaul'. A Roman era copy of this statue is now located in the Capitoline Museum in Rome.

[472] Polyb. 21.17.6; App. *Syr.* 38.

[473] Polyb. 16.25.1–16.26.10.

[474] Livy, 42.5.3.

2

The Experiment

2.1 Eratosthenes' Experiment

No record exists of what point during his career Eratosthenes undertook his experiment to calculate the circumference of the Earth. Newton dates the experiment to 300 BC.[1] However, this is twenty-four years prior to the time in which Eratosthenes is said to have been born (276 BC) and seems highly unlikely. Only a rough time frame is possible: Eratosthenes had to have undertaken his experiment sometime between when he was appointed to the post of Chief Librarian in Alexandria in 245 BC, and the time of his death in *ca.* 196 BC.

There had been prior attempts to calculate the distance around the Earth.[2] Aristotle, for example, refers to observations of such things as the shape of the Earth's shadow on the Moon during an eclipse, and stars that can only be seen in certain locations—both of which he uses to justify the argument that the Earth is a sphere—and calculations that placed the planet's circumference at 400,000 *stadia*.[3] Evans dates this calculation to around 350 BC.[4] Horace also mentions measurements of the Earth taken by Archytas in the fourth century BC, but provides no details as to what these measurements were.[5] A century later, Archimedes stated that some calculations placed the circumference at 300,000 *stadia*, while his own estimates placed it at 3,000,000 *stadia*—which would give the Earth a diameter bigger than that of Jupiter.[6] These determinations were based upon trigonometric calculations utilizing the observations of the angle of elevation of various stars at different locations and the distances between the observing sites (the so-called 'astronomical

---

[1] Newton, 'Sources of Eratosthenes' Measurement', p.379.
[2] Sagan (*Cosmos*, p.145) claims that Eratosthenes was the first person in history to measure the size of a planet—a feat not accomplished again until the time of Christian Huygens. However, there are numerous passages, many referred to here, that attest to attempts to calculate the size of the Earth both prior to Eratosthenes' attempt, and following Eratosthenes but before the time of Huygens.
[3] Arist. *Cael.* 297b31–298b.
[4] J. Evans, *The History and Practice of Ancient Astronomy* (Oxford University Press, Oxford, 1998), p.61.
[5] Hor. *Carm.* 1.28; even earlier (fifth century BC), Aristophanes' *Clouds* (lines 202–205) similarly mentions the measuring of the land. Whether both of these texts refer to attempts to calculate the circumference of the planet, or some other form of geographical calculation, is not specified.
[6] Archim. *Psam.* 1; Dreyer (*History of Astronomy*, pp.173–174) attributes this measurement to Dicaearchus *ca.* 285 BC.

Eratosthenes and the Measurement of the Earth's Circumference (c.230bc). Christopher A. Matthew, Oxford University Press. © Christopher A. Matthew (2023). DOI: 10.1093/oso/9780198874294.003.0002

method'). Eratosthenes took the principles set down in these calculations, rotated them by 180°, and attempted to refine the methodology and the result using a 'geometric method'.[7]

Eratosthenes published the results of this work in a text titled *The Book of Measurements* (*Libri Dimensionum*) according to the writer Macrobius, while Heron of Alexandria calls it *On the Measurement of the Earth* (Περί της αναμετρήσεως της γης).[8] Sadly this work, whatever its name, has not survived the passing of the centuries. However, the details of Eratosthenes' methodology, calculations and results have survived in a description of the experiment found in the later work *On the Circular Motions of Celestial Bodies* (Κυκλικὴ Θεωρία Μετεώρων) by Cleomedes. Cleomedes states:[9]

> Let us suppose, in this case, first that Syene and Alexandria sit under the same meridian [i.e. longitudinal] circle; second, that the distance between the two cites is 5,000 *stadia*; and third, that the rays sent down from different parts of the Sun onto different parts of the Earth are parallel . . . Fourth, let us assume that . . . straight lines falling on parallel straight lines make alternate equal angles . . . Since meridian circles are great circles in the universe . . . the circle on the Earth passing through Syene and Alexandria . . . will be the size of the great circle [i.e. circumference] of the Earth. Eratosthenes states, and it is a fact, that Syene sits under the summer tropic [i.e. the Tropic of Cancer]. Therefore, whenever the Sun, being in the Crab at the Summer Solstice, is exactly in the middle of the sky, the *gnomons* [i.e. the pointers on sundials] necessarily cast no shadows as the Sun in exactly vertical above them[10] . . . However, in Alexandria at the same hour the *gnomons* cast shadows

[7] For an examination of how the prior attempts to calculate the circumference of the Earth may have influenced Eratosthenes' methodology, see: Pinotsis, 'Comparative Study', pp.127–138.

[8] Macr. *In Somn.* 1.20; Hero. *Dioptr.* 35.

[9] The section containing this passage was originally omitted from a collection of astronomical and mathematical texts, possibly from the fourteenth or fifteenth centuries, titled Ἀράτου Φαινόμενα. Κλεομήδους περὶ κυκλικῆς θεωρίας μετεώρων. Νικόμαχου Γερασήνου Πυθαγορικοῦ ἀριθμητικὴν εἰσαγωγή (acquired by the Central Scientific Library of V.N. Karazin Kharkiv National University, Ukraine around 1807), but was added into the margins at an unknown later date. For a discussion of the manuscript tradition of Cleomedes' work, see: R.B. Todd, 'Cleomedes' in V. Brown, P.O. Kristeller, and F.E. Cranz (eds.), *Catalogus Translationum et Commentariorum: Medieval and Renaissance Latin Translations and Commentaries Vol. VII* (Washington DC, Catholic University of America Press, 1992), pp.1–11.

[10] Cleomedes also states (in sections of the passage not cited here—*De motu* 1.10, 2.1) that this lack of shadow occurs in an area with a diameter of 300 *stadia*. As Bowen and Todd (*Cleomedes' Lectures*, p.82) point out, 300 *stadia* is exactly 1/800th of 240,000 *stadia*, and 1/600th of 180,000 *stadia*—the two figures for the circumference of the Earth reportedly calculated by Posidonius (Cleom. *De motu* 1.10, Str. *Geog.* 2.2.2). It has also been suggested that the figure of 300 *stadia* was obtained from a demonstration of water clocks which showed that the Sun is 1/750th the size of its own orbit (Cleom. *De motu* 2.1) and that, based upon Cleomedes' concept of meridian circles being circles of the universe, that a circumference of 240,000 *stadia* divided by 750 equals 320 *stadia*, which was then rounded to an

as Alexandria is further north than Syene ... If we then project straight lines from each of the pointers through the ground, they will meet at the centre of the Earth ... while on the angle ... stands the arc reaching from Syene to Alexandria ... [T]he arc is found to be one-fiftieth of the great circle. And the stated distance is 5,000 *stadia*. Therefore, the complete great circle measures 25 *myriads*.[11]

Cleomedes states that the distance for the circumference was calculated to be 25 *myriads*. A *myriad* was a unit of measure in the ancient Greek world representing a value of ten thousand. As such, the result of Eratosthenes' calculations was 250,000. However, the question is: 250,000 of what? It would seem that the answer to this question is 250,000 *stadia*, based not only on the use of this unit of distance by Cleomedes in earlier passages, but also on other writers who refer to Eratosthenes' result. This fairly valid assumption has made its way into most translations of Cleomedes and into modern examinations of Eratosthenes' calculation of the circumference of the Earth.

In other words (and in simpler terms), Eratosthenes began with several pieces of 'known' knowledge: (a) the cities of Alexandria and Syene (modern Aswan) sit of the same longitudinal line (not correct); (b) the distance between the two cites was 5,000 *stadia* (possibly rounded, but close to accurate—see following); (c) light rays from the Sun are fundamentally parallel when they reach the Earth (not correct); (d) when parallel rays strike two parallel sticks, their shadows will form the same angle (correct);[12] and (e) the city of Syene sits under the Tropic of Cancer (not entirely correct, but not influential in the

---

even 300 *stadia* in the text (see: O. Neugebauer, *A History of Ancient Mathematical Astronomy* (Berlin, Springer, 1975), pp.655–656).

[11] Cleom. *De motu* 1.10: Ὑποκείσθω ἡμῖν πρῶτον μὲν κἀνταῦθα, ὑπὸ τῷ αὐτῷ μεσημβρινῷ κεῖσθαι Συήνην καὶ Ἀλεξάνδρειαν, καὶ δεύτερον, τὸ διάστημα τὸ μεταξὺ τῶν πόλεων πεντακισχιλίων σταδίων εἶναι, καὶ τρίτον, τὰς καταπεμπομένας ἀκτῖνας ἀπὸ διαφόρων μερῶν τοῦ ἡλίου ἐπὶ διάφορα τῆς γῆς μέρη παραλλήλους εἶναι ... Τέταρτον ἐκεῖνο ὑποκείσθω ... τὰς εἰς παραλλήλους ἐμπιπτούσας εὐθείας τὰς ἐναλλὰξ γωνίας ἴσας ποιεῖν ... Ἐπεὶ οὖν μέγιστοι τῶν ἐν τῷ κόσμῳ οἱ μεσημβρινοί, δεῖ καὶ τοὺς ὑποκειμένους τούτοις τῆς γῆς κύκλους μεγίστους εἶναι ἀναγκαίως. Ὥστε ἡλίκον ἄν τὸν διὰ Συήνης καὶ Ἀλεξανδρείας ἥκοντα κύκλον τῆς γῆς ἡ ἔφοδος ἀποδείξει αὕτη, τηλικοῦτος καὶ ὁ μέγιστος ἔσται τῆς γῆς κύκλος. Φησὶ τοίνυν, καὶ ἔχει οὕτως, τὴν Συήνην ὑπὸ τῷ θερινῷ τροπικῷ κεῖσθαι κύκλῳ. Ὁπόταν οὖν ἐν καρκίνῳ γενόμενος ὁ ἥλιος καὶ θερινὰς ποιῶν τροπὰς ἀκριβῶς μεσουρανήσῃ ... κατὰ κάθετον ἀκριβῆ τοῦ ἡλίου ὑπερκειμένου ... Ἐν Ἀλεξανδρείᾳ δὲ τῇ αὐτῇ ὥρᾳ ἀποβάλλουσιν οἱ τῶν ὡρολογίων γνώμονες σκιάν, ἅτε πρὸς ἄρκτῳ μᾶλλον τῆς Συήνης ταύτης τῆς πόλεως κειμένης ... Εἰ οὖν ἑξῆς νοήσαιμεν εὐθείας διὰ τῆς γῆς ἐκβαλλομένας ἀφ' ἑκατέρου τῶν γνωμόνων, πρὸς τῷ κέντρῳ τῆς γῆς συμπεσοῦνται ... ἐπὶ δὲ τῆς πρὸς τῷ κέντρῳ τῆς γῆς ἡ ἀπὸ Συήνης διήκουσα εἰς Ἀλεξάνδρειαν ... Ἡ δέ γε ἐν τῇ σκάφῃ πεντηκοστὸν μέρος εὑρίσκεται τοῦ οἰκείου κύκλου. Δεῖ οὖν ἀναγκαίως καὶ τὸ ἀπὸ Συήνης εἰς Ἀλεξάνδρειαν διάστημα πεντηκοστὸν εἶναι μέρος τοῦ μεγίστου τῆς γῆς κύκλου· καὶ ἔστι τοῦτο σταδίων πεντακισχιλίων. Ὁ ἄρα σύμπας κύκλος γίνεται μυριάδων εἴκοσι πέντε.

[12] See: Euc. *El.* 1.28–30.

result of the experiment—see following). It was on this foundation that Eratosthenes based his calculations. He seems to have known that, at midday, on the Summer Solstice, the *gnomon* on a sundial in Syene would cast no shadow, but a *gnomon* in Alexandria would cast a more significant shadow due to its more northerly location. The angle created by the shadow cast in Alexandria is equal to the angle of intersection between two lines, extended downwards from each of the *gnomons*, at their point of intersection at centre of the Earth. The greater the difference in the shadows, the greater the curvature of the Earth. Eratosthenes measured this angle to be one-fiftieth of a circle ($\approx$7.2°). As such, the reported 5,000 *stade* distance between Alexandria and Syene should be one-fiftieth of the circumference of the Earth—a calculated distance of 250,000 *stadia* (Fig. 4).

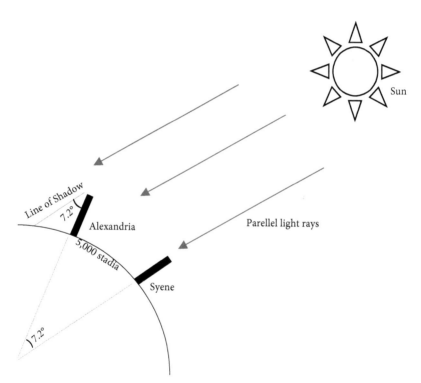

Fig. 4 Eratosthenes' method for calculating the circumference of the Earth (with angles and curvature exaggerated).

Despite the basic concept and principles behind Eratosthenes' experiment, from this point onwards, problems start to arise with our understanding of his calculations and results.

Some modern retellings of Eratosthenes' work, for example, contain different details (usually unreferenced) of the experiment. Some state that the knowledge of a lack of shadows at Syene during the solstice was based upon observations of columns or obelisks, rather than a *gnomon*, and/or that the overhead position of the Sun was determined by observations of it shining directly down a well.[13] The origin of the well story may be based on an incorrect interpretation of a passage found in the works of Strabo. In Book 17 of his *Geography*, Strabo describes the 'Nile-o-meter' (νειλομέτριον) found in Syene. The Nile-o-meter was a circular, vertical shaft (which Strabo describes as being 'like a well' (φρέαρ)) which was connected to the waters of the Nile. The walls inside the Nile-o-meter were inscribed with markings indicating the height of the average flood, as well as levels for greater and lesser inundations.[14] When the waters of the river began to rise with the annual flood, the water level inside the Nile-o-meter would also rise. Officials would be able to descend into the shaft and record the height of the flood waters. This had an important impact on Egyptian society from year to year. The inundation would spread rich silt and other nutrients across the farmland on either side of the Nile. Thus, the height of the flood would determine how much arable land was available for the coming year. This, in turn, was used to determine the tax rate for the next twelve months.[15] A flood that was below average could be an indication of an imminent food shortage in Egypt, depending upon how low the waters were, or would dictate a lower-than-average tax rate for the coming year at the very least. A higher-than-average flood would mean that the waters had spread into the urban areas which were situated further back from the river than the farmland. The government could then prepare for wide-scale rebuilding programmes. In addition, the transportation of the large stone blocks that were used in the construction of many of the temples and other grand monuments along the Nile was only carried out during the time of the flood as the higher waters were used to carry the blocks by boat from quarries in Syene in the south to the respective building sites down river. Furthermore, with their farms temporarily submerged, farmers were not able to tend their

---

[13] For example, see: Fischer, 'Another Look', pp.154–155; O'Neil, *Early Astronomy*, pp.139–140; M. Kline, *Mathematical Thought from Ancient to Modern Times* (Oxford, Oxford University Press, 1990), p.160; A.E. Roy and D. Clarke, *Astronomy Principles and Practice* (Bristol, Institute of Physics Publishing, 2003), pp.10–11; Rawlins, 'Too-Big Earth', p.6; Evans, *Ancient Astronomy*, p.20; Nicastro (*Circumference*, pp.25, 107) suggests, despite any reference to a well being present in the account of Eratosthenes' work, that it was a multilingual library assistant who informed Eratosthenes of the story about the well, and that for his calculations, Eratosthenes had to have assumed that the well was perpendicular.
[14] Str. *Geog.* 17.1.48.
[15] Str. *Geog.* 17.1.48.

fields and these men were then employed in the work crews for many of these construction projects.

A representation of the Nile-o-meter in Syene may be preserved in the Palestrina Nile Mosaic from Hellenistic-period Italy (*ca.* second to first century BC) (Figs. 5 and 6)[16].

Fig. 5 River scene in the Palestrina Nile Mosaic.[a]
Image supplied by kind permission of the Lazio-Rome Regional Museums Directorate, Photographic Archive.

While the remains of the Syene Nile-o-meter have never been found, a similar, well-like Nile-o-meter can be seen in the temple complex at Kom Ombo, north of Syene (Fig. 7).

The Nile-o-meters located at the temples of Karnak in Thebes (modern Luxor), the temple of Horus at Edfu, and on Elephantine Island near Syene were shaped like a descending staircase rather than a well, but operated on the same principle (Fig. 8).

Thus, a careful monitoring of the level of the Nile, in order to determine the time of the arrival of the annual flood, was a crucial aspect of the functioning of ancient Egypt. Importantly, Strabo also says that the Nile-o-meter in Syene

---

[16] For the interpretation of this part of the mosaic being a depiction of the Nile-o-meter at Syene, see: P.G.P. Meyboom, *The Nile Mosaic of Palestrina: Early Evidence of Egyptian Religion in Italy* (Leiden, Brill, 1995), pp.51–53.

Fig. 6 Detail of the Palestrina Nile Mosaic showing the Nile-o-meter at Syene (centre-left).[a]

[a] Image supplied by kind permission of the Lazio-Rome Regional Museums Directorate, Photographic Archive.

was additionally used to determine the time of the Summer Solstice as the Sun would shine down directly into the bottom of the shaft.[17] Strabo's statement bears many similarities to the variations of Eratosthenes' experiment that are offered by some modern scholars, and may be where such claims have come from.

Pliny also says that, as well as no shadows being cast, the Nile-o-meter in Syene was specifically constructed for the observation of the Solstice.[18] It seems that, as part of their duties, the temple officials in Syene were responsible for recording when the Solstice had occurred. Strabo says that those watching for the flood knew when it was coming from 'signs and the days'.[19] This suggests that there were set days which heralded the onset of the flood—most

---

[17] Str. *Geog.* 17.1.48; at 2.5.7, Strabo also says that in Syene, at noon on the Summer Solstice, the *gnomon* of a sundial casts no shadow—echoing the work of Cleomedes.

[18] Plin. *HN* 2.75.

[19] Str. *Geog.* 17.1.48.

Fig. 7 The Nile-o-meter at the temple of Kom Ombo, Egypt.[a]
[a] Author's photo.

likely the Summer Solstice at the end of June—which would have been identified through the use of the Nile-o-meter and the observations of the shadows
cast by sundials (or lack thereof). One of the other 'signs' that Strabo refers to
would have also been the heliacal rising of the star Sirius, which first appears
on the eastern horizon around the time of the Summer Solstice.[20] The peak of
the annual flood would then be due about two weeks following the Solstice
and the heliacal rise of Sirius, and so the population had adequate time to prepare. The Summer Solstice occurred towards the end of the fourth month of
the season of Shemu—the last month of the Egyptian year—and the new year

---

[20] The association of the rise of Sirius with the Summer Solstice, the annual flood of the Nile, and
the beginning of the Egyptian year is attested to in texts such as the *Book of Nut* which outlines the
rising and setting times of several target stars across the course of a year in a series of tables which was
then used to determine the hour of the night. Symons (*Ancient Egyptian Astronomy*, pp.107–108) calls
these texts the 'first formal timekeeping instrument'. These tabulated 'star clocks' date back to around
2150 BC, but their use carried into the New Kingdom period as several examples are inscribed on the
ceilings of the tombs of Rameses VI (*ca.* 1150 BC), Rameses VII (*ca.* 1129 BC), and Rameses IX (*ca.* 1111
BC) in the Valley of the Kings. See: O. Neugebauer and R. Parker, *Egyptian Astronomical Texts Vol. I*
(Providence, Brown University Press, 1960), pp.16–51; O. Neugebauer and R. Parker, *Egyptian Astronomical Texts Vol. II* (Providence, Brown University Press, 1966). Parker ('The Calendars of Ancient
Egypt' *SAOC* 26 (1950), pp.1–83) suggests that the rise of Sirius was the most important stellar event
in the Egyptian year because of its association with the annual flood and the beginning of the new year
(see also: M. Isler, 'The Gnomon in Egyptian Antiquity' *Journal of the American Research Center in
Egypt* 28 (1991), p.167).

Fig. 8 The Nile-o-meter at the temple of Karnak, Egypt.[a]
[a] Author's photo.

started in the first month of the season of Akhet which followed.[21] This would place the flood at the same time as the beginning of the Egyptian year.

The importance of the flood, its connection to the Summer Solstice and the Egyptian calendar, and the need to observe them all carefully, made the Nile-o-meter in Syene one of the most important in ancient Egypt. Syene contained a number of cult centres dedicated to deities associated with the Nile and its annual flood including temples to Khunum (god of the Nile waters), Satis (war

---

[21] *Book of Nut*, section T2.

and fertility), and Anuket (goddess of the flood), and some of their temples date back to *ca.* 3000 BC. As such, the Egyptians must have been undertaking observations of the time of the Solstice and the height of the annual flood for centuries (if not millennia), and when Cleomedes reports that Eratosthenes had found a text referring to no shadows being cast in Syene on the Solstice, it can only be assumed that it was one of the earlier temple records of the observation of the time of the flood that he was consulting.

Pinotsis suggests that Eratosthenes made measurements of shadows cast by a *gnomon* on the Solstice in both Alexandria and Syene.[22] This would imply one of the following: (a) Eratosthenes had an assistant who took one of the measurements for him (for which there is no evidence in the accounts of the experiment); (b) that the measurements were taken by Eratosthenes on the Solstice across a two-year period—with the recording of the shadow length being taken at one location during the Solstice in one year, and at the other location in the next year (again for which there is no evidence); or (c) that one of the measurements was not actually taken on the Solstice (for which there is no evidence). There is also no need to doubt Cleomedes' account of Eratosthenes basing his experiment on a text describing a lack of shadow being cast in Syene on the Solstice—which would then require no observations to be made in Syene at all. Consequently, Pinotsis' claim seems doubtful. Regardless of these changes to the base story, the fundamental principles of the experiment are the same and would not have had too much of an impact on the results if they were true (see following).

However, before an examination of the size of the *stade* that Eratosthenes gave his results in can be undertaken, there are other issues to consider. As Bowen points out, much of the controversy around the understanding of Eratosthenes' calculations stems from the usually unargued assumption that the details provided in Cleomedes' account are correct and, as such, most scholars then attempt to reconcile this account with later ones given in works such as those by Strabo and Pliny.[23] Yet, even here, further problems continue to rise. Only two other ancient writers attribute a figure of 250,000 *stadia* to the result of Eratosthenes' experiment—Arrian in the first century AD, and the sixth-century AD writer Philoponus (who cites Arrian).[24] Several other sources, mostly from the first century AD, also refer to Eratosthenes' conclusions, but provide a figure of 252,000 *stadia*.[25] To add even more confusion to the matter, Strabo states that the figure he provides (252,000) is only

---

[22] Pinotsis, 'Significance and Errors', p.57.
[23] Bowen, 'Cleomedes and the Measurement', p.59.
[24] Arr. *Frag. Phy.* fr. 1; Phlp. *In Mete.* 15.13–15.
[25] Gal. *Inst. log.* 12.2; Plin. *HN* 2.112; Str. *Geog.* 2.5.7, 2.5.34; Theon. *Expos.* 3.3; Vit. *De Arch.* 1.6.9.

'approximate' (περὶ), while Heron of Alexandria, on the other hand, states that Eratosthenes was much more careful than others about his calculations. Similarly, Pliny states that Eratosthenes' calculations, which he says resulted in a figure of 252,000, were so bold and subtle that it would be shameful not to accept them.[26] As such, there is no way of determining what the exact figure for the result of Eratosthenes' calculation was from any of these sources alone.[27] It is also interesting to note that all of the sources that directly refer to the result of Eratosthenes experiment come from several hundred years after the event. This has led to complications in the modern scholarship on Eratosthenes as some scholars base their examinations on a figure of 250,000 *stadia*, while others use the figure of 252,000 *stadia*.[28]

It must be remembered that what Cleomedes composed, was not a work on astronomy per se, but a text of Stoic philosophy. Bowen provides a detailed argument of how Cleomedes' work is a commentary on how knowledge of the unknown is obtained by starting with a premise, or 'hypothesis' (ὑποθέσεις) as Cleomedes calls it, and then proceeding through rigorous enquiry, observation, and experimentation (as per Cleom. *De motu* 2.1).[29] Consequently, it is argued, Cleomedes' figures may have been adjusted to suit the Stoic notion of a spherical universe—although, in passages following his account of Eratosthenes' experiment, Cleomedes discusses how the same calculations had been made in later versions of the determination of the circumference of the Earth and still reached the same result of 250,000 *stadia*.[30] However, this seeming confirmation of a figure of 250,000 *stadia* for the circumference of the Earth did not make it into later texts recounting the calculation, nor into many modern examinations on the topic. Additionally, the later accepted number of 252,000 held a somewhat esoteric position in Greek thought. For example, Pliny states that this was the number of *stadia* that Pythagoras had calculated

---

[26] Plin. *HN* 2.247.

[27] Str. *Geog.* 2.5.7; Hero. *Dioptr.* 35.

[28] For example, for the use of the 250,000 *stade* figure, see: Fischer, 'Another Look', p.153; Sagan, *Cosmos*, p.15; O'Neil, *Early Astronomy*, p.59; Kline, *Mathematical Thought*, p.161; J. North, *Cosmos: An Illustrated History of Astronomy and Cosmology* (Chicago, University of Chicago Press, 2008), p.104; for the use of the 252,000 *stade* figure, see: Viedebantt, 'Eratosthenes', p.211; Pinotsis, 'Significance and Errors', p.58; Diller, 'Ancient Measurements', p.7; Diller, 'Julian of Ascalon', p.24; Dreyer, *Thales to Kepler*, p.175; Rawlins, 'Geodesy', p.260; B.R. Goldstein, 'Eratosthenes on the "Measurement" of the Earth' *HIST MATH* 11 (1984), p.412; T. Heath, *Aristarchus of Samos* (Dover, New York, 2004), p.147; Evans, *Ancient Astronomy*, p.51; Rawlins, 'Too-Big Earth', p.6; Carman and Evans, 'The Two Earths', pp.11–14; Russo, *Forgotten Revolution*, pp.273, 276; Bennett et al. (*The Cosmic Perspective*, p.67) use a *stade* equal to 1/6 km (or 166.67 m) in a statement that the result of Eratosthenes' calculation was a circumference of 42,000 km. This would equate to a distance of roughly 251,995 *stadia*, rounded to 252,000.

[29] For a discussion of the Greek uses of the term *hypothesis*, see Russo, *Forgotten Revolution*, pp.174–175.

[30] Bowen, 'Measurement of the Earth', pp.59–68.

as being the distance from the Moon to the Sun.[31] Furthermore, Plato found the figure 5,040—that is, 252,000 divided by 50—to be the ideal number for use in all matters pertaining to the state.[32] Diller, somewhat similarly, suggests that the value of Eratosthenes' result was increased to 252,000 because it is equal to $2^2 \times 3^2 \times 7 \times 1000$.[33] Consequently, it cannot be ruled out that the figure presented by the later ancient writers, and attributed to Eratosthenes, was not done so in order to meet some preconceived mathematical and/or philosophical ideal.

Indeed, there are other possible reasons for 'mathematical convenience' to be at play in the accounts of Eratosthenes' calculations, and the results, that we have. It has been suggested that the initial figure of 250,000 *stadia* was increased, possibly by Eratosthenes himself, to 252,000 so that one degree of latitude would equal exactly 700 *stadia* (i.e. 252,000 / 360 = 700) and that this revised figure was then taken up by the likes of Pliny and Strabo.[34] The caution to be taken here is that, while Eratosthenes divided the circumference of a circle into sixty segments, *hexacontade* (Strabo 2.5.7), and Greek sundials had been divided into twelve segments as early as the fifth century BC (Hdt. 2.109), the first evidence for the ancient Greeks dividing a circle into 360° comes from a century after Eratosthenes' calculations were made in the works of Hypsicles.[35] Thus, it would seem unlikely that Eratosthenes, or anyone else prior to the end of the first century BC, would adjust the figure to make it fit to an exact degree. However, it cannot be ruled out that the figure was adjusted by later writers to conform with this exact premise. It can also not be discounted that the figure may have been adjusted to fit to an even division by the *hexacontade*—250,000 *stadia* divided by 60 equals 4166.7, while 252,000 divided by 60

[31] Plin. *HN* 2.83.

[32] Pl. *Leg.* 737e–744d; see also: A. Gratwick, 'Alexandria, Syene, Meroe: Symmetry in Eratosthenes' Measurement of the World' in L. Ayres (ed.), *The Passionate Intellect: Essays on the Transformation of Classical Traditions Presented to Professor I.G. Kidd* (New Brunswick, Transaction, 1995), pp.177–202.

[33] Diller, 'Ancient Measurements', p.7.

[34] The two biggest proponents of this idea were Dreyer (*History of Astronomy*, p.175) and Neugebauer (*History of Ancient Mathematical Astronomy*, pp.734–735). See also: C.M. Taisbak, 'Posidonius Vindicated at All Costs? Modern Scholarship Versus the Stoic Earth Measurer' *Centaurus* 18 (1974), p.261; Fischer, 'Another Look', p.154; Carmen and Evans, 'Two Earths', pp.3–4; Evans, *Ancient Astronomy*, p.65; Rawlins, 'Geodesy', p.260; Russo, *Forgotten Revolution*, pp.69, 273; Rawlins ('Too-Big Earth', p.6) references Strabo (2.5.7) in relation to this rounding, but the ancient text does not say this.

[35] Hypsicl. *Anaph.* 5.25–31: 'The circumference of the zodiac circle having been divided into 360 equal arcs, let each of the arcs be called a degree in space, and similarly, if the time in which the zodiac circle returns to any position it has left be divided into 360 equal times, let each of the times be called a degree in time.' [Τοῦ τῶν Ζωδίων κύκλου εἰς τξ περιφερείας ἴσας διῃρημένου, ἑκάστη τῶν περιφερειῶν μοῖρα τοπικὴ καλείσθω. ὁμοίως δὴ καὶ τοῦ χρόνου, ἐν ᾧ ὁ ζωδιακὸς ἀφ' οὗ ἔτυχε σημείου ἐπί τὸ αὐτὸ σημεῖον παραγίγνεται, εἰς τξ χρόνους ἴσους διῃρημένου, ἕκαστος τῶν χρόνων μοῖρα χρονικὴ καλείσθω.]

equals exactly 4,200.[36] Goldstein suggests that the figure has been rounded to the nearest thousand.[37] Either of these may be the root cause for the 'adjustment' from 250,000 *stadia* to 252,000 *stadia*, but it seems unlikely, whatever the reasoning behind it, that the alteration can be attributed to Eratosthenes.

2.2 Eratosthenes' Sundial

A fundamental component of Eratosthenes' experiment to calculate the circumference of the Earth was the sundial. As devices that use a shadow cast by a pointer to determine the time of day, sundials in various forms had been in use in places like Egypt and Greece for hundreds of years prior to the time of Eratosthenes. Indeed, both the Egyptians and the Greeks seem to have had a long tradition of measuring time via solar and stellar observations and this knowledge fed into many aspects of culture such as myth, religious practices, architectural design, monumental construction, and their calendars. However, not all of the instruments that had been used prior to the time of Eratosthenes, and which may have still been in use in his own time, were suitable to the type of experiment that he undertook. A review of the historical background and influence of ancient Egyptian and Greek timekeeping, the importance of the Solstices in these cultures, and an examination of the various types of instruments that were available to Eratosthenes, shows that he used the instrument that was best suited to the methodology that he employed—the hemispherical *skaphe*.

The attempt to measure time is one of humankind's oldest scientific endeavours. In the modern world, time is understood to be the measurement of the rate at which something changes. A car, for example, may travel down a road at a certain number of kilometres *per hour*, a household may use a certain amount of water *per day*, and the interest on our investments may accrue *per month*. However, many of these more modern concepts would not have been a concern for most ancient peoples. Ancient cultures were predominantly agricultural in nature, and the concept of 'time' was, initially, more inherently linked to the cycle of the seasons across a year. This was because a better knowledge of when to plant, when the rains were due, and when to harvest would have been paramount to their survival. As such, early timekeeping in the ancient world was more calendrical—attempting to determine the time of

---

[36] See: Dueck, *Geography*, p.73.
[37] Goldstein, 'Eratosthenes', p.412.

the year—rather than needing to know what hour of the day it was. Waugh suggests that this early form of timekeeping merely involved the counting of days or months—as was marked by the change from day to night, or from season to season, respectively—and there were no obvious subdivisions of the day into smaller units of time.[38] At some stage the day was divided into three parts—morning, afternoon, and night—which were delineated by dawn, noon, and sunset, but the concept of 'time' retained its strong link to the agricultural cycles of nature and the path that the Sun travelled across the sky over the course of a year.

From any point of the Earth, the path that the Sun takes across the sky will shift from one day to the next—moving slightly north or south of where it had been the day before depending upon the time of year. This is because the Sun does not travel at right angles to the axis of the Earth, but follows the plane of the ecliptic, which is inclined to the equatorial plane at about 23.5°, and passes through the twelve signs of the zodiac. This means that, during summer, the Sun is higher in the sky, and remains above the horizon for longer, than it does during the winter months. The dual motion of the solar disk across the celestial sphere and the ecliptic means that that Sun follows a different path across the sky each day. The Sun's path along the ecliptic, and through all of the zodiacal signs, takes one solar year. The time of the year (in terms of modern months) was determined by which sign of the zodiac the Sun was traversing at any given moment.

At some stage, the day was divided into twelve hours of equal length. Pliny attributes the separation of the period of daylight into twelve hours to the ancient Greeks.[39] However, instruments which divided the day into smaller subunits (i.e. 'hours') have been found in Egypt which date back to the fifteenth century BC, and architectural alignment and construction, based upon the elevation of the Sun, may go back to at least 3000 BC (see following). According to Waugh, no one knows when people began counting concepts of time like the hours of the day, or what they used to measure them, but he offers that it was most likely based upon the observation of shadows.[40] As the Sun moves from east to west across the sky during the day, the shadow cast by a stationary object will change in length and direction. Thus, there is a clear correlation between the length of a shadow, the location of the observer, the time of year,

---

[38] A.E. Waugh, *Sundials: Their Theory and Construction* (New York, Dover, 1973), p.1.
[39] Plin. *HN* 2.187.
[40] Waugh, *Sundials*, p.1.

and the time of day. The combined nature of these phenomena led to the compilation of data sets, and the creation of the first instruments, by the ancient Egyptians with the specific purpose of determining both time and place for any location and at any point in the calendar. However, during the early development of such instruments and tables, time was not as clearly defined as it is today—with the day broken down into hours, and hours segmented into minutes and seconds, and with the length of the 'daylight hours' altering from season to season. Rather, it was believed that, across a single day, the Sun traversed the sky at a constant rate, and so each hour simply equalled one-twelfth of that arc.[41] This posed a problem to the early determination of time, as the amount of daylight during summer is greater than the amount in winter. By simply dividing the amount of daylight into twelve equal portions, summer 'hours' were therefore longer than winter ones. This means that any instrument that was designed to account only for the length of an hour during a certain time of year, rather than the 'seasonal hour' of each day across a year, would be out on almost every other day. Later, as these cultures developed, the importance of the rise of a certain star in the evening sky, or the changes in the path of the Sun across a year, became not only signifiers of a change of the seasons, but also intertwined with the agriculture of the people, as well as with the religious and cultural aspects of their lives. Through advances in both technology and thought, these people were able to conduct some of the first recorded experiments relating to time and its correlation to the motions of the Sun and stars.

2.3 Time And The Solstices In Ancient Egypt And Greece

According to Isler, a staff or vertical pole was used to observe the angle of the Sun in ancient Egypt. Given the Greek term *gnomon*, this staff was the first scientific instrument and was used to determine such things as the time of the year, the time of day, and geographic location, based upon the length and direction of the shadow that was cast.[42] Another important aspect of this early observation of time was the Solstices—the moments when the position of the Sun is at its most northerly or southerly. In ancient Greece, the term for the

---

[41] A. Jones, 'Greco-Roman Sundials: Precision and Displacement' in K.J. Miller and S.L. Symons (eds.), *Down to the Hour: Short Time in the Ancient Mediterranean and Near East* (Leiden, Brill, 2020), p.130.
[42] Isler, 'The Gnomon', pp.155–168, 169–177, 178–185.

Solstices was *tropai helioio* (τροπαὶ ἠελίοιο)—meaning 'the turning of the Sun'. Bilić calls the Solstices 'the defining moments in the annual solar motion'.[43] This was because the Solstices could not only be used to separate the year into different 'seasons', but the dates of the Solstices heralded many important events in the cultures of both ancient Egypt and Greece.

According to Vitruvius, in ancient Greece, Eudoxus (408–355 BC), Callippus (370–300 BC), and several others used instruments to determine the rising and setting times of stars and the changes of the seasons.[44] Eusebius states that it was Anaximander (610–546 BC) who was '... the first to construct *gnomons* for identifying the Solstices, lengths of time, the *horai* (ὧραι) and the equinox'.[45] Anaximander's instrument for such determinations was a *gnomon*—an upright pole or staff similar to that which had been used in Egypt—which he had erected in the city-state of Sparta.[46] The word ὧραι that Eusebius uses can mean either 'hours' or 'seasons'. Thibodeau argues that, since there is no reference to hours of the day in Greek literature prior to the writings of Herodotus in the fifth century BC (2.109), and that no Greek sundial marking the hours has been found that is dated to before 350 BC, the word ὧραι in this passage should be taken as 'seasons'.[47] Support for this claim comes from the fact that the word ὧραι was used in the context of referring to the seasons as far back as the writings of Homer and Hesiod.[48]

In fifth-century BC Athens, Meton constructed a *heliotropion* (ἡλιοτρόπιον)—a stele upon which the positions of the solstices were marked (καταγράφω).[49] Meton probably made his observations to mark the stele in 431 BC.[50] The stele was erected on the hill of the Pnyx, below the Acropolis, so that the Sun could be observed rising over Mount Lykabettos on the Solstices.[51] Similar devices were erected elsewhere in the Greek world as well and seem to have come in many different sizes. The *heliotropion* erected by Dionysius of Syracuse (405–367 BC), for example, was so big that he could

---

[43] T. Bilić, 'Apollo, Helios, and the Solstices in the Athenian, Delphian and Delian Calendars' *Numen* 59 (2012), p.509; T. Bilić, 'The Island of the Sun: Spatial Aspect of Solstices in Ancient Greek Thought' *GRBS* 56 (2016), p.195.

[44] Vit. *De Arch.* 9.6.3.

[45] Eus. *PE* 10.14.

[46] Diog. Laert. 2.1; see also Plin. *HN* 2.187 who erroneously attributes this to Anaximenes rather than Anaximander; for a discussion of where in Sparta this *gnomon* may have been set up, see: P. Thibodeau, 'Anaximander's Spartan Sundial' *CQ* 67.2 (2017), pp.374–379.

[47] Thibodeau, 'Spartan Sundial', p.374.

[48] For example, see: Hom. *Il.* 6.148, *Od.* 5.485; Hes. *Op.* 450, 584.

[49] Ael. *VH* 10.7.

[50] Ptol. *Alm.* 3.1; for a detailed discussion of the date, see also: L. Depuydt, 'The Egyptian and Athenian Dates of Meton's Observation of the Summer Solstice (−431)' *AncSoc* 27 (1996), pp.27–45.

[51] *FGrHist* 382 F122.

stand on it.[52] Another *heliotropion* from fourth-century Crete is referred to in an inscription on a pillar (which is possibly its base).[53]

In both Egypt and Greece, the Solstices, in particular the Summer Solstice at the end of June, had a large impact on many aspects of culture. The Egyptian festival for Min—god of fertility who was often associated with the annual flood of the Nile and the cycles of nature—may have been held prior to the Summer Solstice in April. Isler interprets the staff with a bifurcated tip that is depicted in some representations of Min as the *gnomon* used to determine time.[54] The staff seems to have similarly bifurcated supports, which may have been used to ensure that the staff was vertical (Fig. 9).

Isler suggests that the gnomon was set up at the festival of Min in April so that it could then be used to observe the shortening of the shadows as the year moved towards the important day of the Summer Solstice.[55] The Summer Solstice, and the rising of the star Sirius—which first appears above the eastern horizon around the time of the Summer Solstice—were two of the most important events in the Egyptian solar and stellar calendars, as they not only marked the beginning of the Egyptian new year, but also heralded the imminent arrival of the annual flooding of the Nile which was essential to the continuance of agriculture in Egypt, determined the levels of taxation for the coming year, allowed monumental building projects to be undertaken, and had specialized infrastructure—such as the Nile-o-meter—created in order to monitor it.

In Athens, the new year similarly began on *Hekatombaion* 1, a month which started on the first new Moon after the Summer Solstice.[56] Theophrastus also suggested that certain plants bloomed with the arrival of the Solstice—similar to the way the Solstice heralded the annual flood and rejuvenation of farmland in Egypt.[57] The longest day of the year also held a special position in many religious cults in ancient Greece, and several ancient texts record how certain sacrifices were made when a particular star, such as Sirius, first appeared on the eastern horizon around the time of the Summer Solstice.[58] According to the writer Macrobius, the name of the god Apollo may derive from the Greek word for the casting of the Sun's rays (ἀποπάλλειν).[59] Plato, on the other hand, suggested that the α in the name meant 'moving together in

[52] Plut. *Dion.* 29; Ath. *Diep.* 5.41.
[53] *Syll*³ 1264.
[54] Isler, 'The Gnomon', pp.158–161.
[55] Isler, 'The Gnomon', p.168.
[56] Pl. *Leg.* 6.767c.
[57] Theoph. *Caus Pl.* 7.15.1.
[58] Ap. Rhod. *Argon.* 2.516–527; Diod. Sic. 4.82.1–3; *P. Hib.* frag. c.v.77; d.vi.85; Alcm. *Par.* 39–43, 60–63.
[59] Macr. *Sat.* 1.17.7.

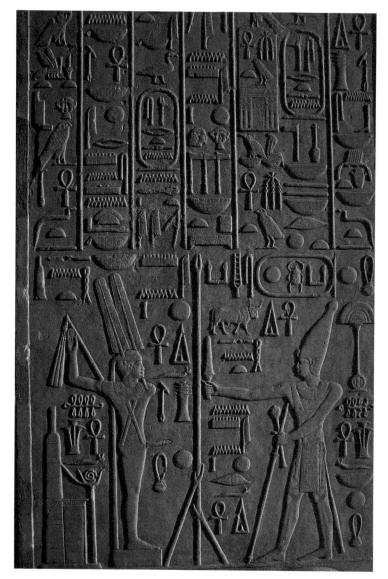

Fig. 9 Representation of Min (left) and bifurcated *gnomon* in the
Shrine of Min, Karnak.[a]

[a] Egypt, Luxor, Column detail, hieroglyphics, God Min, White Chapel, Luxor Temple
by Danita Delimont. Image ID A6RM22, Alamy Stock Photos.

the heavens' (ἀ-πόλησις) around the poles (πόλους).[60] While such a notion
would carry connotations of Apollo being linked with the Sun, the Greeks at
this time had their own divine name for the solar disk—Helios. However, in

[60] Pl. *Cra.* 405c–d.

a different work, Plato refers to the 'common precincts of Apollo and Helios' and with an associated festival dedicated to the Summer Solstice—the θερινὰς τροπὰς or 'turning of the God'.[61] Burkert suggests this indicates that, by Plato's time, Apollo had become synonymous with the Sun.[62] Indeed, there are other ancient texts which support such a conclusion.

The first references to the Solstices in Greek literature are found in Homer's *Odyssey*, and in the works of Hesiod—both *ca.* 800 BC—where Hesiod uses the terms ἠελίοιο τροπαὶ and τροπαὶ ἠελίοιο to designate the Summer Solstice and the Winter Solstice, respectively.[63] However, an important text for understanding the link between Apollo and the Sun is the myth of the god's journey to visit the Hyperboreans—a people whose name translates as 'those who live beyond the North Wind' (Boreas). In the myth, Apollo journeys far to the north at a certain time each year to visit these people, and participate in an annual festival, before returning at a later time.[64] Bilić suggests that this tale may be a metaphor for the annual path of the Sun, similar to the way that Burkert suggests that the language of ancient mythology was designed to 'map reality', to verbalize phenomena, and to give them coherence and sense.[65]

This link between the Solstices and religion, in both ancient Egypt and Greece, also manifested itself physically through the architecture of these two cultures. Magdolen, for example, argues for a solar origin for the so-called 'sacred triangle' that was used by monumental builders in Egypt.[66] The 'sacred triangle' was a right-angle triangle with a ratio of its sides equal to 3:4:5. It is also known as the Pythagorean Triangle. With such a configuration, the angles opposite the right angle of the triangle are 53°07′48″ and 36°52′12″. Magdolen argues that the ratio for the 'sacred triangle' was derived from observations of the angle of the Sun's rays on the Winter and Summer Solstices made using a *gnomon*.[67] At Giza, for example, around 3000 BC, the maximum elevation of the Sun on the Winter Solstice was 36°52′—the same angle as in the sacred triangle. When the Egyptians began dividing the day into hours, two of these— the fourth and ninth hours of the day—were referred to as the 'secret hours' as they were the ones prior to and after the culmination of the Sun when the

---

[61] Pl. *Leg.* 6.767c, 12.945e.

[62] W. Burkert, *Greek Religion* (Cambridge, Harvard University Press, 1985), p.336.

[63] See: Hom. *Od.* 15.404; Hes. *Op.* 479, 564, 663.

[64] Ael. *VH* 11.1.

[65] For a discussion of the religious and literary connections between Helios, Apollo, the Sun, and the Solstices, see: Bilić, 'Apollo, Helios, and the Solstices', pp.509–532; see also: W. Burkert, *Structure and History in Greek Mythology and Ritual* (Berkeley, University of California Press, 1979), pp.23–24.

[66] D. Magdolen, 'The Solar Origin of the "Sacred Triangle" in Ancient Egypt?' *SAK* 28 (2000), pp.207–217.

[67] Magdolen, 'Solar Origin', p.209.

length of shadows cast in these two hours were equal, and were almost the same length as the height of the *gnomon* that cast them (thus forming an isosceles triangle).[68] The Sun passed through an elevation of 53°07' (the other angle in the sacred triangle) during the two 'magic hours' on the day of the Summer Solstice. Many pyramids from the fourth to sixth dynasties—including the great pyramids of Giza—contain a slope (*seked*) based on these proportions.[69] The *Rhind Mathematical Papyrus*, a mathematical text containing problems involving arithmetic, algebra, geometry, and other numerical questions, dated to 1550 BC, outlines the 3:4 ratio of the 'sacred triangle' for use in pyramid building in problems 56, 57, 58, and 59. Problem 57, for example, asks: 'If the *seked* of a pyramid is 5 palms and 1 finger per cubit, and the side of its base is 140 cubits, what is its height?'[70] Christian-Meyer suggests that such ratios can also been seen in the alignment and arrangement of many of the buildings that formed the funerary complexes that were built alongside the pyramids.[71] An example of this can be seen in the positioning of the Great Sphinx at Giza. From the position of the Sphinx, the Sun sets directly between the two great pyramids of Khufu and Khafre on the Summer Solstice.

Something similar seems to have occurred in Greece. Out of a sample of 107 temples scattered across the Greek world that was examined by Dinsmoor, 58 per cent of them were aligned to within 30° of due east, which suggests they were constructed so that their main axis coincided with some form of celestial or solar rising.[72] The problem with this conclusion, as Boutsikas and Ruggles point out, is that 'any architectural alignment must point somewhere, and there is a multitude of possible astronomical targets, so that the mere existence of an astronomical alignment proves nothing: it could have arisen fortuitously through a combination of factors quite unrelated to astronomy'.[73] Thus, other forms of confirmatory evidence are needed. In some cases, the decorations on the temple, which were associated with the cult that was worshipped there, may provide clues, but these are no more confirmatory than the alignment of the building.

---

[68] See: Magdolen, 'Solar Origin', pp.213–216.
[69] Magdolen, 'Solar Origin', pp.207–208; see also: R.J. Gillings, *Mathematics in the Time of the Pharaohs* (New York, Dover, 1982), p.212.
[70] *RMP* 57; see also: Gillings, *Mathematics*, p.187.
[71] W. Christian-Meyer, 'Der "Pythagoras" in Ägypten am Beginn des Alten Reiches' *MDAIK* 43 (1987), pp.195–206.
[72] See: W.B. Dinsmoor, 'Archaeology and Astronomy' *PAPS* 80 (1939), p.115; E. Boutsikas and C. Ruggles, 'Temples, Stars and Ritual Landscapes: The Potential for Archaeoastronomy in Ancient Greece' *AJA* 115.1 (2011), pp.58–66 (total pp.55–68).
[73] Boutsikas and Ruggles, 'Temples, Stars and Ritual', p.59.

When the Greeks embarked on the study of geography—a discipline later taken up by Eratosthenes—the Solstices also played a larger part in how the Greeks saw the world. Strabo, for example, states that Ephorus of Cyme (400–330 BC) used the Solstice points to delineate the four sections of the *oikumene* (the known world).[74] Anaximander may have used a *gnomon* to do something similar for his geographical works.[75] Eudoxus also observed the points on the horizon where the Solstices rose and set.[76] Thus it becomes clear, from a range of evidence encompassing both the literary and the archaeological, and associated with various aspects of culture, that the Solstices played a major role in many facets of the lives of the ancient Egyptians and Greeks. While some of these observations and measurements were undoubtedly taken using a staff-like *gnomon*, one area where the careful and critical observation of the shadow cast by the Sun came to the fore was in the construction of sundials.

2.4 Ancient Egyptian And Greek Sundials To The Time Of Eratosthenes

Waugh states that it is not known when people began using instruments similar to modern sundials, but suggests that, whatever their form, sundials are evidence for a very early understanding of some of the fundamental relationships in astronomy.[77] The earliest written reference to a sundial is found in an account of a battle from the time of the Egyptian pharaoh Thutmose III from the fifteenth century BC (1479–1425 BC).[78] In the account, Thutmose carried a portable sundial with him on a campaign to Syria. This instrument was most likely the Egyptian 'shadow clock'.

THE SHADOW CLOCK
The Egyptians seem to have used shadow measuring instruments to determine time as early as 1500 BC. It is possible that the Egyptians had been using shadows to determine time in the Middle Kingdom Period (*ca.* 2040–1640 BC) as per a passage in the *Prophecy of Neferti*, or even earlier as is suggested by the alignment of some of their monumental buildings from the Old Kingdom

---

[74] Str. *Geog.* 1.2.28; see also: *FGrHist* 70 F30a.
[75] Diog. Laert. 2.1; Eus. *PG* 10.14.11.
[76] Hipparch. *In Ar. et Eud.* 1.9.2.
[77] Waugh, *Sundials*, pp.vii, 3.
[78] See: V.N. Pipunyrov, *Istoriya Chasov s Drevnejshix Vremen do Nashix Dnej [Story Hours from Ancient Times to the Present Day]* (Moscow, Nauka, 1982), p.21.

Period.[79] Floyer says this may have been based on the use of very primitive sundials.[80] This may have involved the use of the *gnomon* often depicted in representation of the god Min (see previous). The Egyptians may have also divided the day into twelve parts as early as the third millennium BC.[81] A funerary text, *The Book of Nut* (otherwise known as *The Fundamentals of the Course of the Stars*), dated to 1290 BC, refers to astronomical events from the nineteenth century BC and also gives instructions for how to make a shadow clock.[82] However, the clock and the astronomical events are not necessarily connected and there is no evidence for the existence of the shadow clock in Egyptian culture in the nineteenth century BC. Symons suggests that such instruments first appear around 1550 BC.[83]

The oldest surviving example of a shadow clock comes from the reign of Thutmose III (fifteenth century BC) and is now housed in the Egyptian Museum in Berlin (inv #19744). Made of green schist, the instrument is in the shape of an L lying on its long side. The base is 230 mm long, 23 mm wide, with the upright 45 mm high and 20 mm deep (Fig. 10).[84]

Symons calls the shadow clock the first 'timepiece' with any form of standardization and a text referring to its construction (*The Book of Nut*).[85] It is believed that a plumbline was used in conjunction with the line marked on the upright of the instrument itself to ensure that the shadow clock was being held horizontally (see the vertical line inscribed on the upright on the right-hand side of Fig. 10). The upright was then pointed towards the Sun and a shadow would be cast by the upright onto the base. Ensuring that the shadow fell squarely onto the base made it possible to confirm that the instrument was pointing correctly towards the solar disk.[86] The base was only marked with four hours, as it was believed that two hours passed in the morning and evening twilight when the Sun's rays would not fall onto the base of the clock.[87]

[79] S. Symons, *Ancient Egyptian Astronomy: Timekeeping and Cosmography in the New Kingdom* (Leicester, University of Leicester—unpublished thesis, 1999), p.128.

[80] E.A. Floyer, 'Primitive Sundials in Upper Egypt' *Athenaeum* (1895) 3545.

[81] See: S. Schechner, 'The Material Culture of Astronomy in Daily Life: Sundials, Science and Social Change' *J. Hist. Astron.* 108 (2001), pp.189–222.

[82] See: J. Fermor, 'Timing the Sun in Egypt and Mesopotamia' *Vist. Astron.* 41 (1997), pp.157–167.

[83] Symons, *Ancient Egyptian Astronomy*, p.128.

[84] For more details and dimensions see: L. Borchardt, 'Altägyptische Sonnenuhren' *ZÄS* 48 (1910), pp.9–17.

[85] Symons, *Ancient Egyptian Astronomy*, pp.128, 134.

[86] Isler, 'The Gnomon', pp.176–180; See also: I. Tupikova and M. Soffel, 'Modelling Sundials: Ancient and Modern Errors' in M. Geller and K. Geus (eds.), *Productive Errors: Scientific Concepts in Antiquity* (Berlin, Max Planck Institute for the History of Science, 2012), pp.18–20.

[87] R.A. Parker, 'Ancient Egyptian Astronomy' *Philos. Trans. R. Soc.* 276 (1974), pp.51–65; Symons (*Ancient Egyptian Astronomy*, p.129), points out that the ratio of the distances between the markings, starting from the base of the upright, is 1:2:3:4:5.

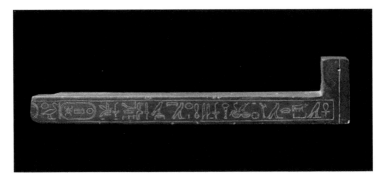

Fig. 10 Egyptian shadow clock from the time of Thutmose III
(Egyptian Museum, Berlin, #19744).[a]

[a] Image taken from the SMB digital Online Collections database: http://www.smb-digital.de/eMuseumPlus?service=direct/1/ResultLightboxView/result.t1.collection_lightbox.$TspTitleImageLink.link&sp=10&sp=Scollection&sp=SelementList&sp=0&sp=0&sp=4&sp=Slightbox_3x4&sp=0&sp=Sdetail&sp=0&sp=F&sp=T&sp=0. Creative Commons image © Egyptian Museum and Papyrus Collection of the National Museums in Berlin—Prussian Cultural Heritage.

Thus, the shadow clock needed to be realigned every time that a measurement was to be taken. It has been suggested that the shadow clock was only aligned along an east–west axis, and then rotated 180° at noon to allow for the time in the latter half of the day to be measured.

The markings on the shadow clock are also of standardized units and do not consider the variances in the elevation and position of the Sun across a year. In 1910, Borchardt theorized that the upright held a cross bar so that the shadow would always fall upon the base no matter what time of year it was (this again assumed that the instrument was only aligned east–west and not pointed towards the Sun).[88] Symons presents a number of arguments against Borchardt's cross-bar hypothesis: (a) there is no representation of a cross-bar found anywhere in the Egyptian artistic record, and no reference to it in any text; (b) New Kingdom Egyptians do not seem to have used a standard unit of time—which then makes a crossbar that accounts for seasonal variances in the height of the Sun superfluous; (c) the hour marks on the base do not correlate to seasonal hours; and (d) the instrument would be too hard to align along three axes.[89] Furthermore, there is no reason to discount the notion that the instrument was just pointed directly towards the Sun whenever

---

[88] Borchardt, 'Altägyptische Sonnenuhren', pp.9–17; See also: E.M. Bruins, 'The Egyptian Shadow Clock' *Janus* 52 (1965), pp.127–137; for an examination of how the cross-bar would have affected the shadow, see: Tupikova and Soffel, 'Modelling Sundials', pp.13–17.

[89] Symons, *Ancient Egyptian Astronomy*, pp.130–144.

a measurement needed to be taken. This would then allow the shadow clock to be used at any time of year—although the length of the shadow would still vary from day to day.

Such devices, with standardized markings for the hours, were very inaccurate and were unlikely to have been constructed based upon any form of detailed observation of the movement of the Sun over the course of a year. For example, as Symons points out, the shadow cast by the upright does not reach the mark closest to the base for half of the year because the Sun does not get high enough in the sky.[90] These sorts of issues were something that the Greeks would later address with the construction of the hemispherical *skaphe* (see following). Despite its inaccuracies, the L-shaped shadow clock remained in use from the New Kingdom Period to Roman times, but it seems to have be an instrument unique to Egypt as no examples of such an instrument have been found elsewhere.[91] However, this does mean that this type of time-measuring instrument was still in use in the time of Eratosthenes.

THE HANGING/VERTICAL SUNDIAL

Another form of an early Egyptian sundial came in the form of a hanging semicircular plate, inscribed with markings that designated the hours, and with a *gnomon* which extended horizontally from the top. An example of most of the vertical face, but missing the *gnomon*, was discovered in the Valley of the Kings near Luxor in 2013. The sundial has been dated to the thirteenth century BC (Fig. 11).

Analysis of the find has shown that the thirteen radial lines inscribed on the face separate the dial into the twelve hours of the day, with the noon line at 90° to the horizontal upper edge of the plate.[92] The sundial would also be oriented so that it faced south at any location north of the Tropic of Cancer.[93] Thus, the hanging sundial marks an improvement over the shadow clock in that it would not have to be turned or realigned. However, the Sun would appear to move faster in the mornings and evenings than it would at midday, but the semi-circular face is divided evenly into segments of 15° each. This would initially suggest that this type of sundial was not overly accurate across a year. Vodolazhskaya, on the other hand, in her modelling of this particular sundial, concludes that it possessed an average error of 0.9°, or about 3.6 minutes.[94] Vodolazhskaya further suggests that the L-shaped shadow clock was used to

---

[90] Symons, *Ancient Egyptian Astronomy*, pp.144–145.
[91] S. Symons and H. Khurana, 'A Catalogue of Ancient Egyptian Sundials' *J. Hist. Astron.* 47 (2016), pp.375–385.
[92] Symons and Khurana, 'A Catalogue', p.378.
[93] Symons and Khurana, 'A Catalogue', p.379.
[94] L.N. Vodolazhskaya, 'Reconstruction of Ancient Egyptian Sundials' *AAATec* 2 (2017), pp.1–18.

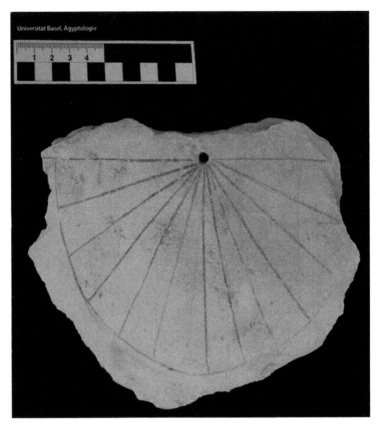

Fig. 11 Vertical sundial from the thirteenth century BC from the
Valley of the Kings, Egypt (University of Basel).[a]

[a] *Ancient-egyptian-sundial*, Image ID J314M8, The History Collection/Alamy Stock
Photo.

define the angles of the radial lines on the face of the hanging sundial, as the
ratio of distances between each line is the same as the distances between the
markings on the base of the shadow clock (Fig. 12).[95]

What is of more importance, however, is that there appears to be markings
on this sundial which designate half-hours. Within each segment, there is a
small dot towards the bottom, and at almost the midpoint across the arc of
the segment (see Fig. 11). These would have denoted the halfway point of the
hour.[96]

[95] Vodolazhskaya, 'Ancient Egyptian Sundials', pp.10, 16.
[96] Vodolazhskaya, 'Ancient Egyptian Sundials', pp.4–8.

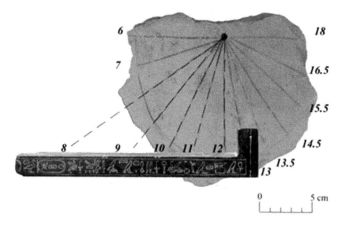

Fig. 12 Use of the shadow clock to mark a vertical sundial as per Vodolazhskaya.[a]

[a] Image taken from: Vodolazhskaya, 'Ancient Egyptian Sundials', p.13.

It is uncertain how common the use of a half-hour was in ancient Egypt, and how common the vertical type of sundial was in the ancient world. Gibbs, in her extensive examination of extant ancient sundials, finds only one example of the vertical sundial—that inscribed on the Tower of the Winds in the Roman-era agora in Athens (although it must be noted that the Egyptian sundial shown in Fig. 11 was discovered after Gibbs' research, which only focuses on Greek and Roman instruments).[97] Vitruvius credits the creation of a variant of this type of sundial, one in which the semi-circular face is cut into a block of stone and is inclined according to the latitude of the location for its use, to Berossus of Chaldea (early third century BC).[98] Using the (albeit limited) sample of two examples of vertical sundials as the start and end points for their usage, a conclusion that is by no means certain, it can still be said that the technological knowledge behind the hanging/vertical sundial existed from *ca.* 1200 BC to at least AD 200. This would mean that such sundials would have still been in use, even if only theoretically, in the time of Eratosthenes.

SLOPING SUNDIALS

Another type of sundial that seems to have been unique to ancient Egypt was the sloping sundial.[99] Symons suggests that these instruments developed out of the shadow clock.[100] The instrument was comprised of a vertical *gnomon*

---

[97] S.L. Gibbs, *Greek and Roman Sundials* (New Haven, Yale University Press, 1976), p.42.
[98] Vit. *De Arch.* 9.8.1.
[99] Symons and Khurana, 'A Catalogue', p.378.
[100] Symons, *Ancient Egyptian Astronomy*, p.148.

at the base of a sloping face. The sundial was turned to face the Sun so that
the shadow cast by the *gnomon* would fall onto the sloping face of the sundial.
A pair of vertical lines running down the sloping surface separated the face
into the three months of the Egyptian year. One edge of the sloping face represented the Summer Solstice, while the opposite edge represented the Winter
Solstice. Obliquely drawn lines on the surface, extending from the Winter Solstice edge to the Summer Solstice, marked the hours of the day at different
times of the year (Fig. 13).

(a)   (b)

Fig. 13 Marble sloping sundial from Thebes, Egypt (fourth
century BC).
Metropolitan Museum of Art #12.181.307.[a]

[a] Public Access images taken from: https://www.metmuseum.org/art/collection/
search/576278.

Such instruments were one of the first to take into account the angular position of the Sun across an entire year. While these sundials demonstrate a shift
in Egyptian sundial design to account for the variances in the lengths of seasonal hours across the year, Symons states that such instruments still have
issues with accurately representing the length of the noon hour.[101] A marble
example of the sloping sundial found in Thebes, measuring 9.3 cm × 5.2 cm ×
14.7 cm, has been dated to 320 BC (Fig. 13).[102] This means that such instruments were in use, in Egypt, around the time that Eratosthenes made his
calculation of the circumference of the Earth.

EARLY GREEK SUNDIALS

The use of lines representative of the seasonal hours to create a sundial that
could be used with accuracy over the course of a year was improved upon

[101] Symons, *Ancient Egyptian Astronomy*, p.150.
[102] N.E. Scott, 'An Egyptian Sundial' *MMAB* 30.4 (1935), pp.88–89.

by the Greeks. The historian Herodotus states that sundials, based upon Babylonian examples, were introduced to Greece by Anaximander of Miletus *ca.* 580 BC.[103] Rather than being the first to divide the day into twelve hours as Pliny suggests, Jones sees an advance in Greek sundial construction in that their design is based upon the concept of the seasonal hour so that all hours across a day were of equal length, and so the seasonal hour became the basic unit of time measurement instead of just dividing the sundial into twelve equal parts.[104] In support of Pliny's claim, it should be noted that early Egyptian sundials such as the shadow clock only marked hours for half of the day and needed to be rotated at noon at the very least. The later sloping sundials similarly only marked half the hours of the day. As such, neither of these instruments actually divide the day into twelve hours. The Egyptian hanging sundials do divide the day into twelve hours, but these are dated to the fourth century BC, after the Greeks had apparently divided the day into that number of hours. Pliny may be referring to this. One of the earliest types of Greek sundial was a vertical *gnomon* on a flat surface similar to what had been used in Egypt.[105] Rohr suggests that the Greeks had been using *gnomons* as an instrument for observation as early as 600 BC.[106] On the flat plane of the base, the tip of the *gnomon*'s shadow traced out a shape called a *pelekinon*—the Greek word for axe—over the course of the year (Fig. 14).

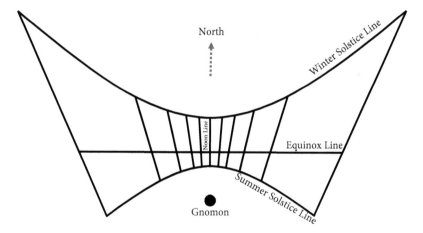

Fig. 14 The *pelekinon* created by a standing *gnomon*-staff sundial.

[103] Hdt. 2.109.

[104] Jones, 'Greco-Roman Sundials', p.130.

[105] W.W. Dolan, 'Early Sundials and the Discovery of Conic Sections' *Math. Mag.* 45.1 (1972), pp.8–12.

[106] R.R.J. Rohr, *Sundials: History, Theory and Practice* (New York, Dover, 1996), p.7.

Both Ptolemy and Vitruvius describe an *analemma* (a diagram that represents the position of the Sun across the year) which serves to determine the shape and direction of the *gnomon* shadow on the face of a planar sundial for given solar positions.[107] This suggests that the markings may have been based on calculations, or at least templates, rather than observations. The use of a simple *gnomon* and plane configuration seems to have been common in Greece for many centuries as evidenced by Anaximander's erection of a *gnomon* in Sparta in the fifth century BC. As Rohr points out, the shadow cast by the *gnomon* does not actually indicate intervals of time, but its length and direction does mark a certain moment in the scale of the year.[108]

Other even simpler methods seem to have been employed to determine the time. In *The Assembly Women* by the Athenian playwright Aristophanes (*ca.* 391 BC), one of the characters determines the time by observing the length of his own shadow.[109] Observing the length of a shadow—cast by either a person or a *gnomon*—seems to have been a method commonly used in rural areas in the ancient Mediterranean well into the fifth century AD. In his work on agriculture (*Opus Agriculturae*), the Roman writer Palladius (fourth to fifth century AD) concludes his examination of agricultural matters pertaining to each month with a table outlining the length of a shadow cast by a standard length staff at different hours of the day during that period.[110] Rohr suggests that the *gnomon* was probably still used in rural areas as it required no specialized knowledge and could be used anywhere.[111] However, in more urbanized areas, the sundial underwent as dramatic transformation with the invention of the instrument that would later play a part in Eratosthenes' calculations—the hemispherical *skaphe*.

THE SKAPHE

The *skaphe* (σκάφη), or *herispherium* as it was known in Latin, was a hollow hemisphere. Variants of this sundial were carved from stone or metal, but no examples of the full hemispherical *skaphe* have survived the passing of the centuries. A passage from Martianus Capella states that the *skaphe* that Eratosthenes used was made of bronze.[112] Ptolemy also refers to an instrument made of metal that has been 'accurately turned, with its faces standing square.'[113]

---

[107] Ptolemy wrote an entire work on this subject called *About the Analemma* (Περὶ ἀναλήμματος). See also: Vit. *De Arch.* 9.1–8.
[108] Rohr, *Sundials*, p.4.
[109] Ar. *Eccl.* 651–652.
[110] Pallad. *Agric.* 2.23, 3.34, 4.16, 5.8, 6.18, 7.13, 8.10, 9.14, 10.19, 11.23, 12.23, 13.7.
[111] Rohr, *Sundials*, p.15.
[112] Mart. Cap. *Phil.* 6.596: 'vessels called *skaphe* were spherical and made of brass' (*scaphia dicuntur rotunda ex aere vasa*).
[113] Ptol. *Alm.* 1.12.

This may account for why no examples of such instruments have survived—the metal would have been broken up, and melted down, for other purposes as locations were conquered and plundered across the ages. Regardless of the material it was made from, the *skaphe* possessed a vertical *gnomon* extending upwards from the bottom and pointing towards the zenith. The 'bowl' of the hemisphere represented a mirror-image of the celestial sphere above, the upper edge of the bowl represented the horizon, and the shadow cast by the *gnomon* would trace, in reverse, the path of the Sun across the sky onto the face of the bowl.[114] A series of lines, called the *arachne*, were inscribed onto the curvature of the bowl which marked the path of the Sun on both the Summer Solstice and Winter Solstice, and the Equinoxes. The area between the Solstice lines was further divided into areas representing the seasonal hours. A later development in the design of the *skaphe* recognized that, when the instrument was correctly aligned to face northwards, the tip of the shadow cast by the *gnomon* would never fall south of the Summer Solstice line. In the surviving examples of the later forms of the *skaphe*, the area of the bowl south of the Summer Solstice line has been removed so that the front edge runs parallel to the Solstice line.[115] From these examples of a partially hemispherical sundial, the basic design of the full *skaphe* can be ascertained (Fig. 15).

Vitruvius credits the creation of the *skaphe* to Aristarchus of Samos (310–230 BC).[116] Gatty calls the *skaphe* the simplest and most ancient form of sundial, claiming that it was the first such instrument, and that it originated in the astronomical schools which flourished in Greek Asia Minor in the fourth century BC.[117] It is uncertain upon what Gatty's confusing claim is based. Clearly there were other devices for measuring time—such as the *gnomon*, the vertical/hanging sundial, and the Egyptian shadow clock—which had been in use for more than nine centuries prior to the time of Aristarchus. Gatty seems to be basing her conclusion on a passage from the writer Apion (30 BC–AD 45). In a commentary on Apion's lost *History of Egypt*, the Jewish writer Josephus

[114] Gibbs, *Greek and Roman Sundials*, p.12.

[115] In her extensive catalogue of extant sundials, Gibbs (*Greek and Roman Sundials*, pp.123–177) lists entries for more than sixty full or partial surviving examples of the trimmed-down spherical sundial from across the Greco-Roman world ranging in age from the third century BC to the third century AD (#1001–#10062). Gibbs also notes (p.12) that the surfaces of eighty-five Greek and Roman 'spherical' sundials exist. Not all of these examples possessed a vertically aligned *gnomon* and, in another variation of the design, some of them had a horizontal *gnomon* which projected out from the upper rim of the 'bowl'. Not only would the trimmed-down variant of the *skaphe* require less material, but an added advantage to this design is that it would not fill with water when it rained—unlike the full *hemispherium*.

[116] Vit. *De Arch.* 9.8.1; Rohr (*Sundials*, pp.9–10, 12) incorrectly credits the creation of the *skaphe*, and the lines upon its face which marked the Solstices, Equinoxes, and hours, to Berossus of Chaldea, and states that Aristarchus merely came up with a refined variant of this type of sundial. However, this is not what is given by Vitruvius.

[117] A. Gatty, *The Book of Sundials* (London, George Bell & Sons, 1900), pp.29–30.

Fig. 15 A modern recreation of the hemispherical *skaphe*.[a]
[a] Author's photo.

refers to a passage in which Apion states that in the city of Heliopolis, the bib-
lical Moses erected pillars, beneath which was a model of a bowl (σκάφη),[118]
and the shadow cast on this basin by the statue on the pillar described a circle
corresponding to the course of the Sun in the heavens.[119] The time of Moses is
roughly dated to around 1200 BC, but there is no other evidence for the Egyptians using sundials of this type in that time period. Josephus doubts Apion's
account on a number of reasons, but none of them has to do with the type of
sundial he describes.[120] Apion may be confusing the use of the *gnomon*-staff,
which was common in Egypt in the thirteenth century BC, with the use of the
hemispherical *skaphe* which was more common in his own time. Regardless,
claims that the *skaphe* was both used in Egypt in the time of Moses, and came
from the fourth century astronomical schools of Asia Minor, do not correlate.
If Aristarchus was the inventor of the *skaphe*, this would mean that it was a
relatively new design, and represented the state of the art for sundials, when
Eratosthenes undertook his experiment.

[118] In some editions of Josephus, the term σκάφη is translated as 'boat'.
[119] Joseph. *Ap.* 2.10–11.
[120] Joseph. *Ap.* 2.12–28.

There were a number of different types of sundial that were in use in Eratosthenes' time, but only one of them would suit his experiment. Cleomedes states
that the Sun on the Summer Solstice cast only a small shadow in Alexandria.[121]
This indicates that the type of sundial that Eratosthenes used possessed a vertically aligned *gnomon*. Certainly, both a *gnomon*-staff and an Egyptian shadow
clock possess a vertical *gnomon* which would have cast a small shadow in
Alexandria on the Summer Solstice, and would have cast no shadow in Syene
at the same time, just as Cleomedes describes. However, while Eratosthenes
would have been able to observe a small shadow length on such instruments
on the Solstice, he would have to determine the ratio of its size to the height of
the upright, determine the angle, and then translate the angle that had been
formed by a shadow cast onto a flat surface to the circumference of a circle. This makes the use of a shadow clock or *gnomon*-staff by Eratosthenes
for anything other than confirming his observations somewhat unlikely. It
is more probable that he used an instrument which was already configured
to produce an angle that was applied to a curved surface—the hemispherical
skaphe—and this accounts for why the specific term for this type of sundial is
attributed to Eratosthenes' experiment by Cleomedes.[122] This was also because
the *arachne* inscribed on the face of the *skaphe* was a fundamental component
of Eratosthenes' calculations.

According to Vitruvius, it was Eudoxus of Cnidus (400–350 BC) who
invented the *arachne*, although Vitruvius also states that some people credit
it to Apollonius of Rhodes (third century BC).[123] Goldstein and Bowen suggest
that, as such markings are found on extant sundials from prior to the time of
Apollonius in the second century BC, Vitruvius' attribution of the invention
of the *arachne to* Apollonius seems unlikely.[124] However, it is Eudoxus who
seems to be an unlikely candidate for the creation of the *arachne* for the spherical sundial as he lived prior to the time of Aristarchus, the apparent inventor
of the *skaphe*, while Aristarchus, Apollonius, and Eratosthenes were relative
contemporaries. Furthermore, the oldest datable example of a *skaphe* with an
arachne in Gibbs' catalogue also comes from the island of Delos in Greece from
the third century BC, which would then correlate with the time of Aristarchus
and Apollonius.[125]

---

[121] Cleom. *De motu* 1.10.
[122] Cleom. *De motu* 1.10.
[123] Vit. *De Arch.* 9.8.1.
[124] Goldstein and Bowen, 'A New View', p.336.
[125] Gibbs, *Greek and Roman Sundials*, p.123, #1001.

Regardless of who created the *arachne*, the series of lines marking the Solstices, Equinoxes, and hours marked an advancement in sundial design over those that had been used before in that it created an instrument that was accurate year-round regarding the location for which it was created. The *arachne* was a projection of the grid of day and hour lines onto the celestial sphere.[126] The geometry of a sundial's grid of day and hour lines is determined by the type of sundial, the shape of its surface, and, for non-planar sundials, the location of the *gnomon*.[127] Most sundial grids were just the day curves for the Solstices and Equinoxes, and the curves for the second to eleventh seasonal hours.[128] All of these curves were within the scope of Greek geometry in the third century BC (Fig. 16).[129]

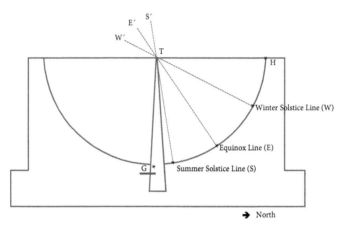

S'
E'
W'
T
H
*Winter Solstice Line (W)
*Equinox Line (E)
G *
Summer Solstice Line (S)

➜ North

Fig. 16 Cross section through the noon line of a hemispherical *skaphe* showing the angle of the Sun (red dotted lines) on the Solstices and Equinoxes.

An inscription from the third century BC, found near Lake Maryut in southwest Alexandria—not far from the site of the *Serapeum* Annex—outlines how sundials by this time not only had lines marking the Equinox and the Solstices, but also had lines indicating the months.[130] The inscription details how the tip

---

[126] Jones, 'Greco-Roman Sundials', p.141.
[127] Jones, 'Greco-Roman Sundials', p.131.
[128] Jones, 'Greco-Roman Sundials', p.132.
[129] It has been suggested that the sloping Amphiareion sundial from the fourth century BC (*ca.* 348 BC) is the 'earliest evidence of the application of the mathematical methods of spherics' (see: K. Schaldach, 'The Arachne of the Amphiareion and the Origin of Gnomonics in Greece' *J. Hist. Astron.* 25 (2004), pp.435–445; Tupikova and Soffel ('Modelling Sundials', pp.8–13) calculate that, if the Amphiareion sundial was constructed for the location where it was found, the measured error in the instrument was about 2°.
[130] *CPI* #106.

of the shadow cast by the pointer would move from one of these monthly lines to the next over a period of thirty days.[131] The same inscription also outlines how the sundial had constellations marked on its face, and how the position of the shadow, in correspondence with these constellations, would indicate the arrival of various seasons throughout the year.[132] However, despite these later advancements in the design and configuration of ancient Greek sundials to indicate both months and seasons, such markings do not seem to have been common in the time of Eratosthenes, nor were they necessary for his calculation of the circumference of the Earth. All that was required was an accurately inscribed line for the Summer Solstice.

As the Sun does not travel at right angles to the axis of the Earth, and is higher in the sky, and remains above the horizon for longer in summer, than it does in winter, this will determine the shape of the *arachne* on a sundial as the position of the lines will vary with the latitude for which the sundial is calibrated. However, due to the design of the *skaphe*, the lines of the *arachne* could be marked on the surface of the sundial without the involvement of any mathematics at all. All that was needed was time and consistent observation. Once the bowl of the sundial had been carved out, and the *gnomon* put in place, the unmarked sundial could simply be set up in a place where the Sun would shine upon it. The arc that the Sun follows on a set day acts as a day curve, indicative of that day of the year, while the fraction of that arc that has been traversed, from one edge of the bowl to the present moment, indicates the seasonal hour. The day curves on a sundial are projections of those arcs for specific days onto the surface of the sundial.[133] The curve for a certain day (e.g. the Summer Solstice) on a hemispherical sundial is formed by the intersection of the curve of the hemisphere with cones generated by rays of sunlight. The axis of the sunlight coincides with the axis of the Earth and the vertex of the cone is the centre of the hemisphere. Thus, the day curves are arcs of parallel circles. The Equinox curve, for example, is half of a great circle.[134] As such, in order to create an accurate *arachne* on the sundial, the location of the tip of the shadow cast by the *gnomon* had to be marked, at periodic intervals across the day, on three days of the year—when the Sun was at its most northerly position on the Summer Solstice (S' in Fig. 16) and the *gnomon*'s shadow was shortest, when the Sun was at its most southerly on the Winter Solstice (W') and the *gnomon*'s shadow was longest, and when the periods of daylight and night were

---

[131] *CPI* #106.
[132] *CPI* #106.
[133] Jones, 'Greco-Roman Sundials', p.130.
[134] Gibbs, *Greek and Roman Sundials*, p.14.

equal on the Equinoxes (E'). These marks could then be connected to create the day curves of the *arachne*. Thus in order to identify the Solstices using the shadow cast by the gnomon, the lines traced by the tip of the shadow had to be followed over the course of a year.[135] Vitruvius discusses just such a method for deriving the *analemmata* for planar sundials, but such a procedure would also apply to the hemispherical *skaphe*.[136] As long as the sundial was not moved from its position once marking the days on the face had commenced, within a twelve-month period, the day lines of the *arachne* would be formed with the Summer Solstice line at the bottom (closest to the *gnomon*), and the Winter Solstice line at the top. As well as the day curves of the *arachne*, the remains of a hemispherical *skaphe* from Delos in Greece (third century BC) also contain inscriptions where each of the day curves were labelled—Equinox (ἰσημηρία), Summer Solstice (τροπαὶ θεριναί), and Winter Solstice (τροπαὶ χειμεριναί).[137]

Indeed, as the face of the *skaphe* is a hemispherical bowl, the blank sundial did not even need to be initially aligned towards the north, or to be totally level, for this process to work, as the regularity of the curvature of the bowl would mean that the day lines would be accurately created regardless of orientation, as long as the sundial was not moved during the process. As Symons and Khurana point out, if the surface of the sundial is not vertical (such as on a hanging sundial), a symmetrical pattern can only be obtained by keeping the cardinal alignment and any rotation of the instrument could only be around the east–west axis.[138] While such movement may make the resultant *arachne* symmetrical, it should be noted that such movement would also make the *arachne* inaccurate for the location depending upon how much, and how often, the sundial was rotated. If the purpose of the sundial was only to act as a clock, then the Equinox and/or Solstice lines did not have to be very accurate (or even exist). These were only needed if the sundial was to act as a calendar as well.

If a levelled, north–south, alignment was required—for aesthetic reasons, for accuracy, so that the instrument could act as a calendar, and/or, importantly, for the later cut-away variants of the *skaphe* that did not possess a full hemispherical bowl—this could have been accomplished in a number of ways. In the first century BC, Diodorus of Alexandria, a student of the astronomer Posidonius, outlined how to find the north–south meridian using three *gnomons*.[139] Whether such a method was known several centuries earlier

[135] Munn, 'Science to Sophistry', p.121.
[136] Vit. *De Arch.* 9.1.1, 9.7.2–7.
[137] F. Dürrback and A. Jardé, 'Fouilles de Délos' *BCH* 29 (1905), pp.250–252; see also: Gibbs, *Greek and Roman Sundials*, p.123, #1001.
[138] Symons and Khurana, 'A Catalogue', p.381.
[139] Papp. *Syn.* 4.22.

when the *skaphe* was first being established as a timekeeping instrument is uncertain. If such a method was unknown, a sundial could still have been aligned to the north at night by aligning the instrument with the star Polaris. Another method would be to mark the place that a star rose on the horizon from a certain location with some form of marker (e.g. a pole or stone), and also mark where that same star set from the same location. Bisecting the resultant angle created by the rising and setting points and the observation location would indicate true north. This may be the process that is mentioned by Diodorus. Regardless of the method employed to align the instrument to the north, once the inscribing of the lines of the *arachne* was complete, the instrument would then be accurate for any location on the same latitude.

The hour lines of the *arachne* should connect the points on all declination arcs (whether represented or not) which correspond to the specific hour.[140] The tip of the *gnomon*'s shadow will travel one-twelfth of its daily path in each of the ancient 'hours'. If the daily path is divided into twelve equal parts, the points of division will mark out thirteen hour lines across a year—with the noon line being one, and the eastern and western horizon edges being another two.[141] Greek sundials do not seem to have possessed marks for half-hours like the example of the vertical sundial from Egypt. Jones, who calls the hour the smallest quantitative unit on the ancient sundial, suggests that the reading of fractional hours could be problematic—especially for the hours close to noon, as the spaces between the curves can be small.[142] However, this would depend on a number of factors such as the size of the sundial and height of the pointer, the thickness of the *gnomon*, and the shape of the tip of the *gnomon*. As Isler states: the taller the *gnomon*, the greater the accuracy, but the less distinct the shadows are.[143] Somewhat contradictorily, Jones also provides examples from ancient texts where the time is given in parts of hours (e.g. two and five-sixth hours, three and four-fifth hours, four and a half hours), and states that determinations could be made to a fraction of an hour with good precision.[144] Jones says that most observers, if asked to indicate where a half-hour mark should be, would pick the point halfway between two of the hour lines. However, this is incorrect for the majority of seasonal hours.[145] It is therefore uncertain how a time measurement to a fraction like five-sixths of an hour could have been determined if there were no other markings—possibly painted and now lost—on the sundials.

[140] Jones, 'Greco-Roman Sundials', p.130.
[141] Gibbs, *Greek and Roman Sundials*, p.15.
[142] Jones, 'Greco-Roman Sundials', pp.126, 133.
[143] Isler, 'The Gnomon', p.157.
[144] Jones, 'Greco-Roman Sundials', pp.128–129.
[145] Jones, 'Greco-Roman Sundials', p.135.

The hour lines are not circular curves except at the horizon and along the noon line. The other lines are determined by the equation:

$$\tan(d) = \tan(\phi) \times \cos(n) \times \sigma \times (\pi/180),$$ [eq. 2.4.1]

where d equals the declination of the Sun, ϕ equals the latitude of the sundial's location, n equals 6/m (where m equals the number of 'hours' from sunrise to noon), and σ equals the number of degrees of the arc that have been traced by the Sun from sunrise to noon.[146] The latitude, or χλῖμα meaning 'inclination', for the sundial's location (ϕ) is equal to the arc length HE (in Fig. 16) because:[147]

$$\text{angle HTE} = \Phi = (\text{HE} \times 180°) / (\pi \times \text{HT}).$$ [eq. 2.4.2]

The distance between the hour marks on a segment of the Solstice curve (λ) also correlates to the latitude that the sundial is calibrated for (ϕ) through the relationship:[148]

$$\tan(\Phi) = -\cos(d) \times \cos[(90° \times 12 \times \lambda)/(\pi \times \text{HT} \times \cos(d))].$$ [eq. 2.4.3]

Despite being non-circular, the hour lines follow very closely circles which pass through the corresponding hour points on the Solstices and Equinoxes at latitudes <45°.[149] As such, the marking of the hour lines did not require careful observations or complicated mathematics to be made. All that was needed was to divide the area into twelve equal parts, which would mark the temporary hours of the day. However, to obtain greater accuracy, if the same division were repeated at multiple times when the Sun was at its highest, and the hour lines were then drawn through these divisions, the unequal hours for every day in the year would be determined.[150]

The noon line is the arc of a great circle that connects the point directly beneath the *gnomon* (G in Fig. 16), and through the points for the Solstices and Equinoxes (SEW) and the edge of the celestial horizon (H). Mathematically,

---

[146] See: T.S. Davies, 'An Enquiry into the Geometrical Character of the Hour Lines upon the Antique Sundials.' *Trans. R. Soc. Edinburgh* 8 (1818), pp.72–122, especially pp.80–86.

[147] Formula derived from: Gibbs, *Greek and Roman Sundials*, p.14. Other ways to determine the latitude of a place are measuring the angle made by the Sun's rays to the vertical at noon on the Solstice, or comparing the ratio of day to night on the Solstice. See: A. Szabó and E. Maula, *Enklima: Untersuchungen zur Frühgeschichte der griechischen Astronomie, Geographie und der Sehnentafeln* (Athens, Akademie Athen—Forschungs-institut für Griechische Philosophie, 1982), pp.155–156); Russo, 'Forgotten Revolution', p.67.

[148] Formula derived from: Gibbs, *Greek and Roman Sundials*, p.16.

[149] Gibbs, *Greek and Roman Sundials*, p.16.

[150] Gatty, *Book of Sundials*, p.29.

the angle, in degrees, between the intersection of the noon line and one of the Solstice lines, and the intersection of the noon line to the Equinox line (for example, angle STE), is equal to the angle between the intersection of the noon line and the other Solstice line, and the intersection of the noon line and the Equinox line (for example, angle ETW). This is because the angles created by the Sun at two different times of the year—W'TE' and E'TS'—are equal. This angle is equal to the obliquity of the plane of the ecliptic (ε).[151] Alternatively, if the line GT is equal to the height of the *gnomon*, which is, in turn, equal to the radius of the hemisphere (r), then:

$$\text{arc WE} = \text{arc ES} = \varepsilon \pi r / 180°. \qquad \text{[eq. 2.4.4]}$$

Thus, on an accurately constructed and aligned sundial, with a central, vertical, *gnomon*, the arc lengths WE and ES should be equal, and the height of the *gnomon* corresponds to:[152]

$$\text{GT} = (180° \times \text{WE}) / (\varepsilon \times \pi) = (180° \times \text{ES}) / (\varepsilon \times \pi). \qquad \text{[eq. 2.4.5]}$$

The Equinox line is equal in length to $\pi \times \text{GT}$ and this directly corresponds to the latitude of the calibrated location, measured along the meridian line of the instrument from the base of the *gnomon*.[153] Both Strabo and Vitruvius say that, on the Equinoxes, the shadow cast by the pointer of a sundial in Alexandria is three-fifths the height of the *gnomon*.[154] Similarly, Pliny says that in Egypt on the Equinoxes the shadow measures 'a little more than half' the height of the *gnomon*. Pliny also says that the readings would alter with every change in latitude of around 300–500 *stadia*.[155]

This indicates that there was some mathematical/geometrical modelling that was understood behind how the lines of the *arachne* were formed, and how a sundial made for use in one place might not be accurate in another. The noon meridian of the *arachne* itself does not depend on the correct location, nor do the first and last hour lines.[156] The Winter Solstice curve, on the other hand, is shorter than the Equinox line and has a length of:[157]

$$\left[(\text{TH} \times \cos(\varepsilon)) \times \pi \times \cos^{-1}(\tan(\Phi) \times \tan(\varepsilon))\right] / 90°. \qquad \text{[eq. 2.4.6]}$$

---

[151] Gibbs, *Greek and Roman Sundials*, p.13.
[152] Formula derived from: Gibbs, *Greek and Roman Sundials*, p.13.
[153] Gibbs, *Greek and Roman Sundials*, p.14.
[154] Str. *Geog.* 2.5.38; Vit. *De Arch.* 9.7.1.
[155] Plin. *HN* 2.315.
[156] Jones, 'Greco-Roman Sundials', pp.140–141.
[157] Formula derived from: Gibbs, *Greek and Roman Sundials*, p.14.

This shows that the shape and position of the Winter Solstice line is very much dependent upon the location of the sundial. Similarly, the Summer Solstice curve is longer than the Equinox line and has a length of:[158]

$$\left[(TH \times \cos(\varepsilon)) \times \pi \times \cos^{-1}(-\tan(\Phi) \times \tan(\varepsilon)) \right] / 90°. \qquad \text{[eq. 2.4.7]}$$

This also demonstrates how the *arachne* of the sundial is specific to the location. However, depending upon how far a sundial is moved from its calibrated location will determine how inaccurate it then becomes in its new position. Pliny, for example, describes inaccuracies found in a sundial moved from Catania (37°30′N) to Rome (41°54′N) in 263 BC.[159] Jones, however, in his examination of this passage, determined that even with a change of 4.25° of latitude, the sundial was out by only 0.07 hours in the summer.[160] Due to the inherent level of accuracy in a properly configured *skaphe*, this made this type of sundial the perfect instrument for Eratosthenes to use in his experiment.

2.5 A Replica Of Eratosthenes' *Skaphe*

In order to engage with the work of Eratosthenes, a replica of the type of *skaphe* that he would have used was created. As no examples of the full hemispherical sundial have survived, the configuration of the instrument had to be designed from scratch. The replica used in this evaluation of Eratosthenes' work was created by Skaphe in Germany (www.skaphe.de) and was then used in a number of experiments to determine the accuracy of the instrument, how Eratosthenes would have used it, and how it influenced his results (the conclusions drawn from these experiments can be found throughout the remainder of this work). The modern replica was turned from a block of solid aluminium (rather than bronze, as per Ptolemy and Capella) measuring 200 mm × 200 mm × 100 mm. The bowl of the *skaphe* was cut with a radius of 75 mm, resulting in a hemisphere with a diameter of 150 mm (Fig. 17).[161]

---

[158] Formula derived from: Gibbs, *Greek and Roman Sundials*, p.14.
[159] Plin. *HN* 2.182, 7.214.
[160] Jones, 'Greco-Roman Sundials', p.142.
[161] This size was chosen for a number of reasons. Firstly, it was twice the size of the sundials that Skaphe.de normally make, and so constituted a customization of their design programmes and manufacturing. Secondly, the size is around the midpoint of examples from Egypt and third-century BC Greece. The find from Delos, for example, has a diameter of 275 mm (see: Gibbs, *Greek and Roman Sundials*, p.124, #1002G). Another example, from Naukratis in Egypt, measures only 75 mm (see: Gibbs, *Greek and Roman Sundials*, p.155, #1040G; R. Thomas and A. Masson, 'Altars, Sundials, Minor Architectural Objects and Models' in A. Villing, M. Bergeron, G. Bourogiannis, A. Johnston, F. Leclère,

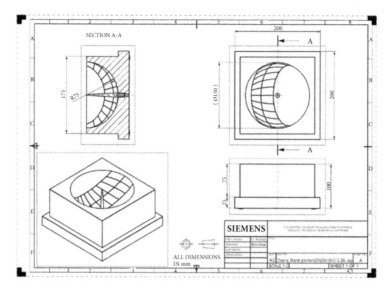

Fig. 17 Design schematics for the replica of Eratosthenes' *skaphe*.[a]

[a] Image courtesy of Werner Schreiner at Skaphe.de.

A *gnomon* of stainless steel was fashioned in the shape of an elongated cone, approximately 5 mm across at the base. This style of *gnomon* was chosen as it was thought that it would provide the sharpest shadow, and the narrow tip would create a more easily readable shadow, than other options that had a small ball on the tip (Fig. 18).

Once the bowl of the *skaphe* had been turned, computer modelling was undertaken to determine the shape and lines of the *arachne* for the sundial—which was set for the location of Alexandria in Egypt, where Eratosthenes had undertaken his experiment in the third century BC (Fig. 19).

The lines of the *arachne* were then etched into the bowl of the *skaphe* with a laser (Fig. 20).

The result was a replica of a hemispherical *skaphe* from the third century BC with an *arachne* perfectly calibrated for the location of Alexandria in Egypt (Fig. 21).

To test the principle that a sundial is accurate within a range of 300–500 *stadia* of the location for which it has been calibrated, a simple experiment was undertaken using the replica of the hemispherical *skaphe*. The sundial was placed on a mount—designed to hold a Dobsonian telescope—which would

A. Masson, and R. Thomas (eds.), *Naukratis: Greeks in Egypt* (London, British Museum, 2015), p.12).
Finally, the cost of manufacture and shipping due to the weight of a larger sundial was considered.

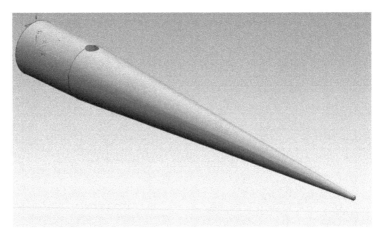

Fig. 18 Computer modelling of the design of the *gnomon* for the replica sundial.[a]

[a] Image courtesy of Werner Schreiner at Skaphe.de.

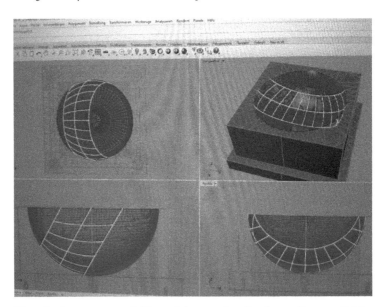

Fig. 19 Computer modelling of the *arachne* for the replica sundial.[a]

[a] Image courtesy of Werner Schreiner at Skaphe.de.

allow the instrument to be rotated and permit the inclination of the base plate upon which it sat to be altered (Fig. 22).

When the Sun was near its zenith point across numerous days, the instrument was positioned so that the tip of the umbra of the shadow cast by the *gnomon* was just touching the intersection point of the noon and Summer Solstice lines on the curved surface of the sundial. This, in effect, replicated

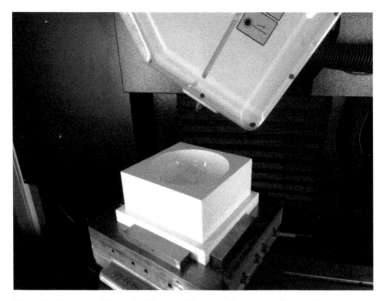

Fig. 20 Laser etching the lines of the *arachne* onto the replica sundial.[a]

[a] Image courtesy of Werner Schreiner at Skaphe.de.

Fig. 21 The complete replica *skaphe*.[a]
[a] Author's photo.

how the shadow would appear at solar noon on the Summer Solstice, in Alexandria, on an instrument specifically calibrated for that location (Fig. 23).

An inclinometer was then used to determine the angle to which the base of the sundial had been aligned, in order to simulate the conditions of noon on

Fig. 22 The replica sundial set onto a Dobsonian mount.[a]
[a] Author's photo.

the Summer Solstice. The angle of the mount was then slowly adjusted until it could be easily determined, via naked-eye observations, that the tip of the shadow was no longer sitting on the intersection of the noon and Summer Solstice lines. The adjusted angle was then recorded using the inclinometer. Across numerous runs of this experiment, the average alteration required to adjust the angle of the sundial so that the tip of the shadow shifted clearly off the markings on the sundial was 0.53°. As the bowl of the *skaphe* is a representation of the 'great circles' of the universe (as the ancient Greeks called them), an angular change in the position of the tip of the shadow corresponds to a change in terrestrial latitude. One degree of latitude is equal to approximately

Fig. 23 The shadow of the *gnomon* when the sundial was set to replicate noon on the Summer Solstice.

111 km. Thus, an angular variation of 0.53° is equal to a change in terrestrial location of (0.53 × 111 km ≈) 59 km. If this distance is converted using the smallest size *stade* used in the ancient Greek world—the Messenian Standard with its 160 m *stade*—this equates to a distance of (59,000 m / 160 m ≈) 369 *stadia*. If converted using the larger Pan-Hellenic system, with its 180 m *stade*, the distance is equal to around 328 *stadia*.[162] These conversions fall within the range provided by Pliny and suggest that the minimum value in his range, a distance of 300 *stadia*—or an angular variation of the orientation of the sundial of around 0.5° based upon conversion using the 180 m *stade*—is the limit of the sundial's calibration.

It was also found that the amount of variance required to shift the tip of the shadow clearly off the intersection of two lines of the *arachne* was the same no matter how the sundial had been initially set up. If, for example, the sundial was aligned so that the tip of the shadow sat on the intersection of the noon and Equinox lines (simulating noon on the Equinox in Alexandria), or on the intersection of the noon and Winter Solstice lines (simulating noon on the Winter Solstice in Alexandria), the amount of change required to shift the tip of the shadow clearly from this point was still around 0.5°. This shows

---

[162] For the different systems of measurement, and size of the *stade*, used in ancient Greece, see section 4.1.

that a *skaphe* that is designed to be used in a specific location will be relatively accurate at any location up to 300 *stadia* (55.5 km) north or south of that location at any time of year.

To test the replica sundial further, and to establish how Eratosthenes used it in his determination of the circumference of the Earth, it was initially planned to take the replica sundial to Egypt and duplicate Eratosthenes' experiment in the forecourt of the New Great Library of Alexandria (the Bibliotheca Alexandrina) on the Summer Solstice in June in the northern hemisphere. However, due to the COVID-19 pandemic restrictions that impacted international travel at the time of writing, an alternative location had to be found. Fortunately, about five hours' drive north of Sydney, Australia lies the coastal town of Crescent Head. This town is on the same southern equivalent latitude (31.2°S) as Alexandria's northern latitude (31.2°N) (Fig. 24).

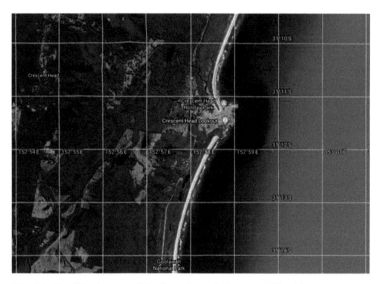

Fig. 24 Satellite image of the location of Crescent Head.[a]
[a] Image taken from Google Earth.

It was thus decided to undertake experimentation with the replica sundial at Crescent Head during the southern Summer Solstice in December 2021. The experiments involved a number of objectives:

1) To examine the potential errors associated with aligning/using a sundial of the style used by Eratosthenes in the third century BC to determine the circumference of the Earth.
2) To determine the accuracy of the replica instrument.
3) To examine potential errors in reading the shadow cast by the *gnomon*.

4) To observe how fast the shadow moves across the face of the sundial at midday on the Summer Solstice.
5) To determine how the length of the shadow could have been used to gauge the distance between two sites.

THE DETERMINATION OF 'SUNDIAL TIME'

A fundamental element of any experimentation involving a sundial—such as gauging the calibration and accuracy of a replica of the kind of *skaphe* that Eratosthenes would have used, and to answer some essential questions in regard to the shadow cast by its *gnomon* at midday—is to know not only what day of the year to conduct the experiment on (in this case, the southern Summer Solstice on 22 December 2021), but also what time the Sun will be at its zenith on that particular day. This is not as simple as it may first appear, as the Sun is not always directly overhead when a clock or watch at the same location reads exactly midday, for a number of reasons. By calculating the factors that influence the Sun's position in relation to local time, the correct time for conducting such an experiment can be determined.

THE EQUATION OF TIME

Due to the elliptical shape of the Earth's orbit and, to a lesser extent, the precession of the Earth's axial tilt and the gravitational influences exerted on the Earth by the Moon and planets, the heliocentric longitude of the Earth does not vary at a constant rate over time. This is because, as is shown by Kepler's Third Law of Planetary Motion, the Earth will move faster in its orbit when it is at perihelion in January, and slower when it is at aphelion in July. Additionally, the Earth's 23° axial tilt makes the Sun appear to cross the meridian at different times at a certain location across a year. The result of this is that the Sun does not follow the path of the ecliptic, nor does its right ascension vary, at a uniform rate. This variance in the Sun's position is then reflected in the difference between the time shown on a sundial, and the time as displayed on a watch or clock in the same location.

A sundial will display 'local apparent time' (LAT) based upon the angle of the Sun. Watches and clocks, on the other hand, show 'local mean time' (LMT), which is based on the movements of a 'mean Sun' and its movement along the ecliptic.[163] This means that, at most times of the year, a clock and a sundial in the same location will not show the same time. In the southern hemisphere, between 1 January and 16 April, the time shown on a sundial will be behind that shown on a clock. From 16 April to 14 June, the sundial will be

---

[163] Waugh, *Sundials*, p.9; Rohr, *Sundials*, p.29.

ahead of the clock. From 14 June to 2 September, the sundial will be behind again, and from 2 September to 25 December, the sundial is again ahead of the clock before falling behind yet again for the last six days of the year.[164] These periods of variance, and the approximate amounts that the two instruments will be out by, known as the *equation of time* (E), can be depicted graphically (Fig. 25).

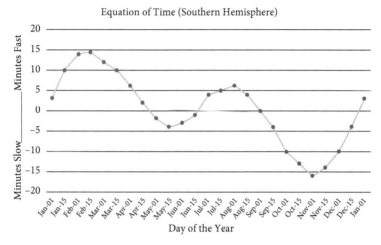

Fig. 25 The *equation of time* for the southern hemisphere.

The value of E for a specific day can be calculated using the following formula:[165]

$$E = L_0 - 0.0057183° - \alpha + \left(\Delta\psi \times \cos\left(\varepsilon\right)\right), \qquad \text{[eq. 2.5.1]}$$

where α equals the apparent right ascension of the Sun in degrees (for 22 December 2021, the RA of the Sun = $18^h04^m48^s$ = 271.2°), $\Delta\psi$ equals the nutation of longitude in degrees (−14.488" = −0.0040244832°), ε equals the obliquity of the ecliptic (23.4°), and L_0 is the mean longitude of the Sun.

The value of L_0 can be derived using the formula:[166]

$$L_0 = 280.4664567 + \left(360,007.698 \times \tau\right) + \left(0.03032028 \times \tau^2\right) + \left(\tau^3/49,931\right)$$
$$- \left(\tau^4/15,300\right) - \left(\tau^5/2,000,000\right), \qquad \text{[eq. 2.5.2]}$$

---

[164] Data derived from Waugh (*Sundials*, p.10), who lists the variances for the northern hemisphere.
[165] See: J. Meeus, *Astronomical Algorithms* (Richmond, Willmann-Bell, 1998), p.183.
[166] See: Meeus, *Astronomical Algorithms*, p.183.

where τ equals time in Julian millennia (units of 365,250 days) from the J2000.0 epoch to the specific day to be calculated. From 1 January 2000 to 22 December 2021, there are 8,026.25 days. Thus:

$$\tau = 8,026.25/365250 = 0.02197467488. \qquad \text{[eq. 2.5.3]}$$

Inserting the value of τ into [eq. 2.5.2] to find L_o yields:

280.4664567	280.4664567
$+(360,007.698 \times 0.02197467488)$	$+7,911.052124$
$+(0.03032028 \times 0.02197467488^2)$	$+0.00001464124$
$+(0.02197467488^3 / 49,931)$	$+2.125186758 \times 10^{-10}$
$-(0.02197467488^4 / 15,300)$	$-1.523986928 \times 10^{-11}$
$-(0.02197467488^5 / 2,000,000)$	-2.56×10^{-15}
	$8,191.518595$

This result then needs to be reduced to a value <360° by subtracting multiples of 360 to find the value of L_o:[167]

$$22 \times 360 = 7,920 \rightarrow 8,191.518595 - 7,920 = 271.518595°. \qquad \text{[eq. 2.5.4]}$$

Placing all of the derived values into [eq. 2.5.1] to find E then calculates how many minutes a sundial will differ in relation to a local clock on 22 December 2021:

$$
\begin{aligned}
E &= L_o - 0.0057183° - \alpha + (\Delta\psi \times \cos(\varepsilon)) \\
&= 271.518595° - 0.0057183° - 271.2° + (-0.0040244832° \times \cos(23.4)) \\
&= 271.518595° - 0.0057183° - 271.2° + (-0.0040244832° \times 0.9177546257) \\
&= 271.518595° - 0.0057183° - 271.2° - 0.0038957142° \\
&= 0.3135259740°
\end{aligned}
$$

This result can then be converted into minutes of time by multiplying it by four:

$$0.313525974 \times 4 \approx 1.25\text{min} \approx 1.3\text{min} = 1^m 18^s. \qquad \text{[eq. 2.5.5]}$$

Another way to calculate the value of E is using the formula:[168]

$$E = 9.87 \times \sin(2B_o) - 7.67 \times \sin(B_o + 78.7), \qquad \text{[eq. 2.5.6]}$$

---

[167] Meeus, *Astronomical Algorithms*, p.184.
[168] Formula derived from the Australian Bureau of Meteorology. See: R. Thompson, *Australian Government Bureau of Meteorology: Space Weather Services—The Equation of Time* (2021), accessed 19 June 2021, https://www.sws.bom.gov.au/Educational/2/1/14.

where B_o = 360 × (N—81) / 365 and N equals the days of the year (for 22 December, N = 356). Thus:

$$B_o = 360 \times (356 - 81)/365$$
$$= 360 \times 275/365$$
$$= 99,000/365$$
$$= 271.2328767$$

and

$$E = 9.87 \times \sin(2x271.2328767) - 7.67 \times \sin(271.2328767 + 78.7)$$
$$= 9.87 \times \sin(542.4657534) - 7.67 \times \sin(349.9328767)$$
$$= 9.87 \times -0.04302223257 - 7.67 \times -0.1748017828$$
$$= -0.4246294355 - 1.340729674$$
$$= 1.29826673 min$$
$$\approx 1.3 min = 1^m 18^s$$

While this is not as detailed as the formula provided by Meeus, Thompson states that this formula is good to within a margin of error of <1 per cent for a century either side of the J2000.0 epoch.[169] The two formulae are also a means of corroborating each other.

In order to convert from LMT to LAT, when the sundial is running behind a local clock—as occurs in the southern hemisphere on the Summer Solstice—the value of E needs to be subtracted from the LMT to determine the LAT: or LAT = LMT − E.[170] Thus, based on the *equation of time* alone, the Sun would be at its zenith on the southern Summer Solstice at: 12:00:00 − 00:01:18 = 11:58:42.

ZONAL TIME

Another thing to consider is that a local clock operates on a 'time zone', which is based upon the mean longitudinal meridian of that zone and not for the meridian that runs through a specific location where a sundial might be set.[171] The east coast of Australia operates on a time zone of UTC plus ten hours based upon the 150°E longitudinal meridian. This is because each hour of difference between two locations is equal to 15° of longitude, so a ten-hour difference between eastern Australia and Greenwich Universal Time equates to 10 × 15° = 150°E. This meridian line runs to the west of the location where the experiment

[169] Thompson, *The Equation of Time*.
[170] Waugh, *Sundials*, p.12.
[171] See: Rohr, *Sundials*, p.30.

was conducted—Crescent Head, New South Wales at a longitude of 152.97°E (or 152°58′E). Thus, there is a difference between the location of Crescent Head and the mean longitudinal meridian of 2.97°. This is almost the exact same longitudinal variance between Alexandria and Syene—the two locations used in Eratosthenes' experiment—and is another reason why Crescent Head was chosen as the site to replicate it.

Each degree of longitude between a specific location and the mean meridian equates to a difference of four minutes of time. Each minute of longitude equals a difference of four seconds of time.[172] The difference between Crescent Head and the 150°E meridian is 2°58′. As such, the difference in time between these two locations is:

$$2 \times 4\text{min} + 58 \times 4\text{sec} = 8\text{min} + 232\text{sec} = 8\text{min} + 3\text{min } 52\text{sec} = 11\text{min } 52\text{sec}$$

In order to convert the local apparent time (LAT) to account for longitude correction (L) and the *equation of time* (E), if the sundial is located east of the meridian, then LAT = LMT − L − E.[173] Thus, at Crescent Head on the 2021 Summer Solstice, if the local apparent time is set as 12:00 noon (i.e. when the Sun should be at its zenith), then the local mean time, as displayed on a watch or clock, should be LMT = 12:00:00 − 00:11:52 − 00:01:18 = 11:46:50.

DAYLIGHT SAVING

A final thing that needed to be considered was that much of eastern Australia was operating under Daylight Saving conditions in December 2021. As such, clocks displaying LMT were an additional hour ahead. To account for this, the time calculated to account for longitude correction and the *equation of time* needed to be increased by one hour. Thus, the LMT when the Sun was at its zenith, and crossing the meridian, at Crescent Head, New South Wales, on the Summer Solstice 2021, was calculated as:

$$11 : 46 : 50 + 01 : 00 : 00 = 12 : 46 : 50. \qquad \text{[eq. 2.5.7]}$$

Such calculations can be confirmed using websites such as suncalc.org, which provides the ephemeris data for the Sun for any location at a set time on a specific day. By feeding in the details for the day and location into the programme, the results showed that on 22 December 2021 at Crescent Head, the Sun culminated at a LAT of 12:46:35 (Fig. 26).

---

[172] Waugh, *Sundials*, p.15.
[173] Waugh, *Sundials*, pp.16–17.

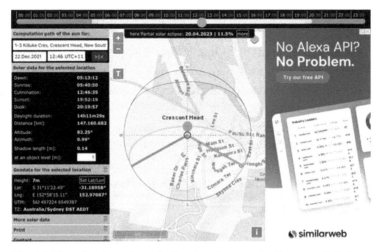

Fig. 26 Data for the solar culmination time on the Summer Solstice at Crescent Head, taken from suncalc.org.

The difference between the calculated time for the solar culmination (12:46.50) and that provided on the suncalc website (12:46:35) is only fifteen seconds, and can be attributed to rounding in the calculations.

Such issues would not have been a consideration for someone like Eratosthenes, who was working before the concepts of time zones and daylight savings existed. The variance between local mean time and local apparent time would also not have existed, and the sundial that he would have used would have accurately placed the location of the intersection of the noon line and the Solstice line onto the face of the instrument based upon the Sun's actual position at the zenith on that day, and the resultant shadow cast by the sundial's *gnomon*—assuming that the sundial had been made correctly for Alexandria and aligned properly. However, for any experiment that attempts to replicate Eratosthenes' work, these modern concepts need to be accounted for, and the local mean time adjusted, so that the correct 'time' to undertake the experiment can be determined.

On the day of the Summer Solstice, the sundial and other equipment needed for the experiment were taken to a local sports field in Crescent Head. This location was chosen as an ideal place to conduct the experiment, as the ground was relatively level and the location would afford unobstructed light onto the sundial at midday. Additionally, as the sundial would be accurate at any location within around 50 km of the site for which it was calibrated, the instrument did not have to be situated at any specific latitude, and this allowed for the most suitable location to be used. The sundial was placed onto the adjustable Dobsonian mount that had been used in previous experiments with the instrument (see Fig. 22) for support. The mount was then levelled, and aligned so that the sundial was facing towards the south using an attachment, also crafted by Skaphe in Germany, which contained two spirit levels and a magnetic compass, which could then be used to align the instrument (Fig. 27).

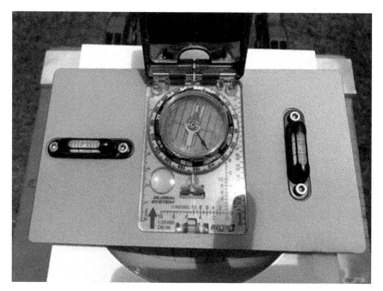

Fig. 27 The compass and spirit levels used to align the sundial.[a]
[a] Author's photo.

In order to get the correct bearing on the compass, the magnetic variation of Crescent Head (+12.6°) had to be considered so that the instrument was aligned to a true north–south orientation. To accomplish this, 12.6° was subtracted from a northerly bearing (0.0°) by rotating the bezel on the compass so that it read around 347.4°, to the best of its ability. The sundial was then rotated until the compass needle pointed to magnetic north, but due to the alteration of the bezel, the instrument itself would be oriented along a true north–south axis.

The correct, level alignment of the sundial along the two horizontal axes was proven when the compass attachment was removed and the bowl of the *skaphe* uncovered. When this was done, and the *gnomon* began casting a shadow, it was immediately observed that the tip of the shadow was tracking right along the Summer Solstice line on the face of the instrument. This would only occur if the sundial had been levelled correctly (Fig. 28).

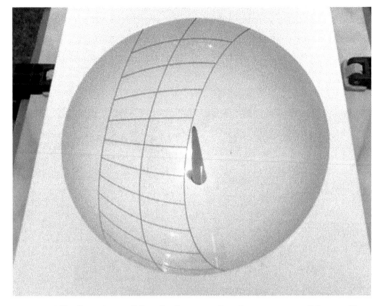

Fig. 28 The location of the shadow cast by the *gnomon* of the replica sundial fifteen minutes prior to noon (LAT) on the Summer Solstice in Crescent Head.[a]

[a] Author's photo.

As the shadow cast by the *gnomon* moved across the face of the sundial, it was observed that the tip of the shadow could first be considered to be sitting on the intersection of the noon and Summer Solstice lines at approximately 12:39:10; the time was recorded and the sundial photographed. The tip of the shadow was also observed to have passed beyond the intersections of these two lines by 12:49:52; the details were again recorded and the sundial photographed. The time of the true Solstice should have been the midpoint between these two recorded times (12:44:31), which differs from the time given for the solar culmination. It was further observed that, at the given time of the solar culmination (12:46:35), the tip of the shadow appeared to be slightly to one side of the intersection of the noon and Summer Solstice lines—indicating a LAT just past the solar culmination—rather than sitting evenly across the intersection (Fig. 29).

The calculated midpoint time for the solar culmination differs from the calculated time and the time given on the suncalc website by approximately two minutes. This is equivalent to a misalignment of the sundial by half a degree,

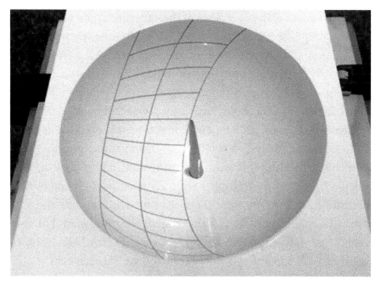

Fig. 29 The shadow cast by the *gnomon* of the replica sundial at noon (LAT) in Crescent Head.[a]

[a] Author's photo.

and can be attributed to one of several possible factors (or a combination thereof):

1) The naked-eye observation of when the tip of the shadow was either first touching or last touching the intersection of the noon and Summer Solstice lines may have been out. Any alteration of the observed times would also have adjusted the calculated midpoint, and hence the 'observed' time of noon.

2) When the sundial was first aligned using the compass, the exact orientation to account for a magnetic variation in Crescent Head of 12.6° may have been out due to attempting naked-eye alignment using a compass with a bezel only 60 mm in diameter, and which displayed only 2° increments using markings only 1 mm in length. The circumference of the bezel is 188.5 mm. This means that each of the segments representing 2° on the compass is just over 1 mm wide, and a variation of 0.5° requires only a misalignment of 0.25 mm, which would be almost impossible to confirm using naked-eye alignment on such a small instrument.

3) Similarly, orientation with a true north–south axis may have been out due to naked-eye alignment of the compass needle with the marking on the bezel, which indicated magnetic north.

Despite such a variance, the objectives of the experiment can be considered to have been accomplished. Firstly, in regard to examining the potential errors in the orientation and use of such a sundial, it is clear that even with a slight misalignment the instrument would still indicate the time with a high level of accuracy, as long as it is positioned so that it is level on the two horizontal axes. This is something that could have easily been accomplished in the time of Eratosthenes. Additionally, as long as the instrument is oriented as accurately as possible with the north–south meridian (again, something that could be achieved easily in the time of Eratosthenes), the time indicated by the tip of the shadow will also be highly accurate—with the potential of being out by only a few minutes from the true time. This demonstrates not only the accuracy with which the replica used in this experiment was produced, but also how accurately the instrument(s) that Eratosthenes would have been using were— assuming they were constructed with a similar level of precision.

One of the main 'errors' in reading the time on such a sundial is due to the small flare of light at the tip of the shadow. While this flare does, in some ways, obscure parts of the shadow itself—so that a clearly defined penumbra, for example, cannot be discerned—the flare also clearly indicates where the tip of the shadow is, and so can actually facilitate the reading of the position of the shadow's apex. This was found to be especially beneficial on the replica sundial, as the lines marking the hours and the Summer Solstice are black, but the flare of light clearly pinpoints the end of the shadow, which might have otherwise blended in with the colour of the line. This again aided in the reading of the instrument as the shadow moved across the face of the sundial.

It was also observed that the tip of the shadow was in a position that could have been considered to be indicating noon, or a time very close to it, for a period of around ten minutes (12:39 to 12:49) at noon on the Summer Solstice. Such a range would not have been a concern in the time of Eratosthenes, when people did not really have to consider the local time down to the minute. Additionally, a duration of ten minutes for 'noon' on the Summer Solstice further indicates that a misalignment of the instrument that would cause a variation of around two minutes in the display of the correct time would not have had a great impact. Lastly, and most importantly for understanding Eratosthenes' calculations, a confirmation that, at noon on the Summer Solstice, the tip of a shadow cast by the *gnomon* of a sundial located in Alexandria would sit right on the intersections of the noon and Summer Solstice lines of the *arachne* of the sundial provides critical insights into how Eratosthenes used these markings to determine the circumference of the Earth—particularly, as will be shown, how he determined that the distance between Alexandria and Syene was equal to one-fiftieth of the circumference of a circle.

3

The Analysis

3.1 The Angle

One of the key aspects of Eratosthenes' experiment and results, as they are reported by Cleomedes, is that he determined that the length of the shadow cast by the *gnomon* of a sundial in Alexandria created an angle with the vertical pointer equal to 1/50 of the circumference of a circle. This value is fundamental to Eratosthenes' calculation of the circumference of the Earth. However, the lack of detail in Cleomedes' account raises the question of how Eratosthenes' managed to determine this angle. And yet, despite how essential this angle is to determining the accuracy of Eratosthenes' calculations, few scholars have suggested ways in which this could have been accomplished in a time before the laws of sines and cosines in trigonometry had been invented. Rather, most prior studies into Eratosthenes either simply accept the 1/50 figure, or alter it in attempts to find some kind of mathematical solution to any questions over Eratosthenes' accuracy, without offering how this value was obtained. However, an examination of the features of the sundial that Eratosthenes would have been using in Alexandria demonstrates that there was a very simple way that the angle of the shadow could have been calculated. Additionally, this method would have not limited Eratosthenes to doing his experiment only on the Summer Solstice, but rather, his calculations could have been carried out at any time of the day, and on any day of the year.

Most scholars, in their examinations of Eratosthenes' calculation of the circumference of the Earth, have dealt with the 1/50th of the circumference of a circle figure that is reported by Cleomedes in some way or another. Some scholars have accepted the 1/50 (7.2°) figure given by Cleomedes in their analyses of Eratosthenes' calculations.[1] Vicdcbantt, on the other hand, suggested that there was an error in the angle reported by Cleomedes, that it was possibly rounded, and offered that the true angle of the shadow represented 1/47.547

---

[1] For example, see: Fischer, 'Another Look', p.156; Engels, 'Length', pp.302–303; Carmen and Evans, 'Two Earths', p.3.

Eratosthenes and the Measurement of the Earth's Circumference (c.230bc). Christopher A. Matthew, Oxford University Press.
© Christopher A. Matthew (2023). DOI: 10.1093/oso/9780198874294.003.0003

of the circumference of the Earth, which is then rounded up to 1/48, or 7.5°.[2] Fischer points out that 1/48th of the sky is also 1/4 of a sign of the zodiac.[3] However, Fischer accepts the 1/50 figure found in Cleomedes and suggests that Eratosthenes would have used fractions and ratios for his measurements.[4] This conclusion is based on a reference found in the works of Ptolemy which states that Eratosthenes had determined the arc length between the two Tropics to be equal to 11/83 of a circle.[5]

Rawlins suggested that there was an approximate 16' error in the reading of the shadow and also that the true angle was close to 1/48—a value that he says would have been easy for Eratosthenes to observe—but states that, as 1/50 worked neatly with the 5,000 *stadia* distance between Alexandria and Syene reported by Cleomedes, and that the rounded figure that was used was based upon someone else's rounding of distances from the Nile Map.[6] Dutka suggests that the real angle was 1/48.2 (7.47°).[7] Newton goes the other way and suggests that the angle was 1/50.6 or 7.11° based upon the difference in the latitudes of Alexandria and Syene.[8] Goldstein suggests that the 1/50 figure was obtained via calculations rather than by being measured directly, that it is rounded, and that the real value was 1/50.4 (7.12°).[9] Goldstein also examines one of the potential, spherical, issues with the angle (Fig. 30).

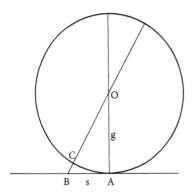

Fig. 30 The spherical problem with Eratosthenes' angle identified by Goldstein.[a]

[a] Image based on that found in Goldstein, 'Eratosthenes', p.413, Fig. 1.

[2] Viedebantt, 'Eratosthenes', pp.211, 215.
[3] Fischer, 'Another Look', p.161.
[4] Fischer, 'Another Look', p.156.
[5] Ptol. *Alm.* 1.12; see also Al-Biruni, *The Determination of the Coordinates of Cities*, 89.1–21.
[6] Rawlins, 'Geodesy', pp.260, 262–263; Rawlins, 'Nile Map', pp.212, 216.
[7] Dutka, 'Eratosthenes' Measurement', p.61.
[8] Newton, 'Sources of Eratosthenes' Measurement', p.384.
[9] Goldstein, 'Eratosthenes', pp.412–413.

Goldstein states that a sundial, with a *gnomon* of height OA (Fig. 30), would cast a shadow on the horizontal line AB of length s when the Sun was in the direction BO. Goldstein then raises the question, but leaves it open as to how the length of the arc AC could be calculated.[10] Goldstein then uses spherical trigonometry to calculate an angle of 1/50.53, which he says was rounded down by ancient mathematicians to 1/50.[11] Gulbekian similarly accepts the 1/50 angle, but provides no explanation on how it may have been obtained.[12] Pinotsis says the angle was actually measured as 7.2° and that this measurement had a 'relative error of about 1%'.[13]

Walkup bases her examination of Eratosthenes' angle on Euclidean geometry. She first discusses how the angle created by the shadow in Alexandria would be equal to the point of intersection of lines extending downwards from the *gnomons* in both Alexandria and Syene, based on Euclid's proposition that 'a straight line falling on parallel straight lines, makes the alternate angles equal to each other' (Fig. 31).[14]

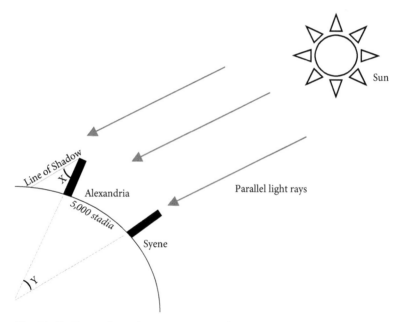

Fig. 31 In Eratosthenes' experiment, angles X and Y are equal.

[10] Goldstein, 'Eratosthenes', p.413.
[11] Goldstein, 'Eratosthenes', p.413.
[12] Gulbekian, 'Stadion Unit', p.362.
[13] Pinotsis, 'Significance and Errors', p.58.
[14] Walkup, 'Mystery of the Stadia' *s.v. Eratosthenes' Argument*; Euc. *El.*, 1.29.

Walkup then looks at another of Euclid's propositions which states that 'in two circles of equal size, angles on equal circumferences are equal to each other'.[15] Walkup then applies this principle to a single circle that is divided by the segment created by the lines of Eratosthenes' experiment, and feeds in some of the known values to a series of ratios to calculate the circumference of the Earth that is found in Cleomedes' account (Fig. 32).

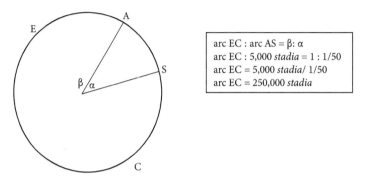

arc EC : arc AS = β: α
arc EC : 5,000 *stadia* = 1 : 1/50
arc EC = 5,000 *stadia*/ 1/50
arc EC = 250,000 *stadia*

Fig. 32 Walkup's method of calculating the circumference of the Earth (with angles exaggerated).[a]

[a] Taken from: Walkup, 'Mystery of the Stadia' *s.v. Eratosthenes' Argument.*

What few of these prior theories have done is examine how Eratosthenes could have arrived at a figure of 1/50 of the circumference of a circle in his result. Walkup, for example, merely accepts the figure and then uses the mathematics that was available at the time to calculate the result given in Cleomedes without offering any suggestion as to how this value may have been obtained. Gulbekian offers even less examination. Scholars who have suggested that the figure is either wrong or rounded, such as Viedebantt, Rawlins, Dutka, Newton and Goldstein, still begin with the 1/50 value without any discussion of how it was derived. Pinotsis, one of the few scholars to offer a suggestion of the figure's origins, suggests that it was measured. However, Pinotsis also says that it was measured at 7.2° which cannot be accurate as the division of a circle into 360° was not used by the Greeks for about a century after the time of Eratosthenes. Similarly, the elements of trigonometry used by Gulbekian in his analysis also did not exist in the time of Eratosthenes.

However, the hemispherical sundial that Eratosthenes most likely used in his experiment does contain all of the information required to determine this

---

[15] Walkup, 'Mystery of the Stadia' *s.v. Eratosthenes' Argument*; Euc. *El.*, 3.27.

fraction to a very high level of accuracy. The key to determining this angle is the location of the line marking the Summer Solstice on sundials that were calibrated for both Alexandria and Syene, and specifically the point on this line where it is intersected by the noon line on each of the sundials. On any sundial, this intersection marks the position of the tip of the shadow cast by the *gnomon* at exactly midday on the Summer Solstice (so long as the sundial is calibrated and aligned correctly). For a sundial that was calibrated for Syene, Eratosthenes would have assumed that the location of this intersection (S_S) was directly beneath the *gnomon* (G_S) as no shadows were believed to be cast at midday on the Solstice (Fig. 33).

However, for a sundial calibrated for Alexandria such as the one Eratosthenes used, the Solstice line, and the point of intersection with the noon line (S_A), would be north of the *gnomon* (G_A) due to the location's more northerly latitude (Fig. 34).

The bowl of the *skaphe* represents one half of the celestial sphere and, as such, a noon line extended from the base of the *gnomon* to the upper rim (which represents the celestial horizon) is equal to one quarter of the celestial sphere or one quarter of the circumference of a circle.

Consequently, in order to determine the difference in the length of the shadows cast by two sundials in Alexandria and Syene at midday on the Summer Solstice (Δ), the distance from the base to the *gnomon* on the Syene sundial (G_S) to the point of intersection of the noon and Summer Solstice line on the Syene sundial (S_S) can be subtracted from the distance from the base of the *gnomon* on the Alexandrian sundial (G_A) to the point of intersection of the noon and Summer Solstice line on the Alexandrian sundial (S_A) (see Figs. 33 and 34). This can be expressed mathematically as $\Delta = (G_A + S_A) - (G_S + S_S)$. For Eratosthenes, assuming that no shadows were cast in Syene on the Solstice, the distance $G_S \rightarrow S_S$ equals zero as the point of intersection between the noon and Solstice lines is directly beneath the *gnomon*. All that Eratosthenes would have then been left with is the distance ($G_A \rightarrow S_A$) on the Alexandrian sundial. The distance from the base of the *gnomon* on the Alexandrian sundial (G_A) to the point where an extended noon line intersects with the celestial horizon (H_A) can then be used to determine the fraction of a full circle that the cast shadow equates to by simply determining how many times the length $G_A \rightarrow H_A$ can be divided by the length $G_A \rightarrow S_A$.

As Goldstein points out, these distances are on the curved surface on the inside of a bowl and are not easy to calculate mathematically. However, there

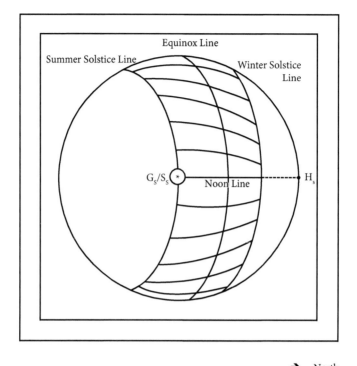

→   North

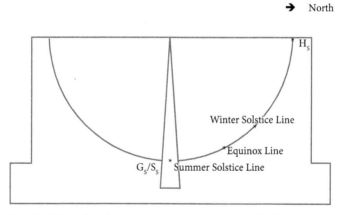

Fig. 33 Plan view (top) and cross section through the noon line (bottom) of a hemispherical *skaphe* calibrated for Syene.

is a very simple method that can be employed in order to determine both distances. A thin, straight-edged, piece of material (such as papyrus, supple leather, or even string) can be pressed into the bowl of the *skaphe* so that one end of the material is positioned at the point directly beneath the *gnomon*.

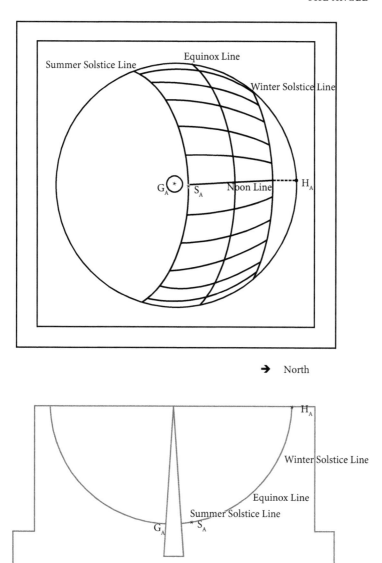

Fig. 34 Plan view (top) and cross section through the noon line (bottom) of a hemispherical *skaphe* calibrated for Alexandria.

The straight edge of the strip of material can then be aligned along the noon line up to the horizon point, and the material pressed to conform with the curvature of the bowl. The thinner and more pliable the material is, the less deformation it experiences when pressed into the curvature of the bowl. The location of the intersection of the noon and Summer Solstice lines, and the position of the horizon, can then be marked on the material.

When this was done on the replica sundial, calibrated for Alexandria, that had been constructed for this research, using a thin strip of paper, it was found that the distance from directly beneath the *gnomon* to the intersection of the noon and Solstice line (G_A → S_A) was 9.5 mm, and the distance from beneath the *gnomon* to the horizon (G_A → H_A) was 118 mm (Fig. 35).

Fig. 35 The strip of paper used to measure the positions of points G_A, S_A, and H_A on the replica sundial.

These distances can also be calculated mathematically (using processes that were not available to Eratosthenes) to demonstrate the accuracy of this method:

DISTANCE GNOMON TO NOON / SOLSTICE INTERSECTION (7.2°)
Radius of sundial bowl = 75mm
Circumference of sphere $(2\pi r)$ = 2 x 3.14 × 75mm = 471.24mm
Distance(G_A → S_A) = 471.47mm/360°× 7.2°= **9.4248mm**

[eq. 3.1.1]

DISTANCE —GNOMON TO HORIZON
Radius of sundial bowl = 75mm
Circumference of sphere $(2\pi r)$ = 2 × 3.14 × 75mm = 471.24mm
Distance(G_A → H_A) = 0.25 × $2\pi r$ = 0.25 × 471.24mm = **117.8mm**

[eq. 3.1.2]

These calculations show that using such a method can obtain an accurate measurement of the distance along the inside of the bowl of the *skaphe* to within about ±0.2 mm. The inherent margin of error would be half of the smallest unit that the distance was measured in (in this case millimetres) and an error of ±0.2 mm falls well within these parameters.

With the two measured values in hand, it is a very simple process of working out how many times the distance $G_A \rightarrow S_A$ fits into the distance $G_A \rightarrow H_A$ and then multiplying the result by four to reach the fraction of a full circle. This could be accomplished by cutting a number of other pieces of material, of length $G_A \rightarrow S_A$, and then seeing how many of them fit side-by-side within the measured distance $G_A \rightarrow H_A$, or it could be calculated mathematically. For example, using the values measured from the replica sundial:

$$G_A \rightarrow S_A = 9.5\text{mm}$$
$$G_A \rightarrow H_A = 118\text{mm}$$
$$(G_A \rightarrow H_A)/(G_A \rightarrow S_A) = 118/9.5 = 12.4 \qquad \text{[eq. 3.1.3]}$$
$$12.4 \times 4 = 48.8$$

As such, the distance from the base of the *gnomon* to the intersection of the noon and Summer Solstice lines ($G_A \rightarrow S_A$)—in other words, the length of the shadow cast by the *gnomon* in Alexandria at midday on the Summer Solstice—created an angle which equals 1/48.8 the circumference of a circle, which is similar to the angle offered by Viedebantt, Rawlins, and Dutka, and which would round to the 1/50 of a circle value reported by Cleomedes.

Using the calculated values yields the following result:

$$G_A \rightarrow S_A = 9.4248\text{mm}$$
$$G_A \rightarrow H_A = 117.8\text{mm}$$
$$(G_A \rightarrow H_A)/(G_A \rightarrow S_A) = 117.8/9.4248 = 12.5 \qquad \text{[eq. 3.1.4]}$$
$$12.5 \times 4 = 50$$

This is exactly the angle fraction as it is reported by Cleomedes. It is also important to remember that Eratosthenes did not work in modern units like millimetres. However, as the calculations are based on ratios, it does not actually matter what units of measure are being used. In Eratosthenes' time, circles were divided into sixty segments (*hexacontades*) which would be equivalent to 6° each. The distance $G_A \rightarrow S_A$ is given by Cleomedes as equalling 1/50 of a circle or 7.2°. This would be equal to 1.2 *hexacontades* (7.2/6 = 1.2), and the distance $G_A \rightarrow H_A$ would be equal to 1/4 of the circumference, or exactly 15 *hexacontades*.[16] Using these unit achieves the same result as the other calculations:

$$G_A \rightarrow S_A = 1.2$$
$$G_A \rightarrow H_A = 15$$
$$(G_A \rightarrow H_A)/(G_A \rightarrow S_A) = 15/1.2 = 12.5 \qquad \text{[eq. 3.1.5]}$$
$$12.5 \times 4 = 50$$

This suggests that Eratosthenes was able to make more precise measurements on the sundial in Alexandria—possibly because it had a larger radius than the replica used in this reconstruction.

Regardless, even accounting for an inherent margin of error, the ratio between $G_A \rightarrow S_A$ and $G_A \rightarrow H_A$ is 1:50, and therefore the value for the fraction of a circle created by the shadow cast in Alexandria, compared to the lack of shadow cast is Syene, is also 1/50—just as it is recounted by Cleomedes.

There would have been several benefits to Eratosthenes using such a method to determine the size of the shadow:

1) The sundial used by Eratosthenes would not have to have the same diameter as the replica used here as the ratio $G_A \rightarrow S_A : G_A \rightarrow H_A$ would be the same regardless of the dimensions of the instrument so long as it was correctly calibrated for Alexandria.
2) The measurement is not dependent upon the use of a specific unit of measurement of a certain size as the ratios are the same regardless.
3) As has been demonstrated, the method is very accurate.
4) There is no need for Eratosthenes to even see a sundial from Syene as he would know that, for any instrument calibrated for a location where no shadows were cast at noon on the Summer Solstice, the intersection of the noon and Solstice lines on the sundial would be directly beneath the *gnomon*.
5) Because no actual shadow is being observed, merely the location of the intersection of the noon and Solstice lines on the Alexandrian sundial, the sundial does not have to be aligned correctly, the measuring can be done indoors. More importantly, the calculations do not actually have to be undertaken on the Summer Solstice, but can be done at any time of the day and on any day of the year.

---

[16] Fischer, in her examination of Eratosthenes ('Another Look', p.157), also places the location of the Summer Solstice line at 1.2 *hexacontades*.

Such a reconstruction of Eratosthenes' possible methodology demonstrates the simplicity of his experiment. All that was required for the completion of the calculations would have been the distance between Alexandria and Syene—a distance that he would have also been able to obtain from the sundial (see section 5.2). Importantly, this reconstruction also demonstrates how accurate parts of Eratosthenes' calculations would have been. However, there were also several errors in the underlying assumed knowledge that Eratosthenes was working from which would affect the overall, comparative, angle of the 'shadow' that was one of the key elements of his determinations. As such, these errors, in turn, have important implications for understanding just how accurate his final result for the size of the world would have been.

3.2 Errors In Eratosthenes' Methodology

Despite the simplicity of the calculations that Eratosthenes would have undertaken, there are several issues with the assumed knowledge that Cleomedes reports Eratosthenes used as a basis for his experiment and calculations. These errors include the following:

- The cities of Alexandria and Syene do not actually sit on the same longitudinal meridian.
- The city of Syene does not actually sit on the tropic of Cancer, and how the precession of the Earth has altered the location of the Tropic over time.
- Light rays coming from the Sun are not actually parallel when they reach the Earth.
- The Earth is not a perfect sphere.

Additionally, there are a few other issues within the premise that Eratosthenes' calculations are based upon which he did not consider:

- the effect of the Earth's rotation on the apparent angle created by the shadow on a sundial;
- the effect of atmospheric refraction on the apparent angle created by the shadow on a sundial;
- any errors resultant from reading the shadow on a sundial due to size and shape of its umbra and penumbra; and

- any of the so-called 'parallax errors' in reading the instrument.

However, as will be shown, all of these issues have a very marginal effect on the premise for the experiment, on the experiment itself, and/or on the results.

3.2.1 The Effect Of Syene And Alexandria Not Sitting On The Same Longitudinal Meridian

A meridian is an imaginary line running around the Earth's circumference which passes through each of the poles. On a sphere, the plane created by any meridian bisects that sphere. The principles of such a concept had been proposed by Euclid.[17] In another proposition, Euclid also states that two lines, perpendicular to the surface of the sphere and on the same meridian, extended downwards, will intersect at the centre of the sphere.[18] The consequence of this for Eratosthenes was that, based upon his assumed knowledge, Alexandria, Syene and the centre of the Earth all sat on the same plane of a 'great circle' that ran through the poles. This assumption was most likely based upon his earlier work on geography in which the main north-south meridian for the map that he compiled ran through Rhodes, Alexandria, and Syene.

However, Alexandria, located at a longitude of 29.92°E, and Syene, at 32.90°E, do not sit on the same meridian, and the two locations are separated by a difference in longitude ($\Delta\lambda$) of 2.98° (Fig. 36).[19]

The latitudinal distance between Alexandria and a point directly south of the city on the same parallel as Syene can be calculated in the following way (Fig. 37):

The longitudinal difference between Alexandria (29.92°E) and Syene (32.90°E) ($\Delta\lambda$) = 2.98° or 2.98 x ($\pi/180$) = 0.05198 radians. On a spherical (or closely spherical) object such as the Earth, the longitudinal distance

---

[17] Euc. *El.* 11.def14; Euc. *Phae.* 1.
[18] Euc. *El.* 3.19.
[19] These values for the longitudes of Syene and the Library of Alexandria are also used by Pinotsis in his examination, in which he gives a difference in the longitudes of approximately 3° ('Significance and Errors', p.60), while Nicastro (*Circumference*, p.119) uses the rounded value to state that Alexandria is 3° west of Syene.

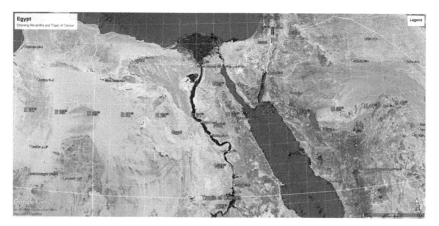

Fig. 36 Egypt, showing the location of Alexandria, Syene (Aswan) and the Tropic of Cancer.[a]

[a] Image created using Google Earth Pro.

between two points is a function of the latitude of those points—with the distance between two longitudinal meridians being closer towards the poles, and further apart towards the equator. The longitudinal distance between Syene and a point on the same meridian as Alexandria, in kilometres (χ), is:

$$0.05198 \times 6,371 \text{km} \times \cos(24.1) \approx 302 \text{km} \qquad \text{[eq. 3.2.1]}$$

This difference in the longitudinal positions of Alexandria and Syene would have had no impact at all on Eratosthenes' calculations or his results. The only difference that would have been resultant from this variance would have been when midday on the Summer Solstice occurred in one location compared to the other. For example, the Sun traverses the 360° of the sky every twenty-four hours (86,400 seconds) as the Earth rotates. The Sun therefore moves 0.00417° every second (360°/86,400 sec = 0. 00417°/sec). If Alexandria and Syene are separated by roughly 3° of longitude, this means that midday, when the Sun is at its zenith point directly overhead, would occur in Alexandria approximately twelve minutes (3°/0.00417°/sec = 720 sec or 12 min) after it had occurred in Syene.[20]

However, the methodology that Eratosthenes would have employed did not require any direct observations to be made in either Alexandria or Syene at

[20] This value was also determined by Tupikova ('Eratosthenes' Measurements', pp.223–224).

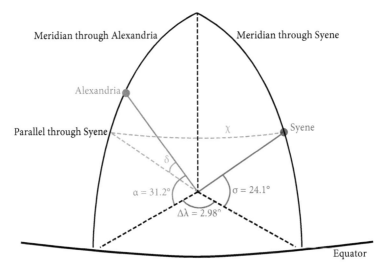

Fig. 37 Determination of the longitudinal distance between
Alexandria and Syene (with the angles and curvature exaggerated).

midday on the Summer Solstice, and the measurements taken from the sundial
in Alexandria could have been done at any time of day and on any day of the
year. Consequently, the difference in the longitudinal positions of Alexandria
and Syene would have had no impact at all on his results. Nicastro suggests that
because Syene and Alexandria are not situated on the same meridian, Eratosthenes cannot have measured the polar circumference of the Earth.[21] However,
as Eratosthenes was basing his determinations on the latitudinal differences
between the two locations, as demonstrated by the location of the intersection of the noon and Summer Solstice lines on sundials calibrated for the two
sites, any longitudinal difference was not a concern and two sites being on
the same meridian was not really a requirement (despite Cleomedes stating
that Eratosthenes was working from a piece of assumed knowledge). Indeed,
Eratosthenes' methodology was so adaptable in its concept that the experiment and calculations could have been done with any two sundials, for any
two locations, regardless of where on the planet the cities they were calibrated
for were located.

Even if a different methodology was used, and observations were made at
both sites, the only impact of a difference in the longitudinal positions of the

---

[21] Nicastro, *Circumference*, p.118.

two locations would have been that the length of a shadow cast by a *gnomon* in Syene (or lack thereof) would have been measured twelve minutes before a similar measurement would have been taken in Alexandria. This, again, would have had no influence on the outcome of the experiment, any observations, or the results of the calculations.

3.2.2 The Effect Of Syene Not Sitting
On The Tropic Of Cancer

One of the key elements of the assumed knowledge that Eratosthenes was basing his calculations on was the assumption that the city of Syene was located on the Tropic of Cancer.[22] However the city, at a latitude of 24.09°N (see Fig. 36 previously), is actually located slightly north of the Tropic. This has led to numerous scholars attempting to determine what impact this error would have on the results of Eratosthenes' calculations. However, as will be shown, the difference in the latitudes of both Syene and the Tropic would have had only a marginal effect on the overall determination of the angle created by the shadow cast by a *gnomon* in Alexandria, and hence the accuracy of Eratosthenes' calculation of the circumference of the Earth.

However, even the location of the Tropic in the time of Eratosthenes is something of a controversial topic among scholars who have suggested numerous values for the angle of the obliquity of the ecliptic (which is equal to the angle of the Tropic) in the third century BC. No ancient text directly records when it was in Eratosthenes' career that he undertook his calculation of the circumference of the Earth. However, the experiment had to have been undertaken sometime after he had been appointed to the position of Chief Librarian in 245 BC and his death in 196 BC. Both Dutka and Nicastro date the experiment to 230 BC, while Tupikova uses a date of 220 BC.[23] While not the mid-point of his career at the Library, a period of around fifteen years into the position would have given Eratosthenes ample time to settle into the post, conduct his research into other topics such his work on geography, and to discover the information that was necessary to formulate his hypothesis, such as the information about

---

[22] Cleom. *De motu* 1.10.
[23] Dutka, 'Eratosthenes' Measurement', p.60; Nicastro, *Circumference*, p.119; Tupikova, 'Eratosthenes' Measurements', p.224.

the city of Syene, the lack of shadows being cast there on the Summer Solstice, and the nature of the Tropic.

In the fourth century BC, Eudemus of Rhodes reported that he had calculated the difference between the pole of the ecliptic and the pole of the equator (i.e. the angle of the Tropic) as being equal to the side of a fifteen-sided polygon that had been inscribed within a circle.[24] Thus, for Eudemus, the angle of the Tropic equalled 1/15 of the circumference of a circle or 24° (360°/15 = 24). Fischer, in her examination of Eratosthenes' work, states that the 'generally accepted' value for the Tropic in this time is the 24° provided by Eudemus—which is equivalent to 4 *hexacontades* in the divisions of a circle that were used by Eratosthenes.[25]

The astronomer Ptolemy (second century AD), on the other hand, reports that Eratosthenes had determined that the arc length between the two Tropics was equivalent to 11/83 of a circle.[26] Thus, for Eratosthenes, the arc length equated to 11/83 x 360° = 47.71°, and the distance from the Equator (0°) to the Tropic of Cancer was therefore 47.71/2 = 23.855° or 23°51'18" which, when rounded, equals the 24° figure given by Eudemus. In a different work Ptolemy reported that the latitude of the location where the longest day occurred (i.e. 13.5 hours of daylight on the Tropic on the Summer Solstice) was 23°51'20", which is almost exactly the same as the value attributed to Eratosthenes.[27] Goldstein and North call the evidence that Eratosthenes measured the obliquity of the ecliptic 'quite unconvincing'.[28] Newton, on the other hand, takes the latter figure of 23°51'20" reported by Ptolemy to be the obliquity of the ecliptic *ca.* 230 BC, and therefore the angle of the Tropic at this time, as well as the latitude of Syene that was used by Eratosthenes.[29] Newton further states that Eratosthenes took this later value to be the latitude of Syene and that he rounded the value to within 30".[30] A value of 23°51'20" for the location of the tropic is also accepted by Rawlins.[31] Tupikova sets the Tropic at 23°43'20" for 220 BC—a value which is claimed to have been taken from the edge of the sub-solar zone (i.e. Syene's actual location) by Eratosthenes with

---

[24] Theon. *Expos.* 198.14–15, 199.6–8.
[25] Fischer, 'Another Look', p.158.
[26] Ptol. *Geog.* 1.12.
[27] Ptol. *Alm.* 2.6.
[28] B.R. Goldstein and J.D. North, 'The Introduction of Dated Observations and Precise Measurement in Greek Astronomy' *Arch. Hist. Exact Sci.* 43.2 (1991), p.95.
[29] Newton, 'Sources of Eratosthenes' Measurement', p.383.
[30] Newton, 'Sources of Eratosthenes' Measurement', p.385.
[31] Rawlins, 'Geodesy', p.260.

a margin of error of about 2ʹ.[32] The problem with such claims is that circles were not divided into degrees, minutes, and seconds, until a century after the time of Eratosthenes, and so such precise positioning for the Tropic seems unlikely. Rawlins states that Eratosthenes' description of the distance between the Tropics equalling 11/83 of a circle was within 1" of the exact value that Eratosthenes wanted to refer to, and that the ancients regularly rounded angles to the nearest 5" and so any rounding to within 1" demonstrated a high level of precision on Eratosthenes' part.[33] Again, such a claim seems unlikely considering that Eratosthenes did not work in values of arc-seconds, and the value can only be seen for what it is—the ratio that Eratosthenes had determined— rather than a rounded value of best fit. Dutka and Nicastro, on the other hand, place the Tropic at 23°44'N in 230 BC, while Pinotsis says the obliquity of the ecliptic was taken as 23°43'N, but was really 8' further north at 23°51'N.

However, what some of these prior hypotheses have failed to account for is that the precession of the Earth—it's 'wobble' about its axis in a 26,000-year cycle—would have changed the position of the Tropic across the more than two millennia between 230 BC and the present day. The obliquity of the ecliptic in 230 BC (ε_{230}) can be calculated using the following formula:[34]

$$\varepsilon_{230} = \varepsilon_0 \quad - 4{,}680.83" \times U \qquad\qquad\qquad \text{[eq. 3.2.2]}$$
$$- 1.55" \times U^2$$
$$+ 1{,}999.23" \times U^3$$
$$- 51.38" \times U^4$$
$$- 249.67" \times U^5$$
$$- 39.05" \times U^6$$
$$+ 7.12" \times U^7$$
$$+ 27.87" \times U^8$$
$$+ 5.79" \times U^9$$
$$+ 2.45" \times U^{10}$$

---

[32] Tupikova, 'Eratosthenes' Measurements', p.224.
[33] Rawlins, 'Geodesy', p.262.
[34] The formula is adapted from J. Laskar, 'Secular Terms of Classical Planetary Theories using the Results of General Theory' *A&AT* 157 (1986), pp.68–70; for other methods of calculation the obliquity of the ecliptic, see: Meeus, *Astronomical Algorithms*, pp.147–148.

where ε_0 equals an initial starting obliquity for the J2000.0 astronomical epoch (23°26'21.448" or 23.4°N—which is the value incorrectly used by Dutka and Nicastro in their examinations), and U equals the amount of time, in 10,000-year blocks, from J2000.0 to the target year for the calculation—in this case the 2,230 years back from AD 2000 to 230 BC—and so U = −2,230/10,000 = −0.223. Solving for each term in [eq. 3.2.2] shows that, in order to determine the angle of the ecliptic in the time of Eratosthenes, the value for the current obliquity of the ecliptic needs to be altered as follows:

$$
\begin{aligned}
\varepsilon_{230} = 23°26'21.448" \quad &+1,043.8407" \\
&+ 0.07707995" \\
&- 22.17081682" \\
&+ 0.1270613754" \\
&+ 0.1376862832" \\
&+ 0.00486231027" \\
&- 0.00019526033" \\
&- 0.0001704417" \\
&- 0.00000789627" \\
&- 0.0000007451"
\end{aligned}
$$

TOTAL   **+ 1,022.0161987555"**

This equates to an addition of approximately 17.03', or 17'2", to the initial angle for the ecliptic. Thus, the angle of the ecliptic (and therefore the latitude of the Tropic of Cancer) in 230 BC was:

$$23°26'21.448" + 17'2" \approx 23°43'23.5" \text{ or } 23.7° \qquad \text{[eq. 3.2.3]}$$

This is close to the value for the Tropic that Ptolemy attributes to Eratosthenes (23.855°) and the value given by Eudemus of Rhodes (24.0°). Whether Eratosthenes found a way to refine Eudemus' figure, and/or whether Eudemus used a rounded value, is uncertain. Laskar states that this method of calculation has a precision of 0.01–0.02" for reckonings out to 1,000 years either side of the J2000.0 epoch, and 'a few seconds of arc' for times beyond 10,000 years either side of the epoch.[35] Even with the highest error, the obliquity of the

---

[35] Laskar, 'Secular Terms', pp.68, 70.

ecliptic in 230 BC would still equal approximately 23.7°. Consequently, Eratosthenes' determination of the location of the Tropic at 23.855° is out by only 0.155° (9'18") to the actual angle of the Tropic—a margin of error of <1 per cent. Newton, who takes the error to be 22', suggests that, if sundials were used to determine the position of the Solstice, the margin of error would be in the range of 4'. Newton further suggests that the possibility of the error being as big as 22' if sundials were used is less than 10^{-8} and therefore claims that Eratosthenes did not actually measure the location of the Tropic or Syene with a sundial (contra what other scholars have suggested).[36] This would conform with Eratosthenes only working with a sundial calibrated for Alexandria, not examining a sundial calibrated for Syene at all, and basing his calculations on the knowledge that the location of the intersection of the noon and Solstice lines on a sundial calibrated for Syene would be located directly beneath the *gnomon*.

Syene (modern Aswan) was located at 24.09°N (see Fig. 36 previously). Thus, the site was 0.39° north of the actual Tropic in 230 BC. This equates to a distance of approximately 43 km. The ancient geographer Strabo seems to acknowledge the more northerly position of Syene in regard to the Tropic when he states that Syene is on the 'boundary' of the Summer Solstice.[37] Similarly, Cleomedes states that the area where no shadows are cast during the Summer Solstice, the so-called 'sub-solar zone', has a size of 300 *stadia* (or approximately 55.5 km depending on the unit of measure).[38] Neugebauer argues that the figure is actually 320 *stadia* based on measurements made by water-clocks and the 240,000 *stade* circumference of the Earth given in a later determination by Posidonius, but that Cleomedes has rounded the figure.[39] Regardless of the figure used, the position of Syene was just inside the boundary of the 'sub-solar zone' as Strabo states, and the Sun would be almost directly overhead in Syene on the Solstice. Nicastro states that, as the angular size of the Sun is 0.5°, the exact centre of the Sun would not have been directly overhead at Syene on the Solstice, but the edge of the Sun would have been.[40]

---

[36] Newton, 'Sources of Eratosthenes' Measurement', p.384.
[37] Str. *Geog.* 2.2.2.
[38] Cleom. *De motu* 2.1.
[39] Neugebauer, *History of Mathematical Astronomy*, pp.655–656; Russo (*Forgotten Revolution*, pp.274–275) suggests that Cleomedes rounded all of his values 'so as to not bother the reader with calculations inessential to an understanding of the method' and that the 300 *stade* figure for the sub-solar zone is evidence of 'experimental averaging'.
[40] Nicastro, *Circumference*, p.120.

The stated size of the sub-solar zone—300 *stadia*—is also the minimum distance that Pliny states a sundial could be moved from its intended location and remain accurate, and the size of the Sun and sub-solar zone explains why a sundial that is calibrated for a certain location is still relatively accurate in locations 55.5 km north or south of that location (see section 2.5). Due to the angular size of the Sun, at the correct location for the instrument, light from the centre of the solar disk would pass over the tip of the *gnomon* to create the shadow. However, at locations north or south of that position, light from nearer to the edge of the solar limb would pass over the *gnomon* to create the shadow. The size and shape of the penumbra would vary with the relocation of the sundial, but the basic shape of the umbra would remain relatively unchanged—assuming that it was large enough to be noticed in the first place. If the sundial is moved to a position beyond the boundaries of the sub-solar zone, then the angle of any incoming light to the sundial is altered—and altered more dramatically the further from its intended location it is moved. This then accounts for why ancient writers such as Pliny gave a stated distance that the sundial could be moved, and yet remain accurate, that was equal to the size of the sub-solar zone. This also accounts for why the city of Syene was assumed to sit on the Tropic of Cancer by the ancient Greeks.

Importantly for the understanding of Eratosthenes' calculation of the circumference of the Earth, the slight northerly position of Syene in relation to the Tropic, means that the *gnomon* on a sundial in this location would have cast a shadow at midday on the Summer Solstice, but it would have been so minimal in size as to be almost undetectable. For example, the angle of the shadow cast by a *gnomon* would increase by about 0.94° for every degree of latitude north of the Tropic on the Summer Solstice. With Syene located at a latitude of 24.09°N, and the Tropic at 23.7°N in the third century BC—a difference of 0.39°—any *gnomon* located in Syene on the Solstice would cast a shadow at an angle of about (0.39° x 0.94° =) 0.37°. A *gnomon* 20 cm in height, for example, would then cast a shadow (tan(0.37) = X/20 =) 0.129 cm, or about 1.3 mm, in length. This would be difficult, if not impossible, to detect on an ancient sundial. On the replica sundial created for this research, the *gnomon* is conical in shape with a height of 75 mm and a radius at the base of just over 4.5 mm. This means that, on this sundial, a shadow cast at midday on the Summer Solstice in Syene would fall onto the sloping surface of the *gnomon* itself and would not project onto the actual face of the sundial. This then accounts for why the information that Eratosthenes was working on included the assumption

that no shadows were cast in Syene at midday on the Solstice. Russo suggests that it was probably suggested that the Tropic ran through Syene as this
was the closest large settlement to the Tropic, but that Eratosthenes desired
to measure the distance from Alexandria to the Tropic and that he took the
trouble to find it precisely and did not just accept someone's say-so that it ran
through Syene.[41] While this would comply with Eratosthenes' determination
of the distance between the Tropics as reported by Ptolemy, no connection is
made with this calculation and the city of Syene. Furthermore, Eratosthenes'
methodology did not require the observation of a sundial in, or from, Syene
at all. Such observations must have been made by temple staff in Syene using
sundials and the Nile-o-meter for centuries, and Eratosthenes is fully justified in his acceptance of this assumed knowledge without any confirmatory
investigation.

However, it must also be conceded that the angle created by the small
shadow cast in Syene, regardless of how imperceivable it was, needs to be
deducted from the 1/50 of a circle (7.2°) angle that is reported by Cleomedes
in order to arrive at an adjusted, 'true', value of the angle. If this were the only
alteration to Eratosthenes' angle that needed to be made, then the adjusted
angle becomes:

$$7.2° - 0.37° = 6.83° \qquad \text{[eq. 3.2.4]}$$

This angle is closer to 1/53 of the circumference of a circle (360°/6.83 = 52.7)
than it is to the 1/50 value recounted by Cleomedes. Based upon Eratosthenes
calculation of the angle between the Tropics being 11/83 of a circle, as per
Ptolemy, the derivation of an angle, even rounded, of 1/53 seems to be something that Eratosthenes would have been capable of. This suggests that an angle
of 1/50 was exactly what Eratosthenes had measured. And yet, because Syene
is slightly north of the Tropic, his reported result is not correct—at least for
Syene. However, the reported angle of 1/50 is accurate for the Tropic of Cancer. This discrepancy in the angles can be accounted for via an understanding
that Eratosthenes was working on the assumption that Syene was located on
the Tropic, and that no shadows were cast there at midday on the Summer
Solstice—which is true for sites on the actual Tropic. Additionally, as Eratosthenes' methodology did not require him to undertake any observation in
Syene itself, he did not make any examination that would have demonstrated

---

[41] Russo, *Forgotten Revolution*, pp.274–275.

that his assumed knowledge was incorrect, and that his calculations may have required minor revisions. As these were the assumptions that Eratosthenes was working on, his calculations are essentially to the Tropic and not really to Syene at all. The result of this is that, by accepting the long-standing notion that Syene was on the Tropic, and undertaking calculations in regard to the features of sundials based on that assumption, Eratosthenes calculated the correct angle for the Tropic. As such, this error in the assumed knowledge that Eratosthenes was working from has no impact on the results of his calculations at all—contrary to what many previous scholars have suggested.

3.2.3   The Effect Of The Oblateness Of The Earth

Another error contained within the assumed knowledge that Eratosthenes was working from is that the Earth is not a perfect sphere as earlier thinkers such as Pythagoras, Plato, Aristotle and Euclid had assumed. On a sphere, all of the polar meridians are perfect circles. The Earth, on the other hand, is a geodic and all of its meridians are elliptical. The level of oblateness of the Earth would therefore affect the angle of intersection between two lines, extending downward from two points on the circumference, which meet in the centre as in Eratosthenes' experiment.

However, the level of oblateness (b) of the Earth is minimal. This can be shown in the calculation $b = (R_e - R_p)/R_e$ where R_e is the Earth's equatorial radius, and R_p is the Earth's polar radius. Giving the Earth a polar radius of 6,357 km, and an equatorial radius of 6,378 km, results in a planetary oblateness as follows:

$$b = (6,378\text{km} - 6,357\text{km})/6,378\text{km}$$
$$b = 21\text{km}/6,378\text{km} \qquad\qquad \text{[eq. 3.2.5]}$$
$$b \approx 0.0033$$

This is very close to spherical (an oblateness of 0.00).

Another way to calculate the effect of planetary oblateness is through the geocentric and geographic latitudes of Alexandria and Syene. Pinotsis defines the geocentric latitude of a location (φ_{gce}) as the angle formed between the radius of the spherical Earth at a specific location and the plane of the equator.[42] The geographic latitude of the same location (φ), on the other hand, is

---

[42] Pinotsis, 'Significance and Errors', pp.59, 61–62.

the angle formed by a line, perpendicular to the geodic surface of the Earth and the plane of the equator[43] (Fig. 38).

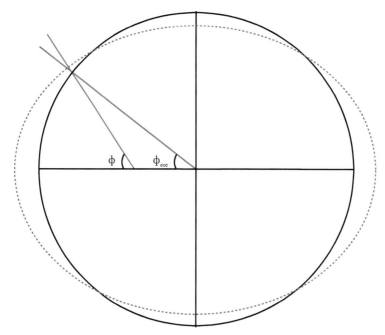

Fig. 38 The relationship between geocentric latitude on a spherical Earth (red line/black circle) and the geographical latitude on a geodic Earth (blue line/dotted ellipse).

The difference between these two angles (γ) is equal to the geographic latitude angle (φ) minus the geocentric latitude angle[44]:

$$\gamma = \varphi - \varphi_{gce} \qquad \text{[eq. 3.2.6]}$$

The variation in the geocentric angle caused by planetary oblateness can be determined using the formula:[45]

$$\tan(\gamma) = (\kappa \times \sin(2\varphi))/1 + (\kappa \times \cos(2\varphi)) \qquad \text{[eq. 3.2.7]}$$

where $\kappa = e^2/2 - e^2$, and e = the eccentricity of the geodic Earth   [eq. 3.2.8]

---

[43] Pinotsis, 'Significance and Errors', pp.59, 61–62.
[44] Pinotsis, 'Significance and Errors', p.60.
[45] Formula adapted from Pinotsis, 'Significance and Errors', p.60.

$$e = \left(1 - \left(\text{polar radius}^2/\text{equatorial radius}^2\right)\right)^{1/2}$$
$$= \left(1 - \left(6,357\text{km}^2/6,378\text{km}^2\right)\right)^{1/2}$$
$$= \left(1 - \left(40,411,449\text{km}/40,678,884\text{km}\right)\right)^{1/2}$$
$$= \left(1 - 0.9934257046\right)^{1/2}$$
$$= 0.0065742954^{1/2}$$
$$= 0.0810$$

thus κ [eq.3.2.8] $= 0.0810^2/2 - 0.0810^2$
$$= 0.006561/2 - 0.006561$$
$$= 0.006561/1.993439$$
$$\approx 0.0033$$

Both Strabo and Vitruvius states that, for Eratosthenes, the ratio of the length of the shadow cast to the intersection of the noon and equinox lines on a sun-dial calibrated for Alexandria, and the height of the *gnomon* of the sundial, was 3:5.[46] The intersection of the noon and equinox line was indicative of the latitude (φ) of the location for which the sundial was calibrated. Consequently, Eratosthenes placed the location of Alexandria as follows:

$$\varphi = \arctan(3/5) = 30°57'49" \approx 30°58" \text{ or } 30.96° \qquad \text{[eq. 3.2.9]}$$

This is not too far removed from the current latitude for Alexandria (31.2°N), and Eratosthenes' value is the same as that given by Ptolemy in his *Almagest*, and is close to the figure given by Ptolemy in his *Geography* of 31°N—which seems to be just a rounded number.[47]

As such, for the Alexandria of Eratosthenes (at 30.96°N), the variation in the geocentric latitude angle is:

$$\tan(\gamma) = 0.0033 \times \sin(2 \times 30.96°)/1 + (0.0033 \times \cos(2 \times 30.96°))$$
$$= 0.0033 \times 0.8823/1 + (0.0033 \times 0.4707)$$
$$= 0.0029/1 + 0.0016$$
$$= 0.0029/1.0016$$
$$= 0.0029$$
$$\gamma = \tan^{-1}(0.0029) = 0.17° = 10.2' = 612" \qquad \text{[eq. 3.2.10]}$$

---

[46] Str. *Geog.* 2.5.38; Vit. *De Arch.* 9.7.1.
[47] Ptol. *Alm.* 5.12; Ptol. *Geog.* 4.5.9.

Pinotsis, who takes κ to be 0.0034, and the latitude of Alexandria as 31.11°N, states that the variation (γ) was 0.18' or 10.8".[48] However, Pinotsis failed to take the inverse of tan(γ) at the end of his calculations, and doing so would have given him a result of 0.17° or 10.2' or 612".[49] The size of the variation in angle can be confirmed using an alternative formula:[50]

$$
\begin{aligned}
\varphi - \varphi_{gce} &= 692.73" \times \sin\left(2 \times \varphi\right) - 1.16" \times \sin\left(4 \times \varphi\right) \\
&= 692.73" \times \sin\left(2 \times 31.2°\right) - 1.16" \times \sin\left(4 \times 31.2\right) \\
&= 692.73" \times \sin\left(62.4\right) - 1.16" \times \sin\left(124.8\right) \\
&= 692.73" \times 0.8862 - 1.16" \times 0.8211 \\
&= 613.90" - 0.9525" \\
&= 612.95" = 10.2' = 0.17°
\end{aligned}
$$

[eq. 3.2.11]

The value γ reaches its maximum value for a geocentric latitude of 45°.[51] For locations at latitudes less than 45°, a conversion from geocentric latitude to geographic latitude requires the calculated adjustment to be added to the geocentric latitude angle. As such, the geocentric angle of Alexandria for Eratosthenes, adjusted to the geographic angle to account for the oblateness of the Earth becomes 31.1° (30.96° + 0.17° = 31.13° = 31.1°) which is the current value for the latitude of Alexandria. In terms of distance, a variance of 0.17° equates to a change in position between a geocentric and geographic location of just under 19 km.

For the northern Tropic (23.7°N), the variation to the geocentric latitude angle is:

$$
\begin{aligned}
\tan\left(\gamma\right) &= 0.0033 \times \sin\left(2 \times 23.7°\right)/1 + \left(0.0033 \times \cos\left(2 \times 23.7°\right)\right) \\
&= 0.0033 \times 0.7361/1 + \left(0.0033 \times 0.6769\right) \\
&= 0.0024/1 + 0.0022 \\
&= 0.0024/1.0022 \\
&= 0.0024 \\
\gamma &= \tan^{-1}\left(0.0024\right) = 0.14° = 8.4' = 504"
\end{aligned}
$$

[eq. 3.2.12]

Here again, the angle remains relatively unchanged due to how close the Earth is to spherical (23.7° + 0.14° = 23.84° = 23.8') and the change in distance is around 15.7 km.

---

[48] Pinotsis, 'Significance and Errors', p.60.
[49] It may be that Pinotsis was relying on the 'small angle approximation', which states that the tan of a small angle is roughly equal to that angle. However, this only holds true for angles in radians, and Pinotsis was working in degrees.
[50] Formula adapted from Meeus, *Astronomical Algorithms*, p.83.
[51] Meeus, *Astronomical Algorithms*, p.83.

These calculations show that the city of Alexandria was, in terms of its geographic latitude, 0.17° further north than where Eratosthenes thought it was. Similarly, the Tropic was 0.14° north of its presumed position. Both of these alterations would have been impossible to determine in the ancient world using sundials calibrated into segments of 6°. As the two alterations are not equal, due to the oblate nature of the shape of the Earth, the difference between these two variations (0.03°) must be added to the angle that Eratosthenes used in his calculations. The 1/50 value reported by Cleomedes equals 7.2°. This value should be adjusted, to account for terrestrial oblateness, by the difference in the two geographic latitudes of Alexandria and Syene, to 7.23°—which would still round to 7.2° or 1/50 of a circle. Thus, Eratosthenes' angle of 1/50 remains true even for a non-spherical Earth.

An additional angle of 0.03° equates to a distance of roughly 3.3 km. This means that the distance between Alexandria and Syene was 3.3 km greater than Eratosthenes thought it was. This constitutes an error of only 0.4 per cent. Such a small change in the geocentric and geographic latitudes of Alexandria and Syene would have had almost no impact on Eratosthenes' experiment and/or calculations except if the values are taken to several decimal places—a mathematical practice that the ancient Greeks did not do.

Nicastro suggests that one of the things that Eratosthenes required for his calculations to work was a spherical Earth.[52] However, this does not seem to be the case. Eratosthenes was working on the assumption of geocentric latitudes for both Alexandria and Syene based upon a spherical Earth, as Eratosthenes and others had believed, and Euclidean spherical geometry. In reality, the angles are based on geographic latitudes on a geodic Earth and need to be adjusted accordingly. However, as can be seen, due to how close the Earth is to spherical, this error is so marginal as to be redundant when the figure is rounded. The consequence of this is that Eratosthenes' figure of 1/50 of the Earth's circumference for the distance between Alexandria and the Tropic remains unaffected by the oblateness of the Earth.

3.2.4 The Effect Of Non-Parallel Rays Of Light From The Sun

The final problem with the assumed knowledge that Eratosthenes was working from is that Cleomedes recounts how Eratosthenes believed that light rays reaching the Earth from the Sun are parallel. However, this is not correct. As the Sun has an angular size of 0.5°, light from both sides of the limb of the

---

[52] Nicastro, *Circumference*, p.37.

Sun will converge and reach a point on the Earth's surface at an angle. Prior to the time of Eratosthenes, Aristarchus had demonstrated that light from the Sun was not parallel when it reached the Earth, and yet Eratosthenes seems to have ignored this idea and believed that it was. This seems to have been based upon a distance between the Earth and the Sun that Eratosthenes himself had calculated and the angular size of the solar disk. Debate has ensued over exactly what Eratosthenes' Earth:Sun distance was due to the varying ways in which the ancient texts which provide this figure can be interpreted. A critical review of the available evidence demonstrates that Eratosthenes had determined the distance to the Sun to within a 2 per cent margin of error. However, Eratosthenes incorrectly believed that the Sun, set at this distance, was four-times smaller than it actually is. These values result in a calculated angle for incoming light that is so small that Eratosthenes would have concluded that the rays were parallel.

Aristotle outlines how, by his time, the distance from the Earth to the Sun had been proven to be many times greater than the distance from the Earth to the Moon.[53] Unfortunately, Aristotle does not provide details of the values that any of these previous examinations had resulted in. Around the time of Eratosthenes, Aristarchus of Samos determined (incorrectly) that the distance from the Earth to the Sun was 'greater than eighteen times, but less than twenty times, the distance from the Moon to the Earth.'[54] The distance from the Earth to the Sun (149,600,000 km) is around 390 times the distance from the Earth to the Moon (384,400 km). Eratosthenes himself would later calculate the distance to the Moon as 780,000 *stadia* and come up with his own value for the distance to the Sun.[55]

However, determining what Eratosthenes' Earth:Sun distance was is no easy task due to the nature of the source material. There are a number of sources that refer to the result of Eratosthenes' calculation, but they all seem to offer different values for the result, incorporate various corruptions in the transmission of the text, and/or have been interpreted in varying ways by modern scholars. All of the values listed in the texts are given in myriads of *stadia*—with a myriad being an ancient Greek unit representing 10,000, and the *stade* being an ancient Greek unit of measure. The actual size of the *stade* is not important (at this stage), and all of the texts are in agreement that Eratosthenes' Earth:Moon distance was 780,000 *stadia*. It is only in the record of Eratosthenes' Earth:Sun distance where problems are encountered.

[53] Arist. *Mete.* 1.8.345a36–b5.
[54] Aristarch. Sam. *De Mag.* proposition 7.
[55] Eus. *PE* 15.53; Stob. *Ecl.* 1.26.

The earliest record of Eratosthenes' Earth:Sun distance is found in the work of Aetius (second century AD).[56] Aetius' work, Περὶ τῶν ἀρεσκόντων φιλοσόφιος φυσικῶν δογμάτων (otherwise known by its Latin name *de Placita Philosophorum* or 'The Opinions of Philosophers') is a doxographical account of the sayings of various philosophers on a range of subjects. The text of Aetius' work has not survived in itself, but has been preserved in fragments which are recounted in other doxographical works by Pseudo-Plutarch (third–fourth century AD), Stobaeus (fifth century AD), and Theodoret (fourth–fifth century AD) who epitomized, redacted and abridged Aetius' text. From such sources, approximately 85 per cent of Aetius' original work can be reconstructed.[57]

However, even the reconstruction of Aetius' text raises issues with the reporting of Eratosthenes' Earth:Sun distance. In his edition of Aetius in the nineteenth century, for example, Diels reconstructed the passage which refers to Eratosthenes' Earth:Sun distance to read Ἐρατοσθένης τὸν ἥλιον ἀπέχειν ἀπὸ τῆς γῆς σταδίων μυριάδας μυριάδων τετρακοσίας καὶ στάδια ὀκτάκις μύρια ('Eratosthenes held that the Sun is distant from the Earth by 400 myriads of myriads of *stadia* and eight myriads of *stadia*').[58] The problematic part of this passage is the word for *stadia* in its genitive case (μυριάδων) meaning 'of myriads'. It is uncertain if this word is an error, a duplication of some kind, whether it should only be applied to the first number of *stadia* (400), or if it should be applied to both values (i.e. both 400 and eight) in the sentence. The separation of the total number into two constituent parts of, for example, 400 myriads and eight myriads, suggests that the accusative-genitive construction 'myriads of myriads' should only be applied to the first value, otherwise simpler terminology such as '408 myriads of myriads', may have been used. However, this is in no way definite and alternate translations of the passage are possible. If this term is applied to only the first number of *stadia* (i.e. '400 myriads of myriads of *stadia*'), then the value for Eratosthenes' Earth:Sun distance is (400 x 10,000 x 10,000) + (8 x 10,000) = 40,000,080,000 *stadia*. If, on the other hand, the genitive term is applied to both values (i.e. '400 myriads of myriads of *stadia* and eight myriads of myriads of *stadia*'), then the Earth:Sun distance becomes (400 x 10,000 x 10,000) + (8 x 10,000 x 10,000) = 40,800,000,000 *stadia*.

[56] Aet. 2.31.
[57] See: E. Jeremiah, 'Not Much Missing: Statistical Explorations of the *Placita* of Aetius' *Philos. Antiq.* 148 (*Aëtiana IV*) (2018), p.295.
[58] See: H. Diels, *Doxographi Graeci* (Berlin, G. Reimeri, 1879), pp.362–363; this is based upon the relevant passages of Stobaeus (*Ecl.* 1.26).

Adding to this uncertainty is that the genitive term is only found elsewhere in the works of Stobaeus—one of the later writers who used Aetius as a source—and in no other text (see following). This has made the interpretation of this passage anything but decisive. In their more recent edition of Aetius, Mansfield and Runia omit the word for *stadia* in the genitive case, and alter the final wording to the accusative word for 80,000 found in Pseudo-Plutarch, so that the reconstructed passage reads Ἐρατοσθένης τὸν ἥλιον ἀπέχειν ἀπὸ τῆς γῆς σταδίων μυριάδας τετρακοσίας καὶ στάδια ὀκτάκισμύριας ('Eratosthenes held that the Sun is distant from the Earth by 400 myriads of *stadia* and 80000 *stadia*').[59] In such a reconstruction of the passage, the Earth:Sun distance becomes (400 x 10,000) + (8 x 10,000) = 4,080,000 *stadia*.

However, no real justification is given for why the word for myriads in its genitive case should be omitted from the text, and such terminology is found in other passages that recount Eratosthenes' Earth:Sun distance (see following). Additionally, no justification is given for why, if a word does need to be omitted, that it has to be the term in the genitive case rather than the word in the accusative (μυριάδας) which precedes it. If the accusative term is deleted, and the passaged altered to read σταδίων μυριάδων τετρακοσίας καὶ στάδια ὀκτάκις μύρια, it can still be translated as meaning 4,080,000 *stadia*, but has the additional, alternative, translation of meaning '400 myriads of *stadia* and eight myriads of myriads of *stadia*'. This would then make the total equal (400 x 10,000) + (8 x 10,000 x 10,000) = 804,000,000 *stadia*. While this value seems excessively large in comparison to the other way of translating the passage, it is found in other texts containing Eratosthenes' Earth:Sun distance.

Unfortunately, these other works do not aid any attempt to understand Eratosthenes' Earth:Sun distance. The work of Pseudo-Galen (second century AD), for example, reads Ἐρατοσθένης τὸν ἥλιον ἀπέχειν τῆς γῆς σταδίων μυριάδας τρεῖς καὶ ὀκτάκις μυριάδας ('Eratosthenes held that the Sun is distant from the Earth by three myriads and eight myriads').[60] This would make the Earth:Sun distance (3 x 10,000) + (8 x 10,000) = 110,000 *stadia*—which is far smaller than any of the ways of interpreting the passage found in Aetius. Mansfield suggests that the word for three in this passage (τρεῖς) is a misreporting of the word for 400 (τετρακοσίας), or one of its variants, that are found in other texts.[61] If this is the case, then the value for the Earth:Sun distance would be the same as is found in their reconstruction of Aetius (4,080,000 *stadia*).

---

[59] See: Aet. 2.31.
[60] Gal. *Phil. Hist.* 72.
[61] J. Mansfield, 'Cosmic Distances: "Aetius" 2.31 Diels and Some Related Texts' *Phronesis* 45.3 (2000), pp.187–188.

Eusebius (third–fourth century AD) has Eratosthenes' Earth:Sun distance being σταδίους μυριάδων τετρακοσίων καὶ ὀκτακισμυρίων (literally '*stadia* of myriads 400 and 80000').[62] Here too problems are encountered with references to myriads being in the genitive case as two readings of the text are possible. In one interpretation, the distance is '400 myriads of *stadia* and eight myriads [of *stadia*]'. This would make the Earth:Sun distance (400 x 10,000) + (8 x 10,000) = 4,080,000 *stadia*; the same as is found in interpretations of some other texts. However, as both terms are in the genitive, another way in which the passage can be translated reads '40 myriads of *stadia* and eight myriads of myriads [of *stadia*]'. This would make the distance (400 x 10,000) + (8 x 10,000 x 10,000) = 804,000,000 *stadia*—which is, again, a way of interpreting the passage found in Aetius. Mansfield and Runia, in their reconstruction of Aetius, take this passage as referring to the smaller value of 4,080,000.[63] Nicastro, on the other hand, suggests that Eratosthenes' Earth:Sun distance was the larger value of 804,000,000 *stadia* and that, if this is converted into a modern equivalent using a *stade* of the Attic Standard of 185 m, the result (148,700,000 km) is close to accurate with a margin of error of <1 per cent.[64] However, Nicastro considers this to be a 'happy coincidence' and, in his examination of the measurement of the circumference of the Earth, advocates the use of a *stade* 157.5 m in length by Eratosthenes.[65] There are a number of issues with Nicastro's conclusions. Firstly, there is no known evidence for a *stade* 157.5 m in length being used in the ancient Greek world, and it is unlikely that Eratosthenes was even using the Messenian Standard *stade* of 160 m, or the Attic Standard of 185 m, in his calculations (see sections 5.1 and 5.2). Such issues also provide no aid in determining the correct interpretation of the passage found in Eusebius.

A work of Pseudo-Plutarch (third–fourth century AD) contains a large lacuna in the passage, which does not aid in understanding any of the earlier texts either, and reads Ἐρατοσθένης τὸν ἥλιον ... ἀπέχειν τῆς γῆς σταδίων μυριάδας ἑβδομήκοντα ὀκτώ ('Eratosthenes held that the Sun ... is 780000 *stadia* from the Earth').[66] Clearly the section of the passage referring to the Earth:Sun distance is missing, and all that remains is the opening of the sentence referring to Eratosthenes' Earth:Sun distance, followed by the value for his Earth:Moon distance. Diels, in his translations of the text, filled in the lacuna by drawing on the works of both Eusebius and Pseudo-Galen, as well as

[62] Eus. *PE* 15.53.3.
[63] See: J. Mansfield and D.T. Runia, '*Liber 2 Caput 31*' *An Edition of the Reconstructed Text of the Placita with a Commentary and a Collection of Related Texts* (Leiden, Brill, 2020), p.1108.
[64] Nicastro, *Circumference*, pp.37, 137.
[65] Nicastro, *Circumference*, p.137.
[66] Ps. Plut. *Epit.* 2.31.32.

drawing from a later version of the text by Joannes Lydus (sixth century AD) to make the passage read Ἐρατοσθένης τὸν ἥλιον <ἀπέχειν ἀπὸ τῆς γῆς σταδίων μυριάδας τετρακοσίας καὶ ὀκτακισμυρίας, τὴν δὲ σελήνην> ἀπέχειν τῆς γῆς σταδίων μυριάδας ἑβδομήκοντα ὀκτώ <σταδίων>.[67] This version is also found in the edition translated by Berardakis at the end of the nineteenth century.[68] In this reconstruction, the Earth:Sun distance would be translated as 4,080,000 *stadia* as the words for myriads are in their accusative case. An Arabic version of the text by Qusta ibn Luqa (ninth century AD) gives the figure as 408,000 *stadia*.[69] Mansfield states that this is an obvious mistake.[70]

Scholia on Ptolemy's *Almagest* provides yet another variant of the text: Ἐρατοσθένης τὸν ἥλιον ἀπέχειν σταδίων ἀπὸ τῆς γῆς μυριάδων τ̄ καὶ ὀκτάκις μυρίων ('Eratosthenes held that the Sun is distant from the Earth by 300 myriads and eight myriads').[71] Here again, with the words for myriad in their genitive case, there are several ways of translation. The first is that the distance is '300 myriads and eight myriads' or (300 x 10,000) + (8 x 10,000) = 3,080,000 *stadia*. The other is '300 myriads and eight myriads of myriads' or (300 x 10,000) + (8 x 10,000 x 10,000) = 803,000,000 *stadia*. The first value is, stylistically, not far removed from the 4,080,000 figure found in some other sources, while the larger figure is close to the 804,000,000 *stade* figure resultant from one method of translating the text of Aetius, and which is favoured by Nicastro. Mansfield suggests that the source for the Scholia is an uncorrupted version of Pseudo-Plutarch because the text of the sentence is complete.[72] However, it is difficult to justify such a claim when both texts provide different values (regardless of how they are translated).

Mansfield also suggests that the value given in the Scholia equals 300,080,000 *stadia*.[73] It may be that Mansfield has taken the letter tau in the sentence to be a representation of thirty (thirty myriads of myriads = 300,000,000). However, the dash above the letter indicates that it is a Byzantine era script for the number 300. Not long later in his examination, Mansfield suggests that the τ̄ in the sentence is a misreporting of the word τετρακοσίας found in other versions, similar to the supposed misreporting of the word τρεῖς in Pseudo-Galen.[74] This would make the value 4,080,000 *stadia*. Somewhat confusingly, in the commentary on their reconstruction of Aetius, Mansfield and

[67] Diels, *Doxographi Graeci*, pp.362–363.
[68] See: Plut. *Mor.* (trans. G.N. Berardakis) (Leipzig, Teubner, 1893).
[69] See: *Aetius Arabus* in H. Daiber (trans.), *Die Vorsokratiker in arabischer Überlieferung* (Wiesbaden, Franz Steiner, 1980).
[70] Mansfield, 'Cosmic Distances', p.186.
[71] *Schol. in Ptol. Alma.* 5.1. See also: Mansfield and Runia, 'Liber 2 Caput 31', p.1104.
[72] Mansfield, 'Cosmic Distances', p.187.
[73] Mansfield, 'Cosmic Distances', p.187.
[74] Mansfield, 'Cosmic Distances', pp.187–188.

Runia suggest that the passage in the Scholia reads as 308 myriads, or 3,080,000 *stadia*, which is the same value in one of the ways of translating the text. Interestingly, the classical Greek symbol for 300 was a capital tau (T), while the symbol for 400 was a capital upsilon (Υ). These were changed in the Byzantine era to lowercase letters with a dash above them. If the Classical version of the numbers had been somehow confused during the handwritten transmission of the text in antiquity, such as the cross-bar on the capital tau being slightly bowed to make it look like an upsilon, the figure 400—which would then make the total equal either 4,080,000 *stadia* or 804,000,000 *stadia* as per other sources—could have been erroneously converted to 300 which would give a different value of either 3,080,000 or 803,000,000.

The work of Stobaeus (fifth century AD) states that Ἐρατοσθένης τὸν ἥλιον ἀπέχειν ἀπὸ τῆς γῆς σταδίων μυριάδας μυριάδων τετρακοσίας καὶ στάδια ὀκτάκις μύρια.[75] Again there are a number of ways in which this passage can be translated due to the additional word for myriads in the genitive case (μυριάδων) in the text. One way would result in an Earth:Sun distance of 40,000,080,000 *stadia* (400 myriads of myriads of *stadia*, or 400 x 10,000 x 10,000 *stadia*, plus eight myriads of *stadia*, or 80,000 *stadia* = 40,000,080,000). The other way results in a value of 4,080,000,0000 *stadia* ((400 x 10,000 x 10,000) + (8 x 10,000 x 10,000)). In his initial examination of the text, Mansfield suggested that Stobaeus' Earth:Sun distance was 4,080,000 *stadia*. This conclusion was based upon the work of Diels who deleted the term μυριάδων from the text to make it read 400 myriads of *stadia*, or 400 x 10,000, in his reconstruction of the passage in Pseudo-Plutarch.[76] This alternate interpretation of the text was also favoured by Carmen and Evans.[77] In their later commentary on Aetius, Mansfield and Runia take the value to be 400 myriads of myriads and eight myriads, or 40,000,080,000 and state that the values given by Stobaeus clearly involve a misreading of Aetius as his source material.[78] Either one of these translations and/or interpretations differs from the more common figure of 4,080,000 *stadia* by several factors of ten.

Heath saw 4,080,000 as being an 'alternative interpretation' of the figures and favoured a value of 804,000,000 *stadia*.[79] As with the passage found in the reconstructions of Aetius (see previous), a result of such a figure can be

---

[75] Stob. *Ecl.* 1.26.
[76] See: Mansfield, 'Cosmic Distances', p.187; oddly, Mansfield suggests ('Cosmic Distances', p.187) that if the term μυριάδων is retained, this then makes the total 'enormously increased' to 400,080,000. This, however, would not be the meaning of the Greek.
[77] Carmen and Evans, 'Two Earths', p.9.
[78] Mansfield and Runia, 'Liber 2 Caput 31', p.1108.
[79] Heath, *Aristarchus of Samos*, p.340.

achieved via the deletion of the first word for myriads in the accusative case, and the retention of the term in the genitive. This then makes the passage read σταδίων μυριάδων τετρακοσίας καὶ στάδια ὀκτάκις μύρια ('400 myriads of *stadia* and eight myriads of myriads of *stadia*').

The Byzantine Era (sixth century AD) antiquarian Joannes Lydus, in his work *de Mensibus* ('On the Months') states that Eratosthenes' distance was τετρακοσίας καὶ ὀκτάκις μυρίας ('400 and eight myriads [of *stadia*]').[80] This makes the value 4,080,000 *stadia*. Mansfield, in his earlier examination, claims that the figure was 408,000—the same as is found in Qusta ibn Luqa's edition of Pseudo-Plutarch—despite citing a different version of the text (τετρακοσίας καὶ ὀκτακισμυρίας) which says 408 myriads—a value given in his later commentary on Aetius.[81] It was also this text that was used by Deils to reconstruct the passage of Aetius (2.31) to read eight myriads of myriads (i.e. 8 x 10,000 x 10,000 = 800,000,000) of *stadia*, plus 400 myriads (400 x 10,000 = 4,000,000) of *stadia*, or 804,000,000 *stadia*—the figure cited by Nicastro.[82] However, Carmen and Evans suggest that this interpretation is based upon a copying error of the original terminology, and that the text is better translated as eight myriads plus 400 myriads (i.e. 8 x 10,000 + 400 x 10,000 = 408 x 10,000) or 4,080,000 *stadia*.[83] As with the passage in Stobaeus, Heath viewed the translation of this passage as 4,080,000 *stadia* to be an alternative interpretation.[84]

Finally, the work of Theodoret (fourth–fifth century AD) provides a generalized statement on the Earth:Sun distance when he says τετρακοσίας ἀριθμοῦσαι καὶ μέτοι καὶ πλείους σταδίων μυριάδας ('it has been determined to be 400 myriads of *stadia* and more').[85] Thus for Theodoret, the Earth:Sun distance was more than 4,000,000 *stadia* which implies similarity with some of the values found in interpretations of other texts. However, in an earlier passage, Theodoret does provide a value for the Earth:Sun distance, but does not attribute the value directly to Eratosthenes. Interestingly, the text of this passage uses terminology found in other references to both Eratosthenes' Earth:Sun distance, and his Earth:Moon distance, to provide a figure of τετρακοσίας καὶ ἑβδομήκοντα σταδίων μυριάδας ('470 myriads of *stadia*') or 4,700,000 *stadia*.[86] Plutarch, whom Mansfield claims may be drawing upon Aetius' *Placita*, also provides a generalized value for the Earth:Sun distance of

[80] Lyd. *Mens.* 3.12.
[81] Mansfield, 'Cosmic Distances', p.187. See also: Mansfield and Runia, 'Liber 2 Caput 31', p.1108.
[82] For comments and discussions on this passage, see: Mansfield, 'Cosmic Distances', pp.186–88; Carmen and Evans, p.10; this value is also found in translations of a similar text: Plut. *Plac. Philos.* 2.31.
[83] Carmen and Evans, 'Two Earths', pp.9–11.
[84] Heath, *Aristarchus of Samos*, p.340.
[85] Thdt. *CAG* 4.24.
[86] Thdt. *CAG* 1.96.

4,030 myriads of *stadia* (40,300,000 *stadia*).[87] Both of these are larger than the value of 4,080,000 *stadia* commonly seen in earlier works.

An examination of these texts then leaves a number of options for Eratosthenes' Earth:Sun distance. Five of the seven texts can be translated as either 4,080,000 *stadia* or 804,000,000 *stadia* depending upon how the terms in the genitive case are applied, removed, or retained. The text by Pseudo-Plutarch reads only 4,080,000 *stadia*, but it has been reconstructed to read that way. Similarly, only one text, that of Pseudo-Galen, provides a firm number of 4,080,000, but only because the number given in the text has been exchanged for one that results in this value (Table 1):

There are also the generalized statements of Theodoret and Plutarch which provide an Earth:Sun distance of 4,000,000+ *stadia* and 40,300,000 *stadia* respectively. Neither of these texts helps to narrow down the true interpretation of the passages. This uncertainty in translation has led to some scholars favouring one value, while others favour the alternative. For example, as can be seen, all of the texts that directly refer to Eratosthenes' Earth:Sun distance can be interpreted to mean a value of 4,080,000 *stadia* in one way or another. This is the value most favoured by Mansfield and Runia, while Carmen and Evans state that 'most recent scholars have inclined towards 4,080,000'.[88] Heath and Nicastro, on the other hand, both favour the 804,000,000 *stade* interpretation.

Some scholars find support for a value of 4,080,000 *stadia* in the various ratios for distances from the Earth to the Sun and Moon that are found in the ancient texts from before the time of Eratosthenes. Aristarchus had said that the Earth:Sun distance was eighteen to twenty times that of the Earth:Moon

Table 1 Sources on Eratosthenes' Earth:Sun distance.

Text	4,080,000 *stadia*	804,000,000 *stadia*
Aetius	✓	✓
Ps. Galen	✓	
Eusebius	✓	✓
Ps. Plutarch	✓	
Scholia on Ptolemy	✓	✓
Stobaeus	✓	✓
Lydus	✓	✓

[87] Plut. *De Facie* 10; see also: Mansfield, 'Cosmic Distances', p.192.
[88] Mansfield, 'Cosmic Distances', p.188; Mansfield and Runia, 'Liber 2 Caput 31', p.1108; Carmen and Evans, 'Two Earths', p.11.

distance. The Pythagoreans also took the ratio to be 1:18.[89] Archimedes, in his work *The Sand-Reckoner*, states that Eudoxus had determined the ratio to be 1:9.[90] Archimedes' father had favoured a ratio of 1:12 for the size of the lunar and solar diameters which, due to their similar angular size, is also the ratio of their distances.[91] Archimedes himself favoured a ratio of 1:30.[92] Thus ancient estimates for the ratio of the Earth:Moon distance to the Earth:Sun distance, from prior to the time of Eratosthenes, range between 1:9 and 1:30. If Eratosthenes' Earth:Moon distance is taken as 780,000 *stadia*, and his Earth:Sun distance is taken at 4,080,000 *stadia*, this gives a ratio of 1:5.23—which is just outside the lower end of the range. Rawlins took the ratio to be even lower at 1:5.[93] However, if Eratosthenes' Earth:Sun distance is taken as 804,000,000 *stadia* as per Heath and Nicastro, then the ratio becomes 1:1,031. This is used by some scholars to dismiss the 804,000,000 *stadia* interpretation of Eratosthenes' Earth:Sun distance.[94] However, while this ratio far exceeds the range of prior estimates, that does not exclude the larger Earth:Sun result from being the accurate interpretation of the texts.

Aristarchus had also estimated the diameter of the Sun to be greater than 19/3 the diameter of the Earth (a ratio of 1:6.33), but less than 43/6 the Earth's size (a ratio of 1:7.17), and stated that both the Sun and Moon covered an area of the sky equal to 1/15 of a sign of the zodiac.[95] This would give the Sun and Moon an angular size of 2° each ($(360°/12)/15 = 2°$). This is four times too large as both the Sun and Moon have an angular size of 0.5°—an angular size that was more correctly determined by both Hipparchus and Ptolemy after the time of Eratosthenes.[96] Walkup, in her examination of Eratosthenes, suggests that Aristarchus gave a value of 6.75 Earth diameters for the size of the Sun.[97] This is the average of the two values given in the ancient texts ($(6.33 + 7.17)/2 = 6.75$). Nicastro, on the other hand, simply states that Aristarchus' value was seven Earth diameters.[98] If the currently accepted value for the diameter of the Earth (12,742 km) is applied to Aristarchus' ratios, the diameter of the Sun calculates as being between 80,657 km and 91,360 km. This is far too small as the

[89] Eus. *PE* 15.53.
[90] Archim. *Psam.* 3.
[91] Archim. *Psam.* 3.
[92] Archim. *Psam.* 3.
[93] Rawlins, 'Too Big Earth', p.7.
[94] For example, see: Mansfield, 'Cosmic Distances', p.188; Carmen and Evans, 'Two Earths', p.11.
[95] Aristarch. Sam. *De Mag.* proposition 15; Archim. *Psam.* 4.
[96] Pappus, in his commentary on Aristarchus (*Syn.* 6.37) states that Hipparchus determined the sizes of the Sun and Moon to be 27' (0.45°), and Ptolemy determined them to be 31' (0.52°) in size.
[97] Walkup, 'Mystery of the Stadia' *s.v. Parallel Light Rays*.
[98] Nicastro, *Circumference*, p.37.

diameter of the Sun is approximately 1,392,700 km. Carmen and Evans point out that, if Eratosthenes' Earth:Sun distance is taken as 4,080,000 *stadia*, and the angular size of the Sun is correctly taken as 0.5°, this results in a size of the Sun that is smaller than that of the Earth.[99]

The Roman-period writer Macrobius states that Eratosthenes determined the size of the Sun to be equal to twenty-seven Earth diameters.[100] This is the same size that was given to the Sun by Anaximander.[101] This also is not too far from the estimates of Archimedes who stated that the Sun was thirty times bigger than the Earth.[102] This would make the Sun's diameter 344,034 km—still considerably short of being accurate as the Sun's diameter is actually 109 times that of the Earth. Carmen and Evans dispute the accuracy of the 'technical competence of this late Latin compiler' and suggest that there is a mathematical error involved in the numbers given by Macrobius. They suggest that, assuming an Earth:Sun distance of 4,080,000 *stadia*, and a Solar angular diameter of 2° as per Aristarchus, that the ratio of the Solar radius to the Earth radius ($R_\odot/R_\oplus$) equals 1.775. As such, the ratio of the surface area of these two bodies ($R_\odot/R_\oplus$)^2 is 3.15. Carmen and Evans suggest that this number was determined in some undisclosed context, was rounded to three, and then cubed at some time in the transmission of the text to read as twenty-seven, and then applied to Eratosthenes' size of the Sun. Based on this assumption Carmen and Evans state that it cannot be discounted that the value found in Macrobius is 'simply nonsense'.[103] On the other hand, such a chain of mathematical and transmission errors as Carmen and Evans suggest seems somewhat unlikely, and there is just as much reason to accept Macrobius' figure as there is to doubt it—particularly when there seems to be little doubt over the validity of the statements made by Anaximander and Archimedes, which provide a similar size.

Unfortunately, we do not know the context of Eratosthenes' determination of either the Earth:Sun distance or the size of the Sun's diameter, and references to these results are passed on third-hand from other writers from much later in antiquity. It is also unknown if these calculations were made before or after his investigation into the circumference of the Earth. Carmen and Evans suggest that Eratosthenes' calculation of the Earth:Sun distance was done in

---

[99] Carmen and Evans, 'Two Earths', p.15.
[100] Macr. *In Somn.* 1.20.
[101] Hippol. *Refut.* 1.5–6.
[102] Archim. *Psam.* 3.
[103] Carmen and Evans, 'Two Earths', p.15.

conjunction with his determination of the circumference of the Earth, indicating that the Earth:Sun distance of 4,080,000 *stadia* found in Aetius is the only value that works numerically with the circumference value of 252,000 *stadia* found in the later sources on Eratosthenes' experiment—which they claim is the upper limit of Eratosthenes' size of the Earth.[104]

Clues as to when Eratosthenes may have made these calculations can be found in the account of the assumed knowledge that he used to determine the circumference of the Earth, and a comparison to attempts made by prior philosophers to examine the relationships between the Earth, Sun, and Moon. Aristarchus' had estimated the distance from the Earth to the Sun to be approximately 180 Earth diameters, while his estimate of the size of the Sun averages (as per Walkup) to 6.75 Earth diameters. Additionally, Aristarchus had previously demonstrated that when a sphere, such as the Earth, is illuminated by a larger sphere, such as the Sun, the part of the smaller sphere that receives the light is greater than the single hemisphere that faces the larger light-source (Fig. 39).[105]

If Eratosthenes had consulted the work of Aristarchus, it would have clearly indicated that, with the Sun placed at a relatively close distance of, for example, 4,080,000 *stadia*, the light that the Earth receives from the Sun could not have been parallel when it reached the surface. This can be demonstrated mathematically using basic trigonometry (which Eratosthenes did not have access to). For example, placing the centres of both the Earth and Sun in a horizontal alignment (similar to Aristarchus' proposition), lines can be extended horizontally from the upper and lower limbs of the Earth to intersect with a diameter of the Sun that is oriented perpendicular to those lines. Similar lines, connecting the upper and lower limbs of both the Earth and Sun then form right-angle triangles (Fig. 40):

Subtracting one Earth diameter from the average Solar diameter of 6.75 Earths leaves 5.75 Earth diameters, and the opposite side of each right right-angled triangle (α) will be (5.75/2 =) 2.875 Earth diameters in size. The angle θ is the difference between the assumed parallel light rays that reach the Earth from the centre of the solar disk and the rays from the outer edge of the Sun at a distance (δ) of 180 Earth diameters ($D_\oplus$). Using basic trigonometry:

$$\theta = \tan^{-1}(2.875/180) \approx 0.915°. \qquad [\text{eq. 3.2.13}]$$

---

[104] Carmen and Evans, 'Two Earths', pp.7–11, 16.
[105] Aristarch. Sam. *De Mag.* proposition 2.

ET DIST. SOL. ET LVNAE.   9

Sphæra enim, cu-
ius centrum B à ma
iori fphera, cuius cé
trû A illuminetur.
Dico partem fphere
illuminatã, cuius cé
trû B dimidia fphæ-
ra maioré eſſe. Qﬁ
enim duas inæqua-
les fpheras idem co-
nus comprehendit,
verticé habés ad mi-
norem fphæram : fit
conus fphæras com-
prehédés; & per axé
planum producatur
faciet illud ſectiones
in fpheris qui dé cir-
culos, in cono autem
triangulum. Itaq; fa
ciat in fphæris circu
los CDE FGH; & in
cono triãgulû CEK.
manifeſtum eſt por-
tioné fphere, quæ eſt 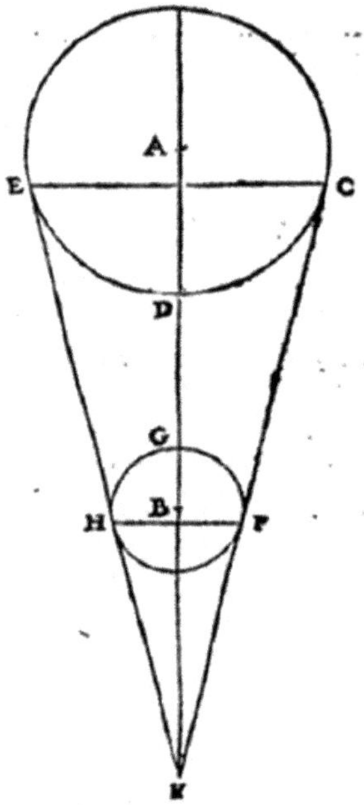
ad FGH circûferentiã, cuius bafis circulus circa dïa
metrû FH, parté eſſe illuminatã à portione, quæ eſt
ad circumferentiam CDE, cuius bafis circulus circa
diametrû CE, rectus exiſtés ad ipſam AB . etenim F
GH circûferétia à circûferentia CDE illuminatur;
quòd extremi radij funt CF EH: atque eſt in pro-
portione FGH centrum fphæræ B. Quare pars fphe
re illuminata, dimidia fphæra maior érit.

B   F E D.

A
E   C
D
G
H B F
A
B
K

Fig. 39 Aristarchus' proof that a larger sphere illuminates more than
a hemisphere of a smaller sphere—from Commandino's 1572 edition
of Aristarchus' *On the Sizes and Distances of the Sun and Moon*.

Walkup calls this error significant, but also suggests that, if it is assumed that
ancient tools of measurement had an inherent margin of error of 1°, then
Eratosthenes' assumption about parallel rays can be accepted as the angle of
the light falls just within the margin.[106] However, even assuming this margin

---

[106] Walkup, 'Mystery of the Stadia' *s.v. Parallel Light Rays*.

Fig. 40 Determination of the angle of sunlight based upon the values of Aristarchus (not to scale).

of error, it should have been clear to Eratosthenes that the light rays from the Sun would not have been parallel.

Additionally, this calculated angle of 0.915° is not correct. The Sun appears in the sky as a disk approximately 0.5° wide and is at a distance of around 147 million kilometres (or about 11,557 times the diameter of the Earth, which is 12,742 km). The Sun's diameter ($\approx$1.39 million km) is approximately 109 times the diameter of the Earth. Thus, using the same basic trigonometry as in [eq. 3.2.13], light rays from the outer limb of the Sun reach the Earth at an angle of (Fig. 41):

Fig. 41 Calculation of the angle at which Sunlight reaches the Earth (not to scale).

$$\theta = \tan^{-1}(54/11,557) \approx 0.27° \text{ (Fig. 41).}[107]$$

Aristarchus had shown that light coming from the Sun should not be parallel. However, according to Cleomedes, Eratosthenes believed that it was. This suggests that Eratosthenes placed the Sun at a greater distance from the Earth than Aristarchus had done. This then provides clues as to how to translate the references to Eratosthenes' Earth:Sun distance that are found in the ancient texts. For example, taking the value for Eratosthenes' circumference of the Earth as 250,000 *stadia* as per Cleomedes gives a terrestrial diameter of approximately 79,578 *stadia*. The Sun's diameter is given as twenty-seven Earth diameters by Macrobius. Subtracting one Earth diameter from this size for the Sun leaves thirteen Earth diameters (or 13 x 79,578 = 1,034,514 *stadia*) on the opposite sides of the triangles created by the lines connecting the limbs of the two bodies (Fig. 42):

If Eratosthenes' distance to the Sun is taken as 4,080,000 *stadia* as some scholars suggest, the angle θ calculates (as per eq. 3.2.13) as

---

[107] Walkup ('Mystery of the Stadia' *s.v. Parallel Light Rays*) suggests that the variance was 0.26°.

Fig. 42 Determination of the angle of sunlight based upon one interpretation of the values of Eratosthenes (not to scale).

$\theta = \tan^{-1}(1{,}034{,}514/4{,}080{,}000) \approx 14.23°$. This angle is too large to fall within any margin of error that modern scholars may attribute to ancient instrumentation. Even if Eratosthenes was working in a time before the laws of trigonometry had been established and the circle was not yet divided into degrees, it would have been obvious that a Sun of that size, and at that distance, could not have produced parallel rays of light.

Even if the Sun is reduced to an accurate angular size of 0.5°, it still produces light at a significant angle when set at a distance of 4,080,000 *stadia* (Fig. 43).

Fig. 43 Determination of the angle of sunlight based upon the values of Eratosthenes and a 0.5° angular size of the Sun (not to scale).

Aristarchus had claimed that the Sun covered 2° of the sky. If Eratosthenes' solar diameter of twenty-seven Earths is also considered to cover this same amount of sky, but then is divided by four to make it 0.5° in size, this makes the Solar diameter 6.75 Earth diameters, and the opposite sides of the right-angle triangles 2.875 Earth diameters or 228,786.75 *stadia*. At a distance of 4,080,000 *stadia*, a Sun of this size emits light at an angle of $\theta = \tan^{-1}(228{,}786.75/4{,}080{,}000) = 3.2°$. Not only is this still outside of the margin of error for ancient instrumentation suggested by some scholars, and runs contrary to Cleomedes' statement that Eratosthenes assumed that light coming from the Sun was parallel, it additionally goes against the size of the Sun attributed to Eratosthenes by Macrobius, and statements that a 0.5° angular size of the Sun was calculated later by Hipparchus and Ptolemy.

However, if the Earth:Sun distance is taken as 804,000,000 *stadia*, as per the interpretations of the ancient texts favoured by Heath and Nicastro, and the Solar diameter is taken as twenty-seven Earths as per Macrobius, this results in an angle more in line with the assumed knowledge regarding parallel light that Cleomedes ascribes to Eratosthenes (Fig. 44).

A Sun of that size, and set at that distance, then makes the angle of the incoming sunlight $\theta = \tan^{-1}(1{,}034{,}514/804{,}000{,}000) \approx 0.07°$ which could easily be

Fig. 44 Determination of the angle of sunlight based upon the alternate values of Eratosthenes (not to scale).

truncated to 0.0° or parallel. Carmen and Evans state that '[t]he obvious impact of a finite Earth-Sun distance is that we can no longer consider the Sun rays falling on the two cities [i.e. Alexandria and Syene] to be parallel'.[108] Similarly, Tupikova states that Eratosthenes' result requires the Sun to be at a distance 'infinitely larger than the radius of the Earth', and the visual diameter of the Sun must be negligible compared to that distance.[109] As Tupikova further points out, the reference to a sub-solar zone at the Tropic during the Summer Solstice must show that the Sun is at a finite distance.[110] However, that distance in itself can be very large compared to the diameter of the Earth due to the larger size of the Sun. Despite the fact that the greater the set distance to the Sun is, the closer to parallel the light becomes, what 'we' are able to consider, as per Carmen and Evans, is something of a moot point. What is of more concern for the understanding of Eratosthenes' work, is trying to understand why, as per Cleomedes, Eratosthenes thought that light was parallel. It seems that Eratosthenes could have determined the angle of the incoming light, but based on the figures attributed to him, that angle would have been so small as to make the light fundamentally parallel anyway.

It must also be considered that Eratosthenes would not have been using modern trigonometry in his calculations. Interestingly, despite advocating an Earth:Sun distance of 4,080,000 *stadia*, Carmen and Evans suggest that Eratosthenes could have reached a similar result for the angle of light for a Sun set at 804,000,000 *stadia* using ratios (Fig. 45).[111]

Carmen and Evans state that setting the Sun as a smaller point-source, the relationship $\beta/\alpha > AU/AE$, and $EU/EA \approx 102$. Since $\alpha = \gamma - \beta$, the angle $\beta > \gamma/(1 + (AE/AU))$. Additionally, as $AU > SU$, the angle $\beta > \gamma/(1 + (AE/SU))$. Furthermore, AE is equal to the radius of the Earth (r), and SU equals the distance from the Earth to the Sun (δ) minus r, then $\beta > \gamma \times (1 - (r/\delta))$. Carmen and Evans suggest that such ratios show that the minimum value of β is 7.1292°

[108] Carmen and Evans, 'Two Earths', p.11.
[109] I. Tupikova, 'Eratosthenes' Measurements of the Earth: Astronomical and Geographical Solutions' *Orb. Terr.* 16 (2018), p.223.
[110] Tupikova, 'Eratosthenes' Measurements', p.223.
[111] See: Carmen and Evans, 'Two Earths', pp.11–14.

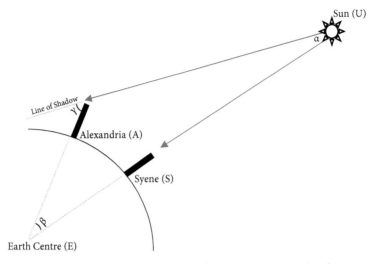

Fig. 45 Carmen and Evan's method of calculating the angle of sunlight using ratios.

when the Sun is placed at a set distance of around 102 Earth diameters, and 7.2° when the Sun is set at an infinite distance.[112] The difference between these two values is 0.0708°—the same value for the angle of the light rays calculated using the larger of the interpretations of Eratosthenes' Earth:Sun distance of 804,000,000 *stadia*.[113]

This implies that Eratosthenes calculated the Earth:Sun distance after he had calculated the size of the Earth. However, one of the precepts that Eratosthenes works from to calculate the size of the Earth is the idea that light is parallel. This then alternatively suggests that he may have calculated the Earth:Sun distance before his determination of the circumference of the Earth. If Eratosthenes determined the Earth:Sun distance first, he must have been working from the results of one of the earlier attempts to calculate the planetary circumference, such as the 300,000 *stadia* results cited by Archimedes, to determine the distance. Doing so, would have still resulted in light that was almost parallel (Fig. 46).

[112] Carmen and Evans, 'Two Earths', p.14.
[113] Carmen and Evans ('Two Earths', pp.11, 14) use this relationship to suggest that the result of Eratosthenes' calculation of the circumference of the Earth was 252,000 *stadia* (contra the 250,000 *stadia* reported by Cleomedes) as 5,000 *stadia* x (360°/7.1292°) = 252,476 *stadia* which they say was then rounded to 252,000. Similarly, Rawlins ('Too Big', pp.6–7) suggests that Eratosthenes' Solar distance was equal to 100 Earth radii, and if the distance was 4,080,000 *stadia*, this made the terrestrial radius 40,800 *stadia* and the circumference 256,353 *stadia*—which is not far removed from the value given by Marcian of Heraclaea. However, neither of these two figures seems to be the correct result of Eratosthenes' calculation of the circumference of the Earth (see section 5).

Fig. 46 Determination of the angle of sunlight based upon Archimedes' value of Earth's circumference (not to scale).

A circumference of 300,000 *stadia* equates to a planetary diameter of 95,494 *stadia*, and the opposite sides of the right-angle triangles formed by connecting the limbs of both the Sun and Earth equal thirteen Earth diameters or 1,241,422 *stadia*. At a distance of 804,000,000 *stadia*, light emitted from the Sun reaches the Earth at an angle of $\theta = \tan^{-1} (1,241,422/804,000,000) \approx 0.09°$ which is still small enough to be truncated to 0.0° or parallel.

Consequently, in the end, it does not matter which order Eratosthenes undertook his experiments and calculations. If Eratosthenes had made initial calculations based upon a terrestrial diameter such as that provided by Archimedes, he would have concluded that incoming light from the Sun was almost parallel—the concept that Cleomedes states Eratosthenes was working from for his calculation of the circumference of the Earth. If, on the other hand, Eratosthenes had merely accepted someone else's concept that light was parallel for his calculation of the circumference of the Earth, his results for the size of the planet, the Sun and the distance between them, would have simply confirmed this idea. Indeed, as will be shown, the result of Eratosthenes' calculation of the Earth was smaller than the value reported by Cleomedes and this would have reduced the calculated angle of the incoming light to around 0.066°.

Eratosthenes seems to have set the Sun at a far greater (and more accurate) distance from the Earth (804,000,000 *stadia*) than Aristarchus had previously done. Eratosthenes also seems to have revised Aristarchus figure for the size of the Sun from thirty Earth diameters to 27. This resulted, for Eratosthenes, in light that reached the Earth at so minimal an angle ($\approx 0.07°$) that it was fundamentally parallel. Eratosthenes had calculated the distance from the Earth to the Sun (144,720,000 km based upon a 180 m *stade*) to within a margin of error of <2 per cent compared to the actual distance of 147,260,000 km. However, Eratosthenes thought that the Sun was approximately four times smaller than it really is (twenty-seven Earth diameters compared to 109 Earth diameters). This explains why he calculated a minimal angle equivalent to about 0.07° compared to a more accurate 0.27° (which is approximately 0.07° x 4). Yet even the small angle that incoming light really creates, and which Eratosthenes seems to have ignored through his miscalculation, would have had an impact of around half a degree either side of the outcome of the experiment (7.2° ± 0.5°)—a percentage error of nearly 7 per cent. Added to this, and possibly negating the error, is the fact that the shadow cast by the *gnomon* on Eratosthenes' experiment should have further highlighted the fact that the Sun was further away than Aristarchus had suggested.

3.2.5 The Effect Of The Components Of The Shadow

Another potential error in Eratosthenes' methodology that needs to be considered is that, due to the Sun not being a point-source of light, any shadow that is cast by an object—such as the upright *gnomon* of a sundial—will have two components: a clearly defined, darker, umbra, and a less-defined, lighter, penumbra around its edges. Some scholars have used this to alter the angle of the shadow that forms a fundamental part of Eratosthenes' calculation of the circumference of the Earth. While Eratosthenes did not actually need to directly observe a shadow in order to undertake his calculation (unless he was checking the calibration of the sundial and/or confirming the 'assumed knowledge' that he was working from), the components of the shadow would affect how the sundial that he would have used would have been calibrated. However, both reconstructive experiments and mathematical modelling demonstrate that the penumbra that was part of a shadow cast at Alexandria on the Summer Solstice would not have been very large, would not have had a large impact on the positioning of the Summer Solstice line on a sundial calibrated for Alexandria like the one Eratosthenes would have used, and, as a result, would have had little to no impact on the results of Eratosthenes' calculations.

The Sun has an angular size of 0.5°. The consequence of this is that the light emitted from the Sun does not come from a single point-source. Rather, light reaches the Earth from across the whole surface of the Sun, and so light that has originated from near the outer limb of the Sun will reach a point on the surface of the Earth at a slightly different angle to light that has originated from the centre of the solar disk (Fig. 47).

Fig. 47 How light from the Sun reaches a point on the Earth at different angles due to it not being a point-source of light.

The umbra (the Latin word for 'shadow') is the area in which all light from the Sun is blocked by an object. The penumbra (derived from the Latin word *paene*, meaning 'almost'), on the other hand, is the area in which only some of the light from the Sun is obstructed due to it being larger than a point-source. The penumbra creates an area of slightly less shaded 'fuzz' on the margins of the shadow—which is graded from full shadow along the edge of the umbra, and which then increases in its lightness towards the boundary with the area of full illumination (Fig. 48).

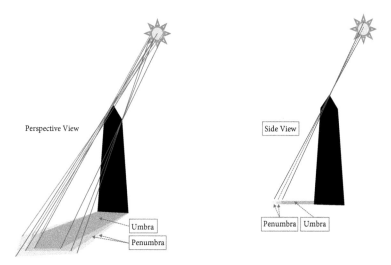

Fig. 48 The creation of the umbra and penumbra of a shadow by the Sun.

As can be seen from Fig. 48, light from the centre of the solar disk (blue lines) will pass over the tip of a *gnomon* and end at the apex in the middle of the penumbra. Light from this location will also pass over all other parts of the object. In the case of light passing over the upper left edge of the *gnomon*, this will also end at the corresponding point in the centre of the penumbra. Light from the upper limb of the solar disk directly above the centre will pass over the top of the *gnomon* at a steeper angle than light from the centre-bottom of the solar disk, and this will create the apex of the umbra and the apex of the penumbra respectively (purple lines in the side view in Fig. 48). Light from the centre-right of the solar disk (red lines in Fig. 48) will pass over the *gnomon* at an angle and project parts of the penumbra to the left. Light emitted from the centre-left of the solar disk (green lines in Fig. 48) will also pass over the *gnomon* at an angle and project parts of the penumbra to the right. As some of these areas of cast shadow overlap, this results in a solid area of umbra that is in line with the axis running from object to solar disk, with a gradation of decreasing shadow extending from the edge of the umbra out to the area of full illumination by the Sun. This creates a 'fuzzy' transitional zone between the areas of umbra and full light. In relation to Eratosthenes' calculation of the circumference of the Earth, the key questions are how big the area of penumbra would have been on a shadow cast by the *gnomon* of a sundial at midday on the Summer Solstice in Alexandria, and how this may (or may not) have affected any resultant calibration of the sundial and/or calculations.

Some previous examinations of the work of Eratosthenes have engaged with this potential error. Newton suggests that the different components of the shadow constitute the largest error in Eratosthenes' methodology because the shadow cast by the *gnomon* would not have had a sharp edge due to the Sun not being a point-source.[114] Dutka also states that because the Sun is not a point source, the shadow would be indistinct and fuzzy, which would make identifying the exact end of the shadow difficult, but conversely suggests that many of the errors associated with the observation of the shadow 'were so small as to be undetectable in the gnomon observations'.[115] Pinotsis simply states that '[t]he measurement of the length of the shadow thrown by the *gnomon* is not accurate because of the existence of the penumbra', but then goes into no further analysis of this potential source of error in Eratosthenes' calculations.[116]

Ptolemy advises against using the shadow cast by a *gnomon* to determine angles on the Equinoxes and the Winter Solstice because, due to the lengthier shadows on these days, the difference between the umbra and penumbra is more pronounced which can lead to errors in calculations (e.g. if one person measured to the outer edge of the penumbra, while another person measured to the edge of the umbra).[117] In another re-creative experiment using the replica sundial, conducted at the time of the Solar culmination (11:59:18) on the Winter Solstice on 21 June 2022, it was found that the shadow cast by the *gnomon* was still well defined, but that the penumbra of the shadow extended beyond the Winter Solstice line on the face of the sundial by approximately 2 mm (Fig. 49):

---

[114] Newton, 'Sources of Eratosthenes' Measurement', p.381.
[115] Dutka, 'Eratosthenes' Measurement', p.61.
[116] Pinotsis, 'Significance and Errors', p.58.
[117] Ptol. *Alm.* 1.12, 2.5.

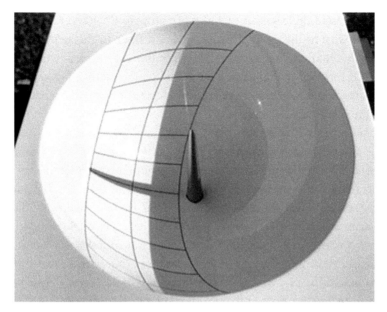

Fig. 49 The shadow cast by the *gnomon* of the replica sundial at noon (LAT) on the Winter Solstice.[a]

[a] Author's photo.

While there was not a large variance between the components of the shadow, the difference between the umbra and penumbra was large enough to be discernible with the naked eye. For Ptolemy, the difference may have been even more apparent if the instrument he was using was larger than the replica used here (Pappus (*in Ptol. Alma* 1.6) states that Ptolemy's armillary sphere had a diameter of around half a metre), or if the *gnomon* he is referring to in his advice against its use of the Winter Solstice is referring to one of the upright staves that had been commonly used in Egypt which would cast its shadow onto a flat surface (e.g. the ground) which would make the difference even more apparent.

However, on the Summer Solstice, when the Sun is more directly overhead in a place like Alexandria, not only would the total shadow be shorter in length, but the difference between the outer edges of the umbra and penumbra would be significantly reduced. Newton suggests that that when using a *gnomon* it is easy to make an error equal to the radius of the sun (i.e. 0.25° or 15') but an error any larger than that is unlikely.[118] Newton argues that it is reasonable to take the standard deviation (σ) for using a *gnomon* to be around 16' ($\approx 0.25°$),

---

[118] Newton, 'Sources of Eratosthenes' Measurement', p.381.

but if a graduated circle is used, the value of σ drops to 4ʹ.[119] However, Newton is unclear by what he means by a 'graduated circle'. In Eratosthenes' time, circles were divided into sixty segments called a *hexacontade*—each of which would equal 6°. If some form of gradation were used on the sundial, for which there is no evidence in the archaeological record, then the margin of error would be half of the smallest calibration, or 3°.

Rawlins accepts Newton's standard deviation of 16ʹ and similarly suggests that the sharp, far edge of the umbra (representing the light from the limb of the Sun) is 16ʹ from the angular centre of the penumbra (representing the true, but unreadable, point corresponding to the centre of the solar disk).[120] Rawlins then uses the assumed standard deviation of the *gnomon*/shadow (−15.8ʹ) to revise an assumed error in Eratosthenes' stated data. The angle recounted by Cleomedes is 1/50 or a circle, or 7°12ʹ. Rawlins compares this to what he calls the real geocentric angle of 7°28.6ʹ. The difference between these two angles is approximately 16.5ʹ. Rawlins subtracts the standard deviation (−15.8ʹ) from this angle to reach a revised margin of error for Eratosthenes' calculations of 0.7ʹ (16.5ʹ−15.8ʹ).[121] This revised margin of error forms parts of the argument made by Rawlins in favour of adjusting the reported distance between Alexandria and Syene to 5,040 *stadia* and a placing of the Tropic at 23°51ʹ15″N (or 23.85°)—which is the correct value for the location of the Tropic in 230 BC.[122]

However, there seems to be a number of issues with modern assumptions regarding the nature of the shadow. When the replica sundial calibrated for Alexandria was used in a simple re-creative experiment, at noon not long after the Summer Solstice, it was observed that the shadow was quite distinct and had quite a clearly defined edge (Fig. 50).

The clearly defined nature of the edges of this shadow suggests that any penumbra present is not overly large. Additionally, it was found that the location of the tip of the shadow was highlighted by a small flare of light which is caused by some of the incoming rays of light reflecting off the curved inner surface of the hemispherical *skaphe* and converging at a point directly opposite the position of the Sun that coincides with the tip of the shadow—similar to the catacaustic curve that is created by light reflecting off the inside of a cylindrical mirror (see Fig. 50). Elements of the flare of light diffuse the tip

---

[119] Newton, 'Sources of Eratosthenes' Measurement', p.384.
[120] Rawlins, 'Geodesy', p.263.
[121] Rawlins, 'Geodesy', p.263.
[122] Rawlins, 'Geodesy', p.264.

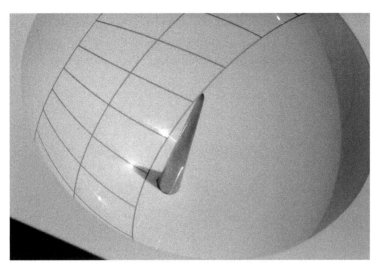

Fig. 50 Result of a re-creative experiment to observe the shadow cast by a sundial at noon not long after the Summer Solstice.[a]

[a] Author's photo.

of the shadow—which was initially confused as being a larger section of penumbra—but this diffusion is offset by the flare of light which pinpoints the tip of the shadow. This flare is most apparent when the angle of the incoming light to the *gnomon* is <45°—as less light is being reflected inside the bowl at greater angles (for example, compare how much of the bowl of the sundial is in shadow at the time of culmination on the Winter Solstice (Fig. 49) compared to around the time of the Summer Solstice (Fig. 50)). Consequently, for the replica sundial, the flare was most intense when the instrument was aligned to simulate noon on the Summer Solstice in Alexandria, but was not as intense when the instrument was aligned for noon on the Winter Solstice. However, even in this case, the shadow was still sharp, and the differentiation in the tip (both umbra and penumbra) was much more easily discernible.

The size of the penumbra created by the *gnomon* of the replica sundial at noon on the Summer Solstice can also be calculated mathematically (Fig. 51).

If it is assumed that the length of the shadow in Cleomedes' account of Eratosthenes' experiment is measured only to the edge of the umbra, the angle created by light from the upper limb of the solar disk, passing over the tip of the *gnomon* (red line in Fig. 51), creates an angle (Θ) of 7.2°. However, light from the lower limb of the solar disk (purple line), passes over the tip of the *gnomon* at an angle (Ψ) equal to 7.2° plus the 0.5° angular size of the Sun, or

7.7°. The height (h) of the *gnomon* of the replica sundial used in this research is 75 mm (±0.1 mm). Thus, on an instrument of this size, at midday on the Summer Solstice, the length of the shadow from the base of the *gnomon* to the tip of the umbra (x) is:

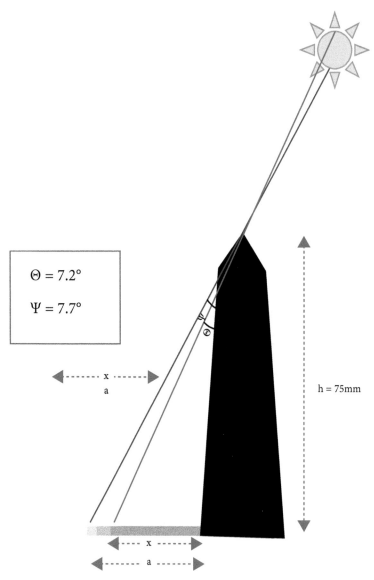

$\Theta = 7.2°$

$\Psi = 7.7°$

h = 75mm

Fig. 51 Calculating the size of the penumbra for the replica sundial at noon on the Summer Solstice.

$$\tan(7.2) = x/75.0\text{mm}$$
$$0.1263293784 = x/75.0\text{mm}$$
$$X = 0.1263293784 \times 75.0\text{mm}$$
$$\approx 9.5\text{mm} \pm 0.1\text{mm}$$

[eq. 3.2.14]

This correlates with the measurements taken from the point directly under the *gnomon* to the Summer Solstice line that is inscribed on the face of the sundial (see section 3.1). At the same time, on the same day, the length of the shadow from the base of the *gnomon* to the tip of the penumbra (a) is:

$$\tan(7.7) = x/75.0\text{mm}$$
$$0.13520563046 = x/75.0\text{mm}$$
$$x = 0.13520563046 \times 75.0\text{mm}$$
$$= 10.1\text{mm} \pm 0.1\text{mm}$$

[eq. 3.2.15]

Thus, the entire width of the penumbra is only (10.1 mm − 9.5 mm =) 0.6 mm (±0.2 mm), and the width from the midpoint of the penumbra (created by light from the centre of the solar disk) to the edge of either the umbra or the fully illuminated area is only 0.3 mm. Such a small penumbra would be almost impossible to determine in naked eye observations—and especially so on a sundial carved from stone or metal, such as the one Eratosthenes would have used, where the surface of the instrument may not be a uniform colour. This runs contrary to the conclusions reached by previous scholars—although it does agree with Dutka's claim that any potential errors may be so small as to be basically undetectable. This also accounts for why the shadow seen in the re-creative experiment appeared to have such a sharply defined edge.

The key aspect of such a small penumbra being cast on the Summer Solstice is how this small transitional area of shadow would have affected either direct observations made by Eratosthenes, and/or how the difference between the umbra and penumbra would have translated to the inscription of the Solstice line on the sundial when it was calibrated for its location. For example, if the Solstice line on the sundial was traced out at the time of the sundial's calibration by someone actually marking the location of the tip of the shadow at regular intervals throughout the day of the Summer Solstice, then the question is whether that person marked the location of the tip of the umbra or penumbra. However, for a sundial being calibrated for a location like Alexandria, where the difference between the umbra and penumbra is very small on the Summer Solstice, such a question becomes something of a moot point. If the person doing the marking was using a reed stylus to do so, for example, and was marking the location of the tip of the shadow by placing a small dot on the surface of the sundial using some form of pigment, that dot would actually be bigger in size than the difference between the umbra and penumbra. When the line had been marked in its entirety, and was then inscribed into the stone, the engraving would similarly be wider than the difference in the components of the shadow. On the replica used in this research, the Solstice line, which was etched into the surface of the sundial using a laser, is 0.5 mm wide. Even with modern engineering and instruments, the lines on the sundial are larger than the difference between the umbra and mid-point of the penumbra.

In regard to the angle that Eratosthenes determined (1/50 of a circle), a measurement to either the umbra or penumbra results in this value. For example, if the angle to the tip of the umbra is taken as 7.2°, this equals (7.2°/360° =) 0.02 of a circle or 1/50—the value recounted by Cleomedes. If the measurement is taken to the penumbra, and the angle is really 7.7°, this then equals (7.7°/360° =) 0.021 of a circle which is still 1/50 when rounded. Thus, no matter which component of the shadow was used, either by those calibrating the sundial for Alexandria, and/or in observations made by Eratosthenes himself, the angle remains the same.

The result of this is that, for Eratosthenes, it did not matter whether he calculated the angle created by the shadow based upon a measurement to the tip of the umbra or to the tip of the penumbra—and even this assumes that Eratosthenes actually undertook some observations as part of his experiment and calculations, which he did not need to do as he could have based his calculations just on the measurement from the *gnomon* to the Solstice line. Even if Eratosthenes did make observations of the shadow, just to confirm the calibration of the sundial he was using, the difference between the umbra and penumbra is so small on the Summer Solstice in Alexandria that measuring to either one would have still placed the tip of the shadow on the Solstice line inscribed on the instrument. Consequently, and a barring any potential error in the north-south alignment or level positioning of the sundial, if an observation was undertaken, the components of the shadow would have had no effect on Eratosthenes' estimations, observations, and calculations at all.

This may additionally explain why Eratosthenes did not accept the distance to the Sun that had been previously offered by Aristarchus, why Eratosthenes placed the Sun at a greater distance from the Earth than Aristarchus had done, and confirm that a value of 804,000,000 *stadia* as the correct way to interpret the ancient passages which recount Eratosthenes' Earth:Sun distance. If, for example, the Sun is given an angular size of 2° and is placed at a distance of 180 Earth diameters as per Aristarchus, the light from the upper limb of the Sun would still create the tip of the umbra with light passing over the tip of the *gnomon* at an angle (Θ) of 7.2°. The light from the lower limb of the Sun, on the other hand, would pass over the *gnomon* at an angle (Ψ) of (7.2° + 2° =) 9.2° (Fig. 52).

Thus, while the length of the distance from the *gnomon* to the tip of the umbra (x) would remain unchanged at ≈9.5 mm, the distance from the *gnomon* to the tip of the penumbra (a) becomes:

$$\tan(9.2) = a/75.0\text{mm}$$
$$0.1619646581 = a / 75.0\text{mm}$$
$$a = 0.1619646581 \times 75.0\text{mm}$$
$$\approx 12.15\text{mm} \pm 0.1\text{mm}$$

[eq. 3.2.16]

The difference between the length of the umbra and penumbra is (12.15 mm − 9.5 mm =) 2.65 mm (±0.2 mm). This is an increase in the size of the penumbra by a factor of ten compared to the areas of shadow cast by a Sun of the correct size and set at the correct distance. An area of penumbra of this size would have been quite noticeable on the shadow cast onto the surface of a sundial. The fact that no such penumbra would have been observable to Eratosthenes would have demonstrated to him that Aristarchus was incorrect.

Instead, if Eratosthenes placed the Sun at a distance of 804,000,000 *stadia*, but gave it an incorrect size of twenty-seven Earth diameters, then the angular size of the solar disk is approximately 0.23° (Fig. 53):

$$\tan(\Theta) = (0.5 \times 1,925,154 stadia) / 804,000,000\ stadia$$
$$= 962,577/804,000,000$$
$$= 0.0019723507$$
$$\Theta - \tan^{-1}(0.0019723507)$$
$$= 0.1130072243^{o}$$

[eq. 3.2.17]

$$Angular size = 2 \times \Theta$$
$$= 2 \times 0.1130072243°$$
$$\approx 0.23°$$

[eq. 3.2.18]

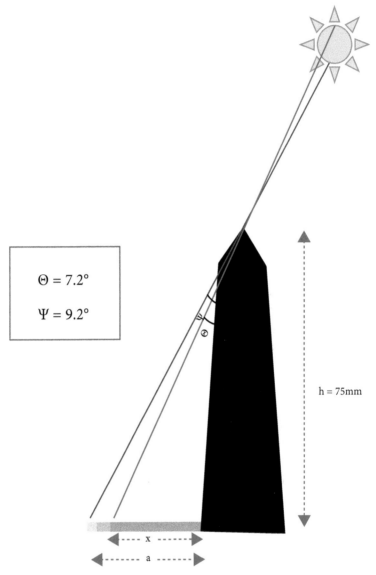

Fig. 52 Calculating the size of the penumbra for the replica sundial at noon on the Summer Solstice with the characteristics of the Sun as per Aristarchus.

With a Sun of this size, and with the light from the upper limb passing over the tip of a *gnomon* at an angle (Θ) of 7.2° to create an umbra ≈9.5 mm in length, the light from the lower limb would pass over the tip of the *gnomon* at

Fig. 53 The angular size of Eratosthenes' Sun at a distance of 804,000,000 *stadia*.

an angle (Ψ) of (7.2° + 0.23° =) 7.43° and the length of the shadow to the tip of the penumbra would be as follows:

$$\tan(7.43) = a/75.0mm$$
$$0.1304097909 = a/75.0mm$$
$$a = 0.1304097909 \times 75.0mm \qquad \text{[eq. 3.2.19]}$$
$$= 9.78mm \pm 0.1mm$$

This creates an area of penumbra (9.78 mm – 9.5 mm =) 0.28 mm (±0.2 mm). This about half the size of the area of penumbra created by a Sun of the correct size at the correct distance (because Eratosthenes assumed the Sun was smaller than it actually is) and, similar to the shadow created by that Sun, would not have been discernible on a shadow falling onto a sundial in Alexandria. However, if the Sun is placed at a closer distance of 4,080,000 *stadia* as per some interpretations of the ancient texts that refer to Eratosthenes' Earth:Sun distance, and the size of the Sun is still set at twenty-seven Earth diameters, then the angular size of the Sun becomes (Fig. 54):

Fig. 54 The angular size of Eratosthenes' Sun at a distance of 4,808,000 *stadia*.

$$\tan(\Theta) = (0.5 \times 1,925,154\ stadia) / 4,080,000\ stadia$$
$$= 962,577/4,080,000$$
$$= 0.2359257353 \qquad \text{[eq. 3.2.20]}$$
$$\Theta = \tan^{-1}(0.2359257353)$$
$$\approx 13.27°$$

$$Angular size = 2 \times \Theta$$
$$= 2 \times 13.27° \qquad \text{[eq. 3.2.21]}$$
$$\gg 26.5°$$

If the Sun was this size, and again with the light from the upper limb passing over the tip of a *gnomon* at an angle (Θ) of 7.2° to create an umbra ≈9.5 mm in length, the light from the lower limb would pass over the tip of the *gnomon* at an angle (Ψ) of (7.2° + 26.5° =) 33.7° and the length of the shadow to the tip of the penumbra would be:

$$\tan (33.7) = a/75.0\text{mm}$$
$$0.6669170964 = a/75.0\text{mm}$$
$$a = 0.6669170964 \times 75.0\text{mm}$$
$$\gg 50\text{mm} \pm 0.1\text{mm}$$

[eq. 3.2.22]

With a Sun of this size and at this close a distance, the width of the penumbra would be approximately (50 mm – 9.5 mm =) 40.5 mm (±0.2 mm) in size. This would have been easily discernible with the naked eye.

Despite Eratosthenes working in a time before many of the elements of modern trigonometry had been determined, this still leaves us with two options for how to interpret the passages which recount Eratosthenes' Earth:Sun distance:

1) That Eratosthenes' Earth:Sun distance was 4,080,000 *stadia*, but he then failed to account for why, with a Sun at this distance, there should have been an easily observable penumbra to every shadow when he would have seen almost no penumbra at all.
2) That Eratosthenes' Earth:Sun distance was 804,000,000 *stadia*, and despite the fact that Eratosthenes' estimate for the diameter of the Sun is four times too small, both a Sun that is too small, but at the correct distance, and a Sun of the correct size at the correct distance, create very minor areas of penumbra that would not have been detectable to the naked eye.

Based upon the description of Eratosthenes' experiment and calculations to determine the circumference of the Earth, the second interpretation, with an Earth:Sun distance of 804,000,000 *stadia*, seems the most likely.

Ptolemy advocated against using sundials to determine angles on the Equinoxes and Winter Solstice due to the lengthy nature of the penumbra of the shadow on those days. However, for Eratosthenes, the key day was the northern Summer Solstice. Re-creative experimentation and mathematical modelling demonstrate that, on this day—and especially at midday—there is a very minimal penumbra to any shadow cast by the *gnomon* of a sundial in Alexandria. The characteristics of this small area of penumbra demonstrate that Eratosthenes had placed the Sun at a much greater distance than Aristarchus had done. Furthermore, contrary to the theories of prior scholars and essential to understanding Eratosthenes' work on calculating the circumference of the Earth, this small area of penumbra would have had no impact on either the calibration of the sundial that Eratosthenes would have used in his experiment, nor affected the outcome of any observations he may have made.

3.2.6 The Effect Of The Earth's Rotation

The rotation of the Earth also has a small effect on the elements of Eratosthenes' experiment to calculate the circumference of the Earth. This effect is caused due to the forces created by the planet's rotation on any plumb line, which may have been used to set the *gnomon* of the sundial to 'vertical'. However, an examination of the total effect of the Earth's rotation on such a plumb line, and a critical understanding of the nature of Eratosthenes' experiment, demonstrates that the amount of variation caused by the Earth's rotation, and how that was incorporated into the design, construction, and alignment of the sundial, would have had almost no effect on Eratosthenes' overall result.

The effect of the Earth's rotation on Eratosthenes' experiment was examined by Pinotsis.[123] In his experiment, Eratosthenes assumed that lines, projected perpendicularly downwards from the *gnomons* of sundials located in both Alexandra and Syene would intersect at the centre of the Earth. This is based upon one of the geometric propositions that had previously been proposed by Euclid, which states 'if a straight line touches a circle, and from the point of contact a straight line be drawn at right angles to the tangent, the centre of the circle will be on the straight line so drawn.'[124]

However, while this holds true for the geometric principles regarding circles and tangents outlined by Euclid, it does not hold for lines projected downwards from a point on the surface of a rotating, mostly spherical, object like a planet. Due to the Earth's rotation, and the effects of centrifugal force, lines that are projected downward from the surface would not actually be perpendicular to the Earth's surface, but would be altered by a very small angle to the direction of the Earth's gravitational field. This would result in a minor variation in the angle of intersection between the two projected lines (Fig. 55).

[123] Pinotsis, 'Significance and Errors', pp.59–60.
[124] See: Euc. *El.* 3.19.

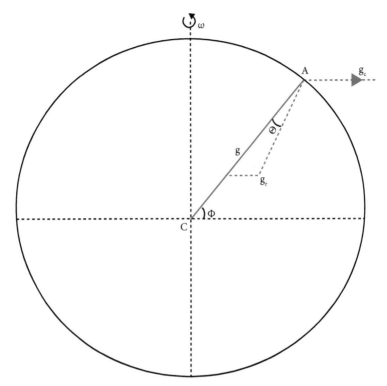

Fig. 55 How planetary rotation affects lines projected downward from the surface.[a]

[a] Image derived from that of Pinotsis, 'Significance and Errors', p.59.

On a spherical Earth, as the ancients believed, the vertical is a line perpendicular to the surface of the Earth as per the proposition of Euclid, and any line extended downward from that point would connect with the centre of the planet (C). At a point of the Earth's surface such as Alexandria (A), which has a set geographic latitude (Φ), a plumb line would thus align in the same direction as the Earth's gravitational field intensity (g)—that is, towards the centre—but only if the Earth was not rotating about its axis. The Earth's rotation results in a variation in the direction and magnitude of how far an instrument such as a plumb line will deviate from 'true vertical'. At any location on the surface, the total intensity of this variation is the sum of the gravitational field strength (g) and the centrifugal force of rotation (g_c) where:

$$g_c = \omega^2 x R_e \; x \cos(\Phi) \qquad \text{[eq. 3.2.23]}$$

and ω equals the angular velocity of the Earth's rotation (7.2921150×10^{-5} radians/second), and R_e is the equatorial radius of the Earth (6,378,000 m).

The result of the force of rotation is that a plumb line, used in the construction and alignment of an instrument such as a sundial, is drawn away from the vertical and any imaginary line extended downwards from the plumb line would not connect with the centre (C), but would form an angle (Θ) to the direction of the gravitational field (g) and pass through a different point along the line representing the gravitational force (g_r) instead.

The line that connects to g_r represents the value of the gravitational field intensity of the Earth, varied by rotation, at a certain point on the Earth's surface. The value of g_r can be expressed mathematically as:

$$g_r = g - \omega^2 x R_e \times \cos^2(\Phi) \qquad \text{[eq. 3.2.24]}$$

where g is the acceleration due to gravity (9.8 m/s^2), ω is the angular velocity of the Earth's rotation (7.2921150×10^{-5} radians/sec), R_e is the equatorial radius of the Earth (6,378,000 m), and Φ is the latitude of the location expressed in radians (degrees x $\pi/180$). Thus, for Alexandria, which Pinotsis places at 31°11'N (or 31.07° or 0.5422 radians):

$$g_r = 9.8\text{m/s} - \left(7.2921150 \times 10^{-5}\text{radians/sec}\right)^2 \times 6,378,000\text{m} \times \cos^2(0.5422)$$
$$= 9.8 - 0.00000000531 \times 6,378,000 \times 0.9999868368$$
$$= 9.766133266\text{m/s}^2 \qquad \text{[eq. 3.2.25]}$$

For the location of Alexandria as per Eratosthenes (30.96°N or 0.5404 radians), the value of g_r is almost the same:

$$g_r = 9.8\text{m/s} - \left(7.2921150 \times 10^{-5}\text{radians/sec}\right)^2 \times 6,378,000\text{m} \times \cos^2(0.5404)$$
$$= 9.8 - 0.00000000531 \times 6,378,000 \times 0.9999870107$$
$$= 9.766133260\text{m/s}^2 \qquad \text{[eq. 3.2.26]}$$

This shows that the intensity of the gravitational field in Alexandria (≈ 9.77 m/s^2) is slightly weaker than the gravitational field towards the Earth's centre (9.8 m/s^2) and this will skew a plumb line in Alexandria away from the vertical. In other words, the plumb line would not actually be perpendicular to the surface of the Earth.

This is given, mathematically, by the trigonometric relationship:

$$\sin(\Theta) / \left(\omega^2 \times R_e \times \cos(\Phi)\right) = \sin(\Phi) / g_r \qquad \text{[eq. 3.2.27]}$$

as the angle created by the variation from true vertical (Θ) is very small, $\sin(\Theta)$ is almost equal to Θ, and so:

$$\sin(\Theta) \gg \Theta = \left(\left(\omega^2 \times R_e\right)/2g_r\right) \times \sin(2\Phi) \qquad \text{[eq. 3.2.28]}$$

For Alexandria (based on a latitude of 31°11', or 0.5422 radians, as per Pinotsis), the angle of deviation is:

$$
\begin{aligned}
\sin(\Theta) &= \left(\left(\omega^2 \times R_e\right)/2g_r\right) \times \sin(2\Phi) \\
&= \left(\left(7.2921150 \times 10^{-5}\,\text{radians/sec}^2 \times 6,378,000\text{m}/2 \right.\right. \\
&\quad \left.\left. \times 9.766133266\text{m/s}^2\right) \times \sin(2 \times 0.5422)\right) \\
&= (0.00000000531 \times 6,378,000/19.53226653) \times \sin(1.0844) \\
&= (0.03386718 \times 19.53226653) \times 0.0189252205 \\
&= 0.00173390937 \times 0.0189252205 \\
&= 0.00003281461 \\
\Theta &= \sin^{-1}(0.00003281461) \\
&= 0.00188013866\,\text{radians} = 0.1077°
\end{aligned}
$$

Pinotsis calculates the angle of variation (Θ) to be 0.0885°.[125] The difference between the angle calculated here (0.1077°) and that determined by Pinotsis (0.0885°) is only 0.0192° and can be attributed to rounding differences in one or both of the calculations. Regardless of the result of the calculation, the variance in the angle away from the vertical would need to be subtracted from the angle that Eratosthenes determined in Alexandria in order to reach a 'true' angle created by a vertical *gnomon*. Based upon this, Pinotsis suggests that the circumference of the Earth was 253,192 *stadia*—a value that is not found in any ancient text unless it is rounded.[126] Pinotsis further claims that this value equates to a distance of 39,868 km which, in turn, assumes the use of a *stade* of 157.5 m in length by Eratosthenes—for which there is no evidence in the literary or archaeological record for the ancient Greek world, but which is close to the 160 m unit from the Messenian Standard.

Despite the claims made by Pinotsis, the question remains as to just how much of an effect the variation caused by the Earth's rotation would have had on Eratosthenes' experiment, the observations, and the calculated results (Fig. 56).

In Fig. 56, the image on the left represents a *gnomon*, skewed away from the vertical due to alignment using a plumb line, which is affected by planetary

[125] Pinotsis, 'Significance and Errors', p.60.
[126] Pinotsis, 'Significance and Errors', p.60.

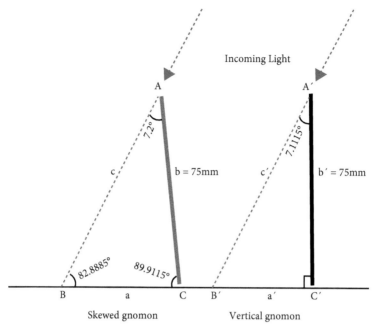

Incoming Light

A

7.2°

c

b = 75mm

82.8885°   89.9115°

B   a   C

Skewed gnomon

A′

7.1115°

c′

b′ = 75mm

B′   a′   C′

Vertical gnomon

Fig. 56 How the deviation of a gnomon away from the vertical affects the length of the shadow that is cast (with angles exaggerated).

rotation. This, according to Pinotsis, is what Eratosthenes would have used in his experiment. Cleomedes tells us that the angle created by the shadow cast by this *gnomon* (A) equalled 7.2° (1/50 of a circle). Angle C is equal to 90° minus the variation due to rotation—which Pinotsis calculates as 0.0885°— or 89.9115°. Angle B is therefore $180°-(7.2° + 89.9115°) = 82.8885°$. If side b represents the height of the *gnomon*, and the height of the replica created for this research is used as an example, then the length of side b equals 75 mm. The length of the shadow, side a, can be determined using the Law of Sines:

$$a/\sin (A) = b/\sin (B)$$ [eq. 3.2.29]

For the *gnomon* in Fig. 55:

$$a/\sin (7.2) = 75.0/\sin (82.8885)$$
$$a/0.1253332336 = 75.0/0.9923071095$$
$$a/0.1253332336 = 75.58143974$$ [eq. 3.2.30]
$$a = 0.1253332336 \times 75.58143974$$
$$= 9.47\text{mm} \pm 0.1\text{mm}$$

If the larger variation of 0.1077° is used, then the length of the shadow also calculates as 9.47 mm ±0.1 mm.

The image on the right of Fig. 56 represents a *gnomon* of the same height set to true vertical. The angle A' is the 7.2° figure reported by Cleomedes minus the variation due to planetary rotation determined by Pinotsis (7.2°–0.0885° = 7.1115°). As the *gnomon* creates a right angle with the surface it is sitting on, the length of the shadow (side a') can be determined using basic trigonometry:

$$\tan (7.1115) = a'/75.0\text{mm}$$
$$0.1247604163 = a'/75.0\text{mm}$$
$$a' = 75.0\text{mm} \times 0.1247604163$$
$$= 9.36\text{mm} \pm 0.1\text{mm} \qquad \text{[eq. 3.2.31]}$$

Thus, if only Pinotsis' values are used, the difference in the length of the shadow caused by planetary rotation is only (9.47–9.36 =) 0.11 mm (±0.2 mm). Using the larger calculated value for the variation (0.1077°) results in a shadow cast by the vertical *gnomon* with a length of 9.33 mm, and a difference between the shadows of (9.47–9.33 =) 0.14 mm (±0.2 mm). Regardless of which values are used, a variation of this size would be undetectable to the naked eye and would be far less than any inherent margin of error for the use of an instrument of measurement that had been designed and manufactured by hand.

Furthermore, as Eratosthenes did not actually have to observe a shadow in order to undertake his calculations, the effect of planetary rotation, and the deviation of any plumb line used, would have only had an influence on the design and construction of the sundial, and the placement of the lines of the *arachne* demarking the Solstices and equinox, when it was first made. Yet even here, the variation resultant from planetary rotation is far less than the thickness of the lines cut into ancient sundials (or even those etched by a laser in the modern replica), and is also less than the difference between the umbra and penumbra of the shadow itself. As such, in regard to the precision of the instrument itself, while there would have been a variation in the shadow length due to planetary rotation, it would have had no effect on the premise of Eratosthenes' experiment or on the instrument he would have used for his calculations.

This can be seen if the adjusted angles are converted into fractions. Using Pinotsis' variation of 0.0885°, the angle becomes 7.2°—0.0885° = 7.1115°. This is 1/50.6 of a circle (360°/7.1115°) which remains the 1/50 value given by Cleomedes if it is truncated to two significant figures. Similarly, the larger variant calculated here (0.1077°) adjusts the angle to 7.0923°, which is 1/50.75 of a circle, which also remains 1/50 if truncated.

Pinotsis, who assumes that Eratosthenes observed shadows in both Alexandria and Syene as part of his experiment, states that no error is assumed to have occurred in Syene because 'Eratosthenes placed the gnomon parallel to the direction of the solar rays and not in the direction of the plumbline'.[127] This is an interesting claim to make as not only is Syene not situated on the equator—which means that there would have been a variation in the gravitational field which would have affected a plumb line used in Syene, but also because a plumb line would have been used to create and calibrate a sundial in Syene and so would have imparted its errors onto the positioning of the *gnomon* and the marking of the lines of the *arachne*. However, because of its more southerly position, the errors would have been even smaller in Syene than they were in Alexandria.

It is also interesting to consider that, in both locations, everything that had been constructed using a plumb line such as temples, libraries, and monuments, would all be skewed in the same manner. This would even extend to the column, pedestal, or table that the sundial itself was sitting on. If the base of the sundial had also been constructed using a plumb line, then the base would incorporate the same variations as the *gnomon*, the lines of the *arachne* would be in their correct locations, and the instrument would contain no inherent errors due to planetary rotation at all. Regardless of whether there was an inherent error incorporated in the sundial, the alignment of its *gnomon* to 'vertical', or in any other piece of architecture that the sundial utilized, the calculations demonstrate that any variation away from true vertical resultant from planetary rotation was so small that it would have had no impact on Eratosthenes' experiment, and the values presented in the account of Cleomedes can be, in regard to this source of potential error, still taken as correct.

3.2.7 The Effect Of Atmospheric Refraction

The atmosphere of the Earth also plays a role in how any shadow is formed. The light that reaches the Earth from the Sun is 'bent' via the process of atmospheric refraction. This not only alters the angle that would be created by the shadow that was cast by, for example, the *gnomon* of the sundial in Eratosthenes' experiment, but also alters where in the sky an object like the Sun is perceived to be. However, modelling the amount of atmospheric refraction that would have occurred in Alexandria at midday on the Summer Solstice in 230 BC indicates that, while a variation in the shadow length would have

---

[127] Pinotsis, 'Significance and Errors', p.60.

occurred, this would have had no impact on the design or calibration of a sun-dial in this location, nor have any effect on the overall results of Eratosthenes' calculations as they are described by Cleomedes.

The process of atmospheric refraction is caused by the bending of rays of light from the Sun (or another source such as a star in a night sky) as they pass through the atmosphere of the Earth. As the light passes through the atmosphere, it encounters layers of ever-increasing density. This causes the ray of light to deviate from its original path and 'bend'. The amount of deviation in the light is dependent upon the angle at which the light encounters these layers of increasing atmospheric density (Fig. 57).

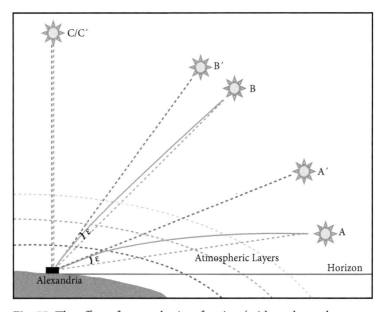

Fig. 57 The effect of atmospheric refraction (with angles and curvature exaggerated).

In Fig. 57, light from a Sun which is low on the horizon (A) at a place like Alexandria passes through the layers of the atmosphere at a relatively shallow angle. Each change in atmospheric density causes the light to bend until it reaches an observer (solid orange line). However, to that observer, the image of the Sun is seen only along the axis of the last bend in the ray of light. This then makes the Sun appear to be higher in the sky than it actually is (A'). The angle of deviation (ε) is the difference between the real (dotted orange line) and apparent (dotted red line) positions of the Sun.

When the Sun is higher in the sky (B) light passes through the atmosphere at a steeper angle, which results in a reduced bending of the light. As a result, the angle of deviation (ε) is smaller, and the difference between the real (B) and apparent (B') positions of the Sun is also smaller, when the Sun is at a higher elevation. When the Sun is at the zenith (C) the light passes directly through the atmosphere with almost no refraction and this makes the angle of deviation essentially zero, and the real (C) and apparent (C') positions of the Sun the same.

Meeus calculates that, when the Sun is at an angle of 45°, the amount of atmospheric refraction is about 1'. When the Sun is on the horizon, the amount of refraction is about 35".[128] The interesting thing about this level of atmospheric refraction at the horizon is that, because the Sun has an angular size of around 30', this means that, at sunrise and sunset, when the bottom limb of the Sun seems to be touching the horizon, this is only the apparent position of the Sun—the whole solar disk is actually below the horizon.

The amount of deviation due to atmospheric refraction also varies due to the latitude of location. Locations at higher latitudes (north or south), for example, experience a greater level of atmospheric refraction due to the angle that the light travels at to reach an observer, than occurs at latitudes closer to the equator (e.g. in Fig. 57, if the location of Alexandria was shifted to the right to a more 'southerly' location towards the equator, the lines connecting the different positions of the Sun to that location would pass through the atmospheric layers at a different angle). This means that, in regard to Eratosthenes' experiment, the angle of deviation would have been larger in Alexandria than it was in Syene. In Syene, at midday on the Summer Solstice, the Sun was almost directly overhead. This can be seen in the relation $\upsilon = \delta + 90° - \varphi$ (where $\upsilon =$ the Sun's altitude above the horizon; $\delta =$ the Sun's declination; $\varphi =$ latitude of Syene).[129] Taking δ to equal the angle of the ecliptic *ca.* 230 BC, or 23.7°, and the latitude of Syene to be 24.09°N, gives:

$$\upsilon = 23.7° + 90° - 24.09°$$
$$= 89.61°$$

[eq. 3.2.32]

If this is taken to be a reference to the centre of the solar disk, then an extra 0.25°, to account for the angular radius of the Sun, needs to be added to this value to give an angle of 89.86°. Thus, as has been discussed previously, the northern limb of the Sun was almost directly overhead (90°) at Syene at midday

---

[128] Meeus, *Astronomical Algorithms*, p.105.
[129] Formula taken from Pinotsis, 'Significance and Errors', p.59.

on the Summer Solstice. This would result in very little atmospheric refraction in Syene. However, with its more northerly location of 31.2°N (or 30.96°N as per Eratosthenes), the centre of the Sun would be at an angle of around 82.5° in Alexandria, and there would have been a more pronounced amount of atmospheric refraction. Some scholars have suggested that this alters the result of Eratosthenes' calculations.

Newton suggests that the refraction error in Alexandria on the Solstice was 7" (0.001944°).[130] Rawlins suggests that 7°12' (i.e. 7.2° or 1/50 of a circle) reportedly observed by Eratosthenes should be 7°25.5'—an adjustment of 16.5' which he attributes to both atmospheric refraction and stellar parallax.[131] Pinotsis, who takes a declination of the Sun on the Solstice of 23°43' (23.71°), suggests that the error in Alexandria is 7.93" or 0.132', and that this should be added to the stated 7°12' value to become 7°12.132'.[132] Tupikova states that the correction due to refraction was 7.6", but this was so small that it would not have had a major impact on Eratosthenes' calculations.[133]

The angle of deviation (ε) can be determined in a number of ways. Pinotsis, for example, uses the formula (ε) = 60.5"/tan(υ) where υ is the Sun's altitude above the horizon.[134] Replacing Pinotsis' value of 23.71° for the declination of the Sun in the formula $\upsilon = \delta + 90° - \varphi$, with the 23.7° value determined previously, gives an angle for the Sun at midday on the Summer Solstice in Alexandria (31.2°N) of 82.5°. Inserting this value into the formula (ε) = 60.5"/tan(υ) gives:

$$(\varepsilon) = 60.5/\tan(82.5)$$
$$\approx 7.96 \qquad\qquad \text{[eq. 3.2.33]}$$

This is not far from the value given by Pinotsis. Nassau states that this formula is only valid for calculations where $\upsilon > 25°$.[135]

Meeus offers that the angle of deviation (ε) due to atmospheric refraction can be determined when the apparent altitude (υ) of an object is >15° using the formula:[136]

$$\varepsilon = 58.294" \times \tan(90°-\upsilon) - 0.0668" \times \tan^3(90°-\upsilon) \qquad \text{[eq. 3.2.34]}$$

[130] Newton, 'Sources of Eratosthenes' Measurement', p.381.
[131] Rawlins, 'Geodesy', p.260.
[132] Pinotsis, 'Significance and Errors', p.59.
[133] Tupikova, 'Eratosthenes' Measurements', pp.224–225.
[134] See: Pinotsis, 'Significance and Errors', p.59; the formula is taken from J.J. Nassau, *Practical Astronomy* (New York, McGraw-Hill, 1948), p.70.
[135] Nassau, *Practical Astronomy*, p.70.
[136] Formula derived from Meeus, *Astronomical Algorithms*, p.106.

Setting υ = 82.5° for the apparent altitude of the Sun in Alexandria on the Summer Solstice yields:

$$\varepsilon = 58.294" \times \tan(7.5) - 0.0668" \times \tan^3(7.5)$$
$$= 28.294" \times 0.1316524976 - 0.0668" \times 0.00228185113$$
$$= 7.674550695 - 0.0015242765 \qquad \text{[eq. 3.2.35]}$$
$$= 7.673026419 \approx 7.67"$$

Meeus states that this formula is only for light in the yellow part of the spectrum—where the eye has maximum sensitivity.[137] Alternatively, Bennett uses the following formula to calculate the angle of deviation:[138]

$$\varepsilon = 1/\tan(\upsilon + (7.31/\upsilon + 4.4)) \qquad \text{[eq. 3.2.36]}$$

For Alexandria, this yields:

$$\varepsilon = 1/\tan(82.5 + (7.31/82.5 + 4.4))$$
$$= 1/\tan(82.5 + (7.31/86.9))$$
$$= 1/\tan(82.5 + 0.08411967779) \qquad \text{[eq. 3.2.37]}$$
$$= 1/\tan(82.58411967779)$$
$$= 7.682906689 \approx 7.68"$$

Thus, it seems that the angle of deviation due to atmospheric refraction in Alexandria on midday on the Summer Solstice of 230 BC was within the range of 7.96" to 7.67" (0.002211° to 0.002131°). This value, as per Pinotsis, would need to be subtracted from the 7.2° reported by Cleomedes. However, doing so results in a value in the range of 7.197789° to 7.197869°. Any of these values would round to 7.2°, and so the reported value given by Cleomedes remains unchanged.

Furthermore, Meeus outlines how the angle of deviation should be adjusted to account for the temperature and air pressure at the specific location by multiplying the calculated value of (ε) by (P/1010) x (283/273+T) where P is the air pressure at the location in millibars, and T is the temperature in Celsius.[139] Alexandria is located on the coast at sea level, and the average air pressure at sea level is 1,013.25 mb. The average temperate for June in Alexandria is 24.9°C.

---

[137] Meeus, *Astronomical Algorithms*, p.106.
[138] Formula derived from G.G. Bennett, 'The Calculation of Astronomical Refraction in Marine Navigation' *J. Inst. Navig.* 35 (1982), pp.255–259.
[139] Formula taken from Meeus, *Astronomical Algorithms*, p.106.

Consequently, the calculated values for (ε) should be adjusted by multiplying it by:

$$(1013.25/1010) \times (283/273 + 24.9)$$
$$= 1.003217822 \times 0.9499832158 \qquad \text{[eq. 3.2.38]}$$
$$= 0.9530400927$$

This makes the range for the angle of deviation, altered to reflect air pressure and temperature, from 0.002107° to 0.002031°, and the range for the altered value of Cleomedes 7.197893° to 7.197696°—both of which would still round to 7.2° or 1/50 of a circle.

Pinotsis suggests that the variation in the angle due to atmospheric refraction would have also resulted in the geographic distances between Alexandria and Syene being calculated as smaller than they really are.[140] If such determinations were made using sundials, then the small variation due to refraction, while present, would have been so small (especially in Syene) that it would have fallen far outside the inherent margin of error for the calibration, alignment and use of an instrument that had been constructed by hand. Even if such a variation was observable to the ancients, it was so small as basically to constitute a nil effect on any subsequent observations or calculations. This means that even the larger amount of atmospheric refraction that occurs in Alexandria would have had no impact at all on the results of Eratosthenes' experiment to determine the circumference of the Earth.

3.2.8 The Effect Of Operational And Instrumentational Error

A final source of potential error comes from the creation and use of the instrumentation that was a part of Eratosthenes' experiment—most notably the sundial. There are numerous sources of potential error that could have affected the creation of the sundial and Eratosthenes' use of it. However, a critical analysis of these sources of error, combined with an understanding of how they may, or may not, have influenced the operation or observation of the sundial, and how this may have affected Eratosthenes' calculations, demonstrates that these potential errors would have not altered the outcome of Eratosthenes' calculations in any way.

Whenever something is measured, there is always a potential for there to be an error inherent in the results of that measurement. The source of some of

---

[140] Pinotsis, 'Significance and Errors', p.59.

these potential errors is the design, construction, and calibration of any instrument involved in the measurement. The source of the other potential errors is based upon how those instruments are used:

Errors with the instrument: These are errors that are created by the misalignment, miscalibration, ageing, or weathering of the instrument.[141] In Eratosthenes' case we do not know much about the sundial that he used in his determination of the circumference of the Earth apart from the singular word that Cleomedes uses to describe it in his account—*skaphe*. As such, while we can be fairly certain as to the type of sundial that Eratosthenes used (see section 2.2), there is no way of determining how well it had been created or set up, or how old it was at the time that Eratosthenes would have been using it. However, based upon examples of other Hellenistic Period sundials that have been found in the archaeological record, it seems clear that the craftsmen who created such instruments in antiquity were able to do so with very high precision. In their examination of the Sundial of Amphiareion (which is configured differently to the bowl-shaped *skaphe* used by Eratosthenes, and has been dated to an earlier time period—350–320 BC) Tupikova and Soffel conclude that 'provided that the latitude of the place where the find was made coincides with the latitude where the sundial was mounted, the measured error [is] about 2° due to the instrument not being mounted correctly.[142] It would be expected, considering the expense the Ptolemies went to in order to secure books for the Library, that such expense was similarly spent on procuring some of the finest instruments for the library as well. This would then suggest that the sundial that Eratosthenes used in his experiment had been expertly crafted, calibrated, and aligned. The seeming accuracy of Eratosthenes' calculations, and of later stellar observations made in Alexandria by the astronomer Ptolemy, would suggest that this was the case. This would, in turn, mean that there was very little error within Eratosthenes' calculations associated with the sundial itself.

Errors with the item being measured: Another issue stems from the item itself that is being measured and issues that would arise if that item was moving or changing in shape and/or size—such as attempting to measure the dimensions of a cube of ice while inside a warm room.[143] For Eratosthenes, the

---

[141] S. Bell, *A Beginner's Guide to Uncertainty of Measurement* (Middlesex, National Physical Laboratory, 1999), p.7.
[142] See: Tupikova and Soffel, 'Modelling Sundials', pp.8–13; see also: Schaldach, 'The Arachne of Amphiareion', pp.435–445.
[143] Bell, *Uncertainty of Measurement*, p.7.

shadow that forms the basis of his calculations was moving, but this would not have posed much of a problem. As the Sun traces its path across the sky, this would constantly alter the angle that light would reach the *gnomon* of a sundial and make its shadow move. However, the Sun moves across the sky at approximately 15° per hour, or 0.25° per minute, or around 0.004° per second. Consequently, at the moment that any reading was going to be taken on a sundial, the shadow would hardly be moving at all, and an accurate reading can be taken. This was observed during the re-creative experiments undertaken with the replica sundial where the tip of the shadow could have been considering to be indicating noon, or very close to it, over a period of ten minutes (see section 2.5). Furthermore, this source of potential error is completely removed if Eratosthenes did not actually observe a shadow at all, but merely based his calculations on the distance from directly under the *gnomon* to the point of intersection of the noon and Solstice lines on the sundial. In this way, there would be no potential error due to the object being measured at all in his experiment.

Errors in the measurement process: These errors arise when the measurement itself is difficult to undertake. Bell uses the example of naturalists attempting to weigh small, fuzzy, animals and how, if they were not cooperative, taking such a measurement may be problematic.[144] Similar to the effect of any potential error based upon the item being measured, the shadow in Eratosthenes' experiment was moving, but it was moving so slowly that this would not have posed any issue within the measurement process itself. Additionally, if Eratosthenes was simply measuring the distance from beneath the *gnomon* to the intersection of the noon and Solstice lines on the face of the sundial, this point would not be moving at all, thus negating this potential source of error. The process of Eratosthenes' measurements would have only been to record the distances between two points on the static surface of a sundial, and then determine the ratio of that distance to the circumference of a full circle.

'Imported' uncertainties: These are errors that are built into the uncertainty of the measurements being made due to an inherent uncertainty in the calibration of the instruments being used.[145] Many of these types of error are based upon the units that the instrument being used is calibrated in. A *uniformly*

---

[144] Bell, *Uncertainty of Measurement*, p.7.
[145] Bell, *Uncertainty of Measurement*, p.7.

distributed uncertainty (α), for example, is half of the smallest gradation of the instrument.[146] A *standard uncertainty* (υ) is $\alpha/\sqrt{3}$.[147] Thus, for an instrument where the smallest gradation is 1.0 mm, such as a ruler, the *uniformly distributed uncertainty* is α = 0.5 mm, and the *standard uncertainty* is $\upsilon = \alpha/\sqrt{3}$ = 0.5/1.732 ≈ 0.29 mm. When a measurement using such an instrument is taken, the *uniformly distributed uncertainty*, for example, can be accounted for by writing the result as 'value ± α'.

Such a potential source of error raises an interesting question: what was the smallest gradation on the sundial and, therefore, the *uniformly distributed uncertainty* in Eratosthenes' experiment? Circles in Eratosthenes' time were divided into sixty segments called *hexacontades*. This would make each segment 6° in size. While the division of a circle in this way was not marked on the curved face of sundials in antiquity, if one *hexacontade* is taken as the smallest gradation, then Eratosthenes' *uniformly distributed uncertainty* is half of that value, or 3°. This seems overly large for a margin of error when using such an instrument.

There were gradations marked on the sundial, but these produce even a larger uncertainty if taken as the smallest. The lines of the *arachne* that mark the hours, for example, while a set of gradations that separate the face of the sundial into the required parts of each day, are separated by around 18° near the noon line of the Summer Solstice, and are separated by smaller distances for the hours at the beginning and end of each day. As such, there is no uniformity in their positioning. Consequently, no value of α can be determined based upon a unit of common size. However, does this then mean that Eratosthenes' *uniformly distributed uncertainty* is half an hour of time? Again, this would seem overly large for an instrument that could be used to a very high level of accuracy.

The final option for a smallest gradation of the sundial (and that is assuming that there even is one) is the thickness of the lines of the *arachne* itself. Such a conclusion would make sense. The time of noon, for example, was determined when the apex of the shadow cast by the *gnomon* of the sundial sat upon the noon line on a particular day. If the apex of the shadow were not sitting on the noon line, then the observer would know that it was either before or after noon by how far the apex was from the noon line. If the tip of the shadow were closer to the line for the previous hour, the user would know that it was some time after 11 a.m., but not yet noon, and the

---

[146] Bell, *Uncertainty of Measurement*, p.19.
[147] Bell, *Uncertainty of Measurement*, pp.12–13, 19.

amount past the previous hour would be determinable by the shadow's distance from the line marking 11 a.m. The lines of the *arachne* on the replica sundial made for this research are approximately 0.5 mm wide. If taken as the smallest gradation of the sundial, the *uniformly distributed uncertainty* of that instrument is 0.25 mm. The circumference of a whole circle with a radius of the size of the replica *gnomon* (75 mm) is 471.25 mm ($2\pi r$). This makes a *uniformly distributed uncertainty* of 0.25 mm for this instrument very small (0.05 per cent). A variance of only 0.05 per cent would be almost imperceivable to the naked eye and would correlate with the perceived accuracy of such a sundial.

Additionally, if Eratosthenes was not observing a shadow at all, this then negates this small potential error. However, there would be the potential for error in the measurement of the distance from beneath the *gnomon* to the point of intersection of the noon and Summer Solstice line, and from beneath the *gnomon* to the horizon of the bowl of the sundial, that Eratosthenes seems to have employed. For Eratosthenes, there is no measuring equipment per se— just some marks placed onto a thin strip of material to record distances and a calculation of ratios. While there would be a small error in this based upon the correct placement of these marks on the material, and the thickness of the mark itself, the fact that Eratosthenes was able to determine the ratio to a high level of mathematical accuracy suggests that such errors were negligible.

Errors due to operator skill: Related to the 'imported uncertainties' are any errors that were resultant from Eratosthenes himself and his ability to correctly and accurately place, for example, the marks on the material for his measurements that were used in his ratio calculations. The main one of these potential errors is a so-called *parallax error*. With any naked-eye observation of distance, if the observer is not directly in line with the thing being measured, when such markings are made, they may be out by a small amount.[148] This is true regardless of whether the thing being measured is a linear distance or an angle. Newton suggests that if the sundial was graduated and aligned accurately, the error in the reading may be less than one arc-minute, but any other reading error may be as high as 6.25 per cent.[149] Applying Newton's largest 6.25 per cent margin of error to the length of the shadow cast by the replica sundial ($\approx$9.5 mm) gives a resultant value of 9.5 mm x 0.0625 = 0.6 mm. Thus the shadow is 9.5 mm ±0.6 mm. Interestingly, 0.6 mm is also the width of the penumbra of the shadow cast by the *gnomon* on the replica sundial (see

[148] Bell, *Uncertainty of Measurement*, p.8.
[149] Newton, 'Sources of Eratosthenes' Measurement', p.381.

section 3.2.5), and the approximate width of the lines of the *arachne*, and either of these may be where Newton has obtained this value for potential error from. Regardless, the value for this potential error is so small, and Eratosthenes' calculation mathematically correct, which suggests that *parallax errors* did not have a significant impact on any of his results.

Sampling errors: Bell suggests that, in order to limit the impact of any potential source of error, how the measurement is taken must also be considered. Bell uses the example of someone measuring samples from a production line, but only taking samples that were made on one particular morning. This would then not be representative of the total output of the production line as a whole.[150] For accuracy, the measurements taken must be representative of what is being assessed. Such a source of potential error does not apply to Eratosthenes. The intersection of the noon and Summer Solstice lines on the sundial is a precise, and unmoving, point of reference that is a clear representation of what he needed to measure.

Environmental errors: The environment in which a measurement is undertaken will also influence the result.[151] For Eratosthenes, a cloudy day on the Summer Solstice, for example, would have affected his ability to observe a shadow cast by the *gnomon* of a sundial in Alexandria. This, in part, is most likely why Eratosthenes employed a method that did not require the shadow to be observed at precisely the right time on the correct day of the year. Rather, by simply determining the ratio of the distance from *gnomon* to intersection of the noon and Solstice lines, compared to the distance from *gnomon* to horizon, Eratosthenes could undertake his measurements and calculations at any time, on any day of the year, and in any conditions, and so negate this source of potential error.

Any errors in Eratosthenes' experiment would have been 'systematic' (i.e. the same influences affect any repeat measurements), rather than 'random' (i.e. repeated measurements include random sources of error each time).[152] If Eratosthenes did not observe a shadow, but only based his calculations on the location of the noon and Summer Solstice line on the sundial, then there is very little error at all (assuming the sundial is made accurately). Regardless of the angle that the sundial was observed at, the point of intersection of these lines of the *arachne* does not alter, and the only potential error would be the angle when a slip of paper (or other material) was used to measure the distance from the *gnomon* to the intersection. However, due to how accurate Eratosthenes'

---

[150] Bell, *Uncertainty of Measurement*, p.8.
[151] Bell, *Uncertainty of Measurement*, p.8.
[152] Bell, *Uncertainty of Measurement*, p.9.

results from his measurement of the sundial in Alexandria were, this then suggests that very little (or no) operational and instrumentational error impacted his calculations.

3.2.9 The Total Effect Of The Errors

There were many errors in the assumed knowledge that Eratosthenes was working from in his calculation of the circumference of the Earth, and the potential for errors in his methodology. However, a critical review of these errors and potential errors demonstrates that all of them would have had a negligible impact on the results (Table 2):

As can be seen from Table 2, all of the errors, or potential errors, within either Eratosthenes' assumed knowledge or his methodology, have little to no impact on the overall result at all. In the case of the cities of Alexandria and Syene not sitting on the same longitudinal meridian, the only impact that this has is that midday would occur twelve minutes later in Alexandria than it did in Syene. In regard to Syene not sitting on the Tropic of Cancer, this would have altered Eratosthenes' angle to approximately 1/53 of the circumference of a circle, but Eratosthenes' was working on the assumption that Syene was on the Tropic, and that no shadows were cast there at midday on the Summer Solstice, and so his determination of an angle of 1/50 of a circle is fundamentally correct as this is the angle that would have been calculated for the Tropic itself.

Three of the errors—the effect of planetary oblateness, the effect of non-parallel light rays reaching the Earth from the Sun, and the effect of the Earth's rotation—while resulting in small variances to the angle of the shadow cast in Alexandria on their own, essentially cancel each other out, with a resultant, overall, variance to the angle from the sum of these three errors of $-0.0077°$. This variance is too small to have any impact at all on any naked-eye observations that Eratosthenes may have made, and/or the calibration of the sundial that he was using as part of his experiment.

The difference between the umbra and penumbra of the shadow cast in Alexandria at midday on the Summer Solstice is also small (0.6 mm). This too is unlikely to have had a significant impact on either any naked-eye observations made by Eratosthenes, or the configuration of the instrument he was using. All of the potential operational or instrumentational errors associated with Eratosthenes' experiment similarly yield little influence on the overall result.

This leaves only the alteration of the angle of incoming light due to refraction of the Earth's atmosphere to have an impact on Eratosthenes' determination

Table 2 The cumulative effect of errors in Eratosthenes' calculations.

Error	Size of error (and % of 7.2°)	Effect on result
Alexandria and Syene not sitting on the same longitudinal meridian	Noon occurs in Alexandria 12 min after Syene.	None
Syene not sitting of the Tropic of Cancer	Angle to Syene is 1/53. Eratosthenes assumes Syene is on the Tropic so angle remains 1/50.	None
The effect of the oblateness of the Earth	+0.03° adjustment to the angle (+0.42%)	None (adjustment negated by others)
The effect of non-parallel Light rays	+0.07° adjustment to the angle (+0.97%)	None (adjustment negated by others)
The effect of the components of the shadow	+0.0° umbra/penumbra difference is too small (0.00%)	None (angle to umbra = 7.2°)
The effect of the Earth's rotation	−0.1077° adjustment to the angle (−1.5%)	None (adjustment negated by others)
The effect of atmospheric refraction	−0.002031° adjustment to the angle (−0.03%)	None (adjustment is too small)
The effect of operational and instrumentational errors	Negligible errors	None
TOTAL (combined)	**−0.009731° (−0.13%)**	**None (adjustment is too small)**

of the angle of the shadow cast by the *gnomon* in Alexandria. This variance is also small: an adjustment of the 7.2° angle by −0.002031°—which would have also been too small to have an individual impact on the results of Eratosthenes. Even when combined with the remainder that has been left over from the nega-tion of other errors (−0.0077°), this results in a total variance to Eratosthenes' angle of −0.009731°. This could not have been applied to calculations based upon naked-eye observations and measurements, and even if it had been, the resultant angle would have been (7.2° − 0.009731° =) 7.190269° which would round to 7.2°—the same as the 1/50 of a circle value reported by Cleomedes in his account of Eratosthenes' calculations. Additionally, all of the 'errors' were unknown to Eratosthenes at the time, and are only determined with the benefit of hindsight—looking into the potential sources of variance in Eratosthenes' readings and results. However, even if they were known at the time, their combined impact is so small as to not contribute to any significant variance to the outcome of Eratosthenes' calculations in Alexandria.

If Eratosthenes had been working in a more northerly location—such as Athens, or at the rival library in Pergamon—some of these errors may have been large enough for Eratosthenes to have noticed them and/or had an impact on his overall result. However, as luck would have it, Eratosthenes was working in just the right place for the various errors inherent in the design of his instrument, and in his calculations, to have had such a minimal effect on his premise, experiment and results as to make them essentially zero. Consequently, the values reported by Cleomedes must be taken as accurate. This means that the only way to gauge comprehensively the accuracy of Eratosthenes' calculation of the circumference of the Earth in the third century BC is to determine the size of the unit of measure in which he gave his results.

4

The *Stade*

4.1 The Size Of The *Stade*

The key to understanding the accuracy of Eratosthenes' calculations lies in correctly converting the unit of measure that he used, the *stade* (στάδιον), into a
modern equivalent. The quest to successfully do this has been the foundation
of a scholarly debate which has been ongoing almost since the time that the
results were announced—but the debate really gained momentum from the
early twentieth century onwards. A review of the available evidence demonstrates that many of the earlier attempts to determine the size of Eratosthenes'
circumference of the Earth—and therefore the accuracy of his calculations—
have been based upon incorrect conversions of the *stade* into modern units
of measure. This is partially due to the fact that there were several systems of
measurement in use across the ancient Greek world which incorporated a unit
called the *stade.*

As with modern systems of measurement, whether imperial or metric, measurements in the ancient Greek world were based upon a series of units of
ever-increasing size. The historian Herodotus, writing in the fifth century BC,
states that the smallest unit of measure was the *daktylos* (δάκτυλος) which represented the width of a finger.[1] In a later passage, Herodotus outlines how 4 x
daktyloi equalled 1 x 'palm' (παλαστή), 4 x 'palms' equalled 1 x 'foot' (πούς),
and 6 x 'palms' equalled 1 x 'cubit' (πῆχυς)—which represented the distance
from the tip of the elbow to the tip of the fingers.[2] A distance of 600 'feet' made
1 x *stade*,[3] as did 400 'cubits'[4] (Table 3).

The *stade* represented the distance between the start and finish lines that
competitors would have to run during a sprint race held within a 'stadium'
(hence the name of the unit). Stadiums in ancient Greece were elongated
oblongs with semi-circular ends on their short sides. Some of the best examples
of this style of stadium are found at the archaeological sites at Olympia, Delphi
and Epidaurus in Greece, and Aphrodisias in Turkey (see following).

---

[1] Hdt. 1.60; see also: *AP* 12.5.
[2] Hdt. 2.49; on the cubit see also: *Suda s.v.* Πῆχυς; *Suda s.v.* Στάδιον.
[3] *Suda s.v.* Μίλιον; *Suda s.v.* Στάδιον.
[4] *Suda s.v.* Πλέθρον.

Eratosthenes and the Measurement of the Earth's Circumference (c.230BC). Christopher A. Matthew, Oxford University Press.
© Christopher A. Matthew (2023). DOI: 10.1093/oso/9780198874294.003.0004

Table 3 Ancient Greek units of measure.

1 x *daktylos*	= width of a finger
4 x *daktyloi*	= 1 x palm
4 x palms	= 1 x foot
6 x palms/1.5 feet/24 *daktyloi*	= 1 x cubit
600 feet/400 cubits	= 1 x *stade*

Sprint races were run along the long-axis of the stadium and could be of single or multiple lengths in distance.[5] Runners would begin at a set of 'blocks' that stretched across one end of the stadium—usually at the point where the shape of the end of the stadium began to curve—and run to a finish line at the other end—also usually at the point where the stadium began to curve to ensure that all competitors had to traverse the same distance. In multiple length races, the competitors went around a 'turning pole' at each end of the stadium (Fig. 58).

Other examples of ancient Greek stadia, such as those in Athens and Messene (see following) had one end squared off. In these venues, the competitors would race from the squared end towards a finish line at the rounded

Fig. 58 The stadium at Olympia, Greece. The line of stones at the end of the 'track' in the foreground are the starter's 'blocks' for the running races. The finish line, one *stade* length away, is just beyond the location of the people in the distance.[a]

[a] Author's photo.

[5] *Suda s.v. Στάδιον.*

end. It was this length that the runners raced across upon which the length of the *stade* was based.

Despite the relatively straightforward system of measurements outlined in the ancient textual evidence, and the seemingly basic concept upon which the length of the *stade* was based, it is from this point on that problems are encountered when trying to convert the results of Eratosthenes' calculations into a modern equivalent. Walkup, for example, states that the question 'what is a *stade*?' has no easy answer.[6] This is because of a number of inter-related factors.

Firstly, the city-states (*poleis*) of ancient Greece were not unified into a single country. Rather there was approximately 160 independent 'states' in the ancient Greek world—some large like Attica (the city-state containing Athens) and Laconia (containing Sparta), and some small like the island of Keos which, while only being 130 km² in size, was divided up into four separate *poleis*.[7] Each *polis* could have its own form of government, its own religious calendar and customs, its own dialect, its own coinage, and its own system of standard weights and measures. Across Greek history, as states merged, became allies, or were conquered, things such as units of measure could have been 'standardized' across all connected territories. Such a thing happened in the mid-fifth century BC when the city-state of Athens initiated the so-called 'coinage degree', which standardized all monies and measures across the members of its growing empire.[8] The result of this isolated existence for many of the regions and city-states of ancient Greece was that there were many different systems of measurement in use, at different times across Greek history, all of which incorporated a unit called a *stade*, but that unit may be of a different length compared to a unit of the same name that was in use elsewhere in Greece at the same time. Additionally, depending upon the region or *polis*, the units of measure in use may also be different depending upon which specific time in that city's history is being examined. This then has a direct impact on any attempt to determine the accuracy of Eratosthenes' calculations.

---

[6] Walkup, 'Mystery of the Stadia' *s.v. How Long is a Stade?*

[7] See: J.V.A. Fine, *The Ancient Greeks: A Critical History* (Cambridge, Cambridge University Press, 1983), p.52.

[8] Ar. *Av.* 1040–1041. See also: inscription IG I³ 1453 in R. Meiggs and D. Lewis, *A Selection of Greek Historical Inscriptions to the End of the Fifth Century BC* (Oxford, Oxford University Press, 2004) #45; the date of the coinage decree has been a topic of considerable scholarly debate.it was initially thought to have been issued around 430 BC. However, stylistic considerations of a fragmentary inscription from the island of Kos have seen this date be revised to between 449 BC and 445 BC—which corresponds to the commencement of the building program in Athens under the leadership of Perikles. However, another fragmentary inscription from Hamaxitos, when it became part of the Athenian Empire in 427 BC, has forced some scholars to revise the date of the 'Coinage Decree' to 425 BC. For example, see: Fine, *The Ancient Greeks*, p.367; H.B. Mattingly, 'The Athenian Coinage Decree' *Historia* 10 (1961), pp.148–188; H.B. Mattingly, 'Epigraphy and the Athenian Empire' *Historia* 41 (1992), pp.129–138; H.B. Mattingly, 'New Light on the Athenian Standards Decree' *Klio* 75 (1993), pp.99–102; Matthew, *An Invincible Beast*, p.711.

O'Neil, for example, states that there were three units called a *stade* in use in the time of Eratosthenes, but which one he used is unclear.[9] Dicks, on the other hand, outlines claims for the existence of *stadia* of least six different lengths.[10] This confusion is compounded by the fact that, due to different states using units of measure of different sizes, and other various reasons such as the available topography, the very structure that the *stade* is named after, the stadium, also came in different sizes.

The most famous stadium in Greece was that at Olympia in the western Peloponnesian state of Elis—site of the religious and athletic festival held every four years in honour of the god Zeus: The Olympic Games. Every four years, competitors and spectators from all over Greece would travel to Olympia for the games. The 'blue ribbon' event of the festival was the *stadion* race, where entrants from many different *poleis* ran the length of the stadium. As such, most of the Greek world would have been familiar with the distance that had to be covered from one end of the stadium to the other. Eratosthenes himself would have most likely been familiar with the size of the stadium at Olympia considering he wrote a work on Olympic victors. In the case of the stadium at Olympia, the distance between the start and finish lines for the foot race was 191 m[11] (Fig. 59).

Fig. 59 Satellite image of the stadium at Olympia, Greece, with the length of the *stade* between the start and finish lines.[a]

[a] Image created using Google Earth.

[9] O'Neil, *Early Astronomy*, p.59.
[10] D.R. Dicks, *The Geographical Fragments of Hipparchus* (London, Athlone Press, 1960), p.43.
[11] M. Andronikos, *Olympia* (Athens, Ekdotike Athenon, 1999), p.41.

Thus, the 'Olympic *stade*' for their system of measurements can also be considered to be 191 m. This unit was used in other parts of the Greek world as well. In Athens, the modern Kallimarmaro (or Panathenaic Stadium), built in 1896 for the first modern Olympic Games, was constructed over an earlier ancient stadium from the fourth century BC (which was then built over by the Romans in the second century AD). If the modern stadium was erected on the same layout as the earlier ones, this too was built using the 'Olympic *stade*' of 191 m (Fig. 60).

Fig. 60 Satellite image of the Panathenaic stadium at Athens, Greece, with the length of the *stade* between the start and finish lines.[a]

[a] Image created using Google Earth.

Similarly, the same unit of measure can be seen in the stadium at Aphrodisias in Western Turkey, an area that was once a colony of Athens (Fig. 61).

It was not only certain stadia that were built based upon a measurement of this size. The temples on the Acropolis of Athens, for example—the construction of which was begun around 450 BC—were also built using a system of measurements that incorporates a cubit of around 48 cm (and therefore a *stade* of 191 m).[12] Similarly, at Isthmia near Corinth, temples there were

---

[12] See: W.B. Dinsmoor, 'The Basis of Greek Temple Design in Asia Minor, Greece and Italy' *Atti VII—Congresso Internazionale di Archologia Classica I* (Rome, L'Erma di Bretschneider, 1961), pp.358–361. For the building program of Pericles, see: Plut. *Per.* 12.

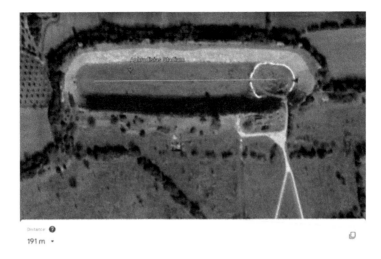

Distance
191 m ▾

Fig. 61 Satellite image of the stadium at Aphrodisias, Turkey, with the length of the *stade* between the start and finish lines.[a]

[a] Image created using Google Earth.

also constructed using a 48 cm cubit/191 m *stade* system of measurements.[13] This shows that the 'Olympic Standard' of measurements was in use by many different states across the Greek world.

Elsewhere in the Greek world, stadiums were built to other lengths. The one at Delphi, for example, built in the fifth century BC and site of the four-yearly Pythian Games in honour of Apollo, is based upon a *stade* length of 177 m (Fig. 62).[14]

The stadium at Epidaurus in central Greece, from the fourth century BC, a stadium with a squared running area and clearly defined start and finish lines, is also built on a 177 m *stade* (Fig. 63).[15]

---

[13] O. Broneer, *Isthmia Vol.I: Temple of Poseidon* (Princeton, American School of Classical Studies in Athens, 1971), pp.175–180.

[14] One of the other four-yearly festivals was the Nemean games, held in honour of Poseidon, at Nemea in the northern-eastern Peloponnese. Part of the remains of the stadium at Nemea have been excavated, and the length of the excavated part is 130 m in length. Unfortunately, the remainder of the stadium is covered with modern farmland, and has not been excavated, so the exact size of the stadium is unknown. The last of the four-yearly festivals was the Isthmian games held in Corinth (also in honour of Poseidon). No remains of the stadium there have been found due to the presence of modern buildings and farms. For a discussion on the stadium in Athens, see W.B. Dinsmoor (*The Architecture of Ancient Greece: An Account of its Historic Development* (New York, Biblo and Tannen, 1973), p.250) who describes the fourth-century stadium as being 850 feet ($\approx$259 m) in length overall.

[15] Charitonidou (*Delphi* (Aharnes, Hesperos, 1978), p.38) outlines how the entire stadium was 181.08 m in length.

Fig. 62 Satellite image of the stadium at Delphi, Greece, with the length of the *stade* between the start and finish lines.[a]

[a] Image created using Google Earth.

Fig. 63 Satellite image of the stadium at Epidaurus, Greece, with the length of the *stade* between the start and finish lines.[a]

[a] Image created using Google Earth.

While the stadium at Messene, in the *polis* directly south of Elis and the site of Olympia, is 160 m in length (Fig. 64).

Fig. 64 Satellite image of the stadium at Messene, Greece, with the length of the *stade* between the start and finish lines.[a]
[a] Image created using Google Earth.

This shows that, even in adjoining states, their system of measurements could be very different. Even within the limited corpus of archaeological evidence presented here, there are clearly three different lengths for the *stade*:

The Olympic/Athenian *stade*	191 m
The Delphic/Epidauran *stade*	177 m
The Messenian *stade*	160 m

There are, however, other sources that provide details of yet other *stadia* of different length that were also in use at different times in the ancient Greek world. A metrological relief found on the island of Salamis in 1985 (and now housed in the Piraeus Museum—#5352), for example, contains representations of a *daktylos* of 2.0 cm, a 'rule' of 32.2 cm, and a cubit of 48.7 cm (Fig. 65).[16]

Such representations would correspond to a system of measurements that incorporated a *stade* of around 191 m—the same that was used in the stadiums

[16] I. Dekoulakou-Sideris, 'A Metrological Relief from Salamis' *AJA* 94.3 (1990), pp.445–451. See also: B. Slapšak, 'The 302 mm foot measure on Salamis?' *DHA* 19.2 (1993), pp.119–136; M.W. Jones, 'Doric Measure and Architectural Design 1: The Evidence of the Relief from Salamis' *AJA* 104.1 (2000), pp.73–93.

(a)

(b)

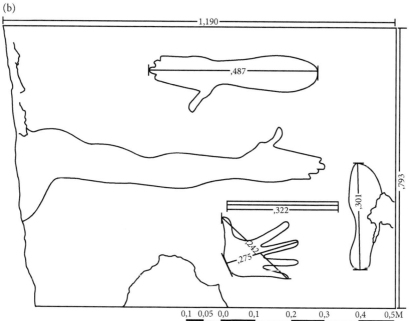

Fig. 65 A photograph (top) and line drawing (bottom) of the metrological relief found on Salamis (now. Piraeus Museum #5352).[a]

[a] Images taken from Dekoulakou-Sideris, 'Metrological Relief', pp.446–447.

at Olympia, Aphrodisias and Athens. A *daktylos* of 2.0 cm should result in a cubit of exactly 48 cm and Dekoulakou-Sideris suggests that any variances in the measurements on the relief, which are only in millimetres, are the result of a lack of precision stone-cutting by the craftsmen who made it.[17]

A similar relief, now in the Ashmolean Museum at Oxford, contains a 'foot' of 29.6 cm—corresponding to a *daktylos* of 1.8 cm, a cubit of 44.4 cm, and a *stade* of around 177 m—the system used in stadiums at Delphi and Epidaurus (Fig. 66).[18]

Fig. 66   Photograph of the metrological relief in the Ashmolean Museum, Oxford.[a]

[a] Image taken from Dekoulakou-Sideris, 'Metrological Relief', p.448.

Michaelis refers to the foot in this relief as 'Samian' and dates the relief to when the island of Samos became part of the Athenian Empire around 440 BC.[19] However, it is interesting to note that the Athenian Empire never seems to have used a system of measurements incorporating a 177 m *stade*. It is most likely that Athens may have operated on a different system—incorporating a cubit of 45 cm and a *stade* of 180 m—prior to the 'coinage decree', and then 'standardized' all of the measurements in its territories to a 48 cm cubit/191 m *stade* 'Olympic Standard' with its inception—and this system was then used in such things as the building of the temples on the Acropolis.[20]

Elsewhere in the ancient Greek world, and at various other times, a *stade* of different lengths was also in use. The metrological relief from Salamis, for example, also contains the representation of a 'foot' of 30.2 cm.[21] This equates to a *stade* of around 180 m in length, and so seems to be from a different set

[17] Dekoulakou-Sideris, 'Metrological Relief', p.449.
[18] See: A. Michaelis, 'The Metrological Relief at Oxford' *JHS* 4 (1883), pp.335–350; E. Fernie, 'The Greek Metrological Relief in Oxford' *AntJ* 61 (1981), pp.255–261; H. Ben-Menahem and N.S. Hecht, 'A Modest Addendum to the Greek Metrological Relief in Oxford' *AntJ* 65 (1985), pp.139–140.
[19] Michaelis, 'The Metrological Relief', p.339; for the annexation of Samos by Athens, see: Thuc. 1.76.
[20] This would then suggest that the Athenian 'coinage decree' occurred around 450 BC—contrary to the hypotheses of some scholars.
[21] Dekoulakou-Sideris, 'Metrological Relief', p.449.

of measures compared to the others given in the relief. In an examination of ancient construction methods, Broneer also identified a measurement of around 30.2 cm which he termed a 'Hellenistic Foot' because he believed it came into use in the time of Alexander the Great in the late fourth century BC.[22] However, other evidence indicates that this system of measurements was much older.

For example, the heavily armed infantryman of Classical Greece (the *hoplite*) carried a round shield (the *aspis*) approximately 90 cm in diameter. Formations of hoplites could be deployed in one of a number of different 'intervals'. The writer Asclepiodotus states that formations could be arranged in either a 'close order', with interlocked shields, with each man one cubit (*pēchus*), from those around him on all sides (τὸ πυκνότατον, καθ' ὃ συνησπικὼς ἕκαστος ἀπὸ τῶν ἄλλων πανταχόθεν διέστηκεν πηχυαῖον διάστημα); an 'intermediate order', or 'compact formation', with each man separated by 2 *pecheis* on all sides (τό τε μέσον, ὃ καὶ πύκνωσιν ἐπονομάζουσιν, ᾧ διεστήκασι πανταχόθεν δύο πήχεις ἀπ' ἀλλήλων); and an 'open order' with each man separated by 4 *pecheis* (τό τε ἀραιότατον, καθ' ὃ ἀλλήλων ἀπέχουσι κατά τε μῆκος καὶ βάθος ἕκαστοι πήχεις τέσσαρας).[23]

Importantly, for the understanding of measurements, due to the way that the *aspis* was configured and carried into battle, for hoplites to create a 'close-order' formation with interlocking shields as is described in the ancient texts, each man must have occupied a space approximately 45 cm by 45 cm. The more widely spaced 'intermediate order' was double that of the 'close order': 90 cm, or the width of the shield, while the more dispersed 'open order' was double that again, 180 cm per man.[24] Consequently, military formations in ancient Greece were based upon the use of a system of measurements that included a 45 cm cubit—and therefore a 180 m *stade*. Interestingly, the evidence we have for ancient Greek warfare shows that the soldiers from almost all city-states would have used this approximate spacing for their formations regardless of what the 'standard' was in their respective regions for things such as construction and other measurements. Thus, a system that used a 45 cm cubit/180 m *stade* was the 'Pan-Hellenic' system in use across many of the Greek states during the Classical Period. As this system would incorporate a 'foot' of around 30 cm, this then suggests that the so-called 'Hellenistic foot'

[22] Broneer, *Isthmia*, pp.173–180.
[23] Asclep. *Tact.* 4.1.
[24] For a detailed examination of how formations of Greek *hoplites* in the Classical Period were based on a cubit of 45 cm—especially the interlocked 'shield wall', and how the other formations were based on this same system of measurements (which includes a 180 m *stade*), see: Matthew, *Storm of Spears*, pp.53–57, 179–196.

identified by Broneer, and the representations on the metrological reliefs from Salamis and in the Ashmolean Museum, are actually part of this 'Pan-Hellenic' Classical Standard. As such, Broneer seems to have misidentified the unit, and therefore misdated the relief, while such a system would correlate with the dating of the so-called 'Samian' foot by Michaelis.

Many scholars argue for, or use in their examinations of the later Hellenistic Period, the continued use of the 45 cm cubit Pan-Hellenic system of measures (also sometimes referred to as the 'Early-Attic Standard') with its associated 180 m *stade* into the fourth century.[25] However, such a Pan-Hellenic system of measurements, and many of the more regional ones that were in use, appear to have been altered, in favour of the 'Olympic standard' with its 48 cm cubit/191 m *stade* for things like construction, following the conquest of Greece by Macedon in the fourth century BC, while the 180 m Pan-Hellenic system continued to be used in other areas. Yet even the identification of the standard units of the Hellenistic Period was not come by easily.

In 1882, Hultsch suggested that the Macedonian cubit was 46.2 cm—making the Macedonian *stade* 184.8 m.[26] In the same year, Dörpfeld suggested that the Macedonian cubit was 44.3 cm—making the *stade* 177.2 m—the same as had been used in the stadiums at Delphi and Epidaurus prior to the Hellenistic Period.[27] Half a century later, in 1930, Tarn suggested that the Hellenistic cubit was 33 cm in length.[28] This would make the Hellenistic *stade* only 132 m. Tarn's conclusions were based on an examination of the heights that are given in the ancient texts for some of Alexander the Great's opponents, and this lead Tarn to the assumption that the Macedonian cubit was three-quarters of the 45 cm Attic cubit. However, Tarn's reasoning seems to be in error—the heights of Alexander's opponents seem to be exaggerated for literary effect, and there is no evidence of a small cubit being in use in anytime

---

[25] For example, see: W.F. Richardson, *Numbering and Measuring in the Classical World* (Bristol, Bristol Phoenix Press, 2004), pp.29–32; J. Champion, *Pyrrhus of Epirus* (Barnsley, Pen & Sword, 2009), p.24; R. Gabriel, *Philip II of Macedonia – Greater than Alexander* (Washington DC, Potomac Press, 2010), p.65; S. English, *The Army of Alexander the Great* (Barnsley, Pen & Sword, 2009), pp.17, 23–34; W.W. Tarn, *Hellenistic Military and Naval Developments* (Cambridge, Cambridge University Press, 1930), pp.14, 28; W. Heckel and R. Jones, *Macedonian Warrior – Alexander's Elite Infantryman* (Oxford, Osprey, 2006), pp.6, 13; A.M. Snodgrass, *Arms and Armour of the Greeks* (Baltimore, The Johns Hopkins University Press, 1999), p.118; P. Connolly, *Greece and Rome at War* (London, Greenhill Books, 1998), p.69; J. Warry, *Warfare in the Classical World* (Norman, University of Oklahoma Press, 1995), pp.72–73; R. Sheppard (ed.), *Alexander the Great at War* (Oxford, Osprey, 2008), pp.54, 82.

[26] Hultsch, atage degree', which standardiswhich abbreviated title would you prefer to use throughout *etrologie*, pp.30–34, 697.

[27] W. Dörpfeld, 'Beiträge zur antiken Metrologie 1: Das solonisch-attische Syetem' *Ath. Mitt. VII* (Athens, 1882), p.277.

[28] Tarn, *Military and Naval Developments*, p.15; W.W. Tarn, *Alexander the Great Vol.II* (Chicago, Ares, 1981), pp.169–171.

across Greek history.[29] Consequently, Tarn's 'Macedonian Cubit' is closer to the Olympic/Peloponnesian/Macedonian 'foot' (with its accompanying *stade* of 191 m in length) which had already been in use in several parts of Greece long before the advent of the Hellenistic Period.—and there is considerable evidence for the use of this system of measures across the Hellenistic World.

For example, many ancient sources describe the length of the main infantry weapon of the Hellenistic period (a long pike called the *sarissa*). While the length of this weapon changed across the Hellenistic Period, the lengths are given in cubits by most ancient writers.[30] An examination of these sources has shown that all of the measurements are given in the units of the 48 cm cubit/191 m *stade* system.[31] Additionally, many ancient writers of works on Hellenistic tactics describe many of the same intervals for formations as were applicable to the *hoplites* of Classical Greece.[32] However, due to the way that Macedonian pike-men deployed in these formations, they could not conform to the small 45 cm interval of the Classical Period, and the minimum interval for a Macedonian formation was 48 cm per man—which was part of the bigger 'Olympic Standard' of measurements.[33]

Furthermore, the ancient writer Asclepiodotus, in his work on tactics, describes how the Macedonian shield (the *peltē*) was two 'feet' in diameter.[34] A mould for the manufacture of the metallic facing for a Macedonian shield (now in the Allard Pierson Museum, Amsterdam #7879) would cover a shield 65 cm in diameter.[35] The remains of a Macedonian shield found in Pergamon in western Turkey is also 65 cm in diameter.[36] This shows that the Hellenistic foot used in the creation of these shields was about 32 cm in size—the Olympic standard—and would be part of a system that included a 191 m *stade*. Elsewhere there is evidence for other sizes of the *stade*. For example, a metrological relief found at the site of Leptis Magna in Libya shows that the Egyptian Ptolemaic Era (fourth century BC onwards) 'Royal Cubit' was around 52.2 cm in size—equating to a 'Ptolemaic Royal *Stade*' of 210 m.[37]

---

[29] For an analysis of Tarn's errors, see: Matthew, *An Invincible Beast*, pp.71–76.

[30] For example, see: Asclep. *Tact*. 5.1; Ael. Tact. *Tact*. 12, 14; Theoph. *Caus Pl*. 3.12; Polyaenus *Strat*. 2.29.2; Polyb. 18.29.2.

[31] Matthew, *Invincible Beast*, pp.66–69.

[32] Asclep. *Tact*. 4.1; Ael. Tact. *Tact*. 14; Polyb. 18.29; Arr. *Tact*. 11.3.

[33] Matthew, *Invincible Beast*, pp.35, 76, 79–80, 148–153, 159–160.

[34] Asclep. *Tact*. 5.1.

[35] E.M. Moorman, *Ancient Sculpture in the Allard Pierson Museum Amsterdam* (Amsterdam, Allard Pierson Series, 2000), pp.187–188; K. Liampi, *Der makedonische Schild* (Bonn, Rudolf Habelt, 1998), pp.59–61.

[36] U. Peltz, 'Der Makedonische Schild aus Pergamon der Antikensammlung Berlin' *Jahrb. Berlin Museen* 43 (2001), pp.331–343.

[37] G. Ioppolo, 'La ta vola délie unità di mi su re ad mercato augusteo di Leptis Magna' *QAL* 5 (1967), pp.89–98.

The result of all of this archaeological and literary evidence is that there were at least six different systems of measurement in place across Greece from the Classical Period and into the Hellenistic Period all of which incorporated a *stade* of a different size (Table 4).

Applying these values to Eratosthenes' figure for the circumference of the Earth—250,000 *stadia* as is reported by Cleomedes—yields the following values, and margins of error when that result is compared to the current value of the polar circumference of the Earth of 40,007 km (Table 5):

Such comparisons would initially suggest that Eratosthenes was working in units based on the Messenian Standard 160 m in length. The reason for this initial assumption is the simplicity of Eratosthenes' experiment. If he was basing his calculations on something as simple as taking the angle of a shadow cast by an upright object, determining what part of the circumference of a circle that angle constitutes, and then applying that ratio to a known distance, then

Table 4 The varying size of the *stade* across Classical and Hellenistic Greece.

Standard	*Stade* Size
The Messenian *stade*	160 m
The Delphic/Epidauran *stade*	177 m
The Early Attic/Pan-Hellenic *stade*	180 m
Attic	185 m
The Olympic/Late Attic/Macedonian *stade*	191 m
Ptolemaic *stade*	210 m

Table 5 The conversion of a 250,000 *stade* circumference of the Earth by known sizes of the *stade*, with margins of error.

Stade size	Calculated Circumference (based on 250,000 *stadia*)	Margin of Error (circum. vs 40,007 km)
160 m (Messenian)	40,000 km	−0.02%
177 m (Delphic)	44,250 km	+9.59%
180 m (Pan-Hellenic)	45,000 km	+11.10%
185 m (Attic)	46,250 km	+13.50%
191 m (Olympic)	47,750 km	+16.22%
210 m (Ptolemaic)	52,500 km	+23.80%

it can only be expected that his results would be highly accurate—regardless of whether any possible rounding of the figures for the angle of the shadow, or the known distance, are taken into consideration or not. This then suggests that any conversion of Eratosthenes' result into a modern unit of measure must likewise have the same high level of accuracy. This would initially suggest the use of a 160 m *stade* by Eratosthenes due to the close correlation between the calculated and actual distances. However, the use of a 160 m *stade* does not conform to the other evidence for his work that we have (see the following section 'Eratosthenes and the 180 m *Stade*'), nor was it a unit that seems to have been in common use across the ancient Greek world.

Alternatively, the evidence for the use of a 191 m *stade* in the Hellenistic Period would suggest that, as the army of Alexander the Great spread Hellenistic culture throughout his conquered territories, that the system of measurements used by the Macedonians at the time—with a *stade* of 191 m—was introduced to these cultures; including into Egypt (which was conquered by Alexander in 332 BC and then later ruled by one of his generals, Ptolemy, and his descendants for the next three centuries). This could imply that Eratosthenes' results were given in 191 m units. However, the high margin of error when the value for the Earth's circumference given in Cleomedes is converted using this unit (+16.22 per cent) would suggest that this, and many of the other units used in the ancient Greek world, was not the unit that Eratosthenes was working in. Despite the inaccuracy of converting the results of Eratosthenes' calculations into any of the previously noted modern units of measure, many of these measurements, and some that are not accounted for in the ancient records, have subsequently been used in examinations of Eratosthenes' calculation of the circumference of the Earth.

Having established the sizes of the different *stadia* across the systems of measurement used in the ancient Greek world, and the corresponding calculation of the circumference of the Earth using these units, does not aid with understanding which unit of measure Eratosthenes was using, nor clarify the accuracy of his determinations. However, despite the inaccuracy of converting the results of Eratosthenes' calculations into any of the previously noted modern units of measure, many of these measurements, and some that are not accounted for in the ancient records, have subsequently been used in examinations of Eratosthenes' calculation of the circumference of the Earth. It is essential to understand these prior analyses of Eratosthenes' calculations in order to gauge their validity, and subsequently re-examine his work (which will be undertaken in the following sections).

4.2 Previous Estimates Of The Size
Of Eratosthenes' *Stade*

In the attempt to determine the accuracy of Eratosthenes' work, previous scholars have assigned the use of a *stade* of different sizes to Eratosthenes in their examinations of his calculation of the circumference of the Earth. However, many of these theories seem to have altered the values provided in the ancient texts for the sake of numerical convenience, failed to engage with some sources of evidence, and/or selectively used some values over others in their examinations. All of them have failed to present a comprehensive argument in favour of one *stade* over another that conforms to all of the available evidence, and some scholars offer the use of a *stade* by Eratosthenes for which there is no evidence of its use in the ancient Greek world. This has led to a scholarly debate over the size of Eratosthenes' *stade*, and subsequently the accuracy of his work, that has been ongoing for centuries.

Much of this ongoing debate is centred around the interpretation of a few ancient passages that discuss the length of the *stade* in antiquity. Pliny, for example, states that 'by the calculation of Eratosthenes, a *schoenus* equals 40 *stadia*, or 5 [Roman] miles, while others make it 32 *stadia*'.[38] In this passage, the size of the *stade* is given in reference to two other units of measure: the Roman mile and the Egyptian *schoenus*. The *schoenus* was equal to 12,000 cubits, and each Egyptian cubit measured 52.5 cm. Thus, one *schoenus* was equal to around 6,300 m. From this information, each of Eratosthenes' *stadia* would be equal to 157.5 m (6,300/40) in length. This is close to the Messenian Standard *stade* of 160 m. However, this passage also states that 8 *stadia* were equal to one Roman mile for Eratosthenes (i.e. 40 *stadia*/5 Roman miles = 8). The Roman mile was around 1.48 km in length, and this would make Eratosthenes' *stade* the Attic Standard of 185 m in length ((5 x 1.48 x 1,000)/40 = 185 m) or 1/8 of a Roman mile.[39] Another source from the first century AD, a metrological table in the writings of Heron of Alexandria, states that a *schoenus* was equal to

---

[38] Plin. (E) *HN* 12.53: *schoenus patet Eratosthenis ratione stadia XL, hoc est p. V, aliqui XXXII stadia singulis schoenis dedere*; Diller ('Ancient Measurements', p.8) claims that this passage from Pliny is the only direct evidence for the size of Eratosthenes' *stade*. This is clearly incorrect as there are numerous passages which directly attribute a *stade* to Eratosthenes (see following), but this is the only one that directly compares the *stade* of Eratosthenes to another unit of measure—which may be what Diller is implying.

[39] In their examinations of these passages, Hultsch (*metrologie*, p.88) used a Roman mile of 1,480 m, Nicastro (*Circumference*, p.125) used a Roman mile of 1,481 m, Diller ('Ancient Measurements', p.8) used a Roman mile of 1,488 m, and Walkup ('Mystery of the Stadia' *s.v. How Long is a Stade?*) used a Roman mile of 1,479 m.

4 Roman miles—meaning that there would be 10 *stadia* to the mile—making each *stade* approximately 148 m in length (1,480 m/10 = 148 m).[40]

However, in other passages, Pliny only adds to the confusion over the size of the *stade*. For example, earlier in his work, Pliny converts the 252,000 *stade* circumference for the Earth that is attributed to Eratosthenes by later writers into 31,500 Roman miles.[41] This again would make Eratosthenes' *stade* the Attic Standard of 185 m in length ((31,500 x 1.48 x 1,000)/252,000 = 185 m). In another passage, Pliny converts 1 *stade* into 625 Roman feet.[42] A Roman foot was approximately 29.59 cm in size, which again makes the *stade* around 185 m (625 x 29.59 cm = 184.9 m). Elsewhere, Pliny states that there were 30 *stadia* to a *schoenus*.[43] This equates to a *stade* of the Ptolemaic Standard of 210 m. Strabo similarly refers to a 30:1 conversion between *stadia* and *schoeni*, but also outlines how the size of the *schoenus* varied from region to region within Egypt, and that there could be 30, 40, 60, or even 120 *stadia* to one *schoenus*.[44] Both Herodotus and Diodorus state that a *schoenus* was equal to 60 stadia.[45] This would equate to a *stade* of only 105 m in length (6,300 m/60 = 105 m). Diller outlines how most ancient sources have 1 Roman mile equal to 7.5, 8.0, 8.3, or 9.0 *stadia*.[46]

Within these equivalents in Roman miles outlined by Diller can be seen what seem to be conversions into *stadia* from different systems of measurement. For example, converting a Roman mile of 1,480 m in length by 7.5 *stadia* gives a *stade* length of ≈197 m. This is only slightly too big to be a conversion into the Olympic Standard with its 191 m *stade* (or possibly a bigger error in converting into the Ptolemaic Standard with its 210 m *stade*). Converting the Roman mile by 8 gives a *stade* of ≈185 m—the Attic Standard, by 8.3 gives a *stade* of ≈179 m—the Pan Hellenic Standard, and by 9 gives a *stade* of ≈164 m—which is close to the Messenian Standard. Thus, it is clear that the ancient sources are citing measurements—in miles, *stadia*, and/or *schoeni*—which are using different units of measure. Previous scholars examining the work of Eratosthenes have interpreted these passages in various ways and, in

---

[40] Hero. *Met.*; the table is found in one main manuscript of the text *Codex Constantinopolitanus, Palatii veteris*. See: E.M. Bruins (ed.), *Codex Constantinopolitanus, Palatii veteris Vol.3* (Leiden, Brill, 1964).

[41] Plin. (E) *HN.* 2.112.

[42] Plin. (E) *HN* 2.21.

[43] Plin. (E) *HN* 5.11.

[44] Str. *Geog.* 17.1.24.

[45] Hdt. 2.6; Diod. Sic. 1.15.5.

[46] Diller, 'Ancient Measurements', p.8.

combination with other elements of their hypotheses, this has led to many different ideas about the size of Eratosthenes' *stade* being proposed.

One of the first examinations into the size of Eratosthenes' *stade* was made by d'Anville in 1759. Following the passage of Diodorus, d'Anville divided the *schoenus* of around 5,920 m by 60 to reach a *stade* length of only 98.67 m.[47] Then d'Anville compared the length of Egypt's coastline (which d'Anville states is given as 1,970 *stadia* by Diodorus (1.31.6)) and multiplying that distance by the size of the Attic *stade* (185 m)—which d'Anville mislabelled as the Olympic Standard—and then again divided the result (364.45 km) by 60 (as per Diodorus and Herodotus) to reach a *schoenus* of ≈101.1 m. The two different sizes for the *stade* (98.67 m and 101.1 m) were then averaged to result in a final value for the size of the *stade of* 99.8 m.[48] As demonstrated by Engels, there are numerous problems with d'Anville's work.[49] Firstly, it assumes a constant size of the *schoenus*. Secondly, there is no evidence for a *stade* of around 100 m being used at any time in the ancient Greek world. Thirdly, d'Anville had based his calculations on a conversion factor of 5 Roman miles per *schoenus* and had not engaged with any of the texts that suggest 4 miles per *schoenus*. Lastly, Diodorus actually says that Egypt's coastline is 2,000 *stadia* and not the 1,970 *stade* value used by d'Anville.[50] Herodotus (2.7), one of the other sources that d'Anville relies upon, states that Egypt's coast was 3,600 *stadia* across, so it is clear that the two ancient writers were using units of different sizes and any conversion into modern units based upon a combination of their two figures was bound to reach an incorrect result.

Another of the early examinations which attempted to define the size of the *stade* used by Eratosthenes had been conducted earlier in another work by d'Anville in 1741.[51] He accepted the values given in the metrological tables of Heron of Alexandria that 1 *schoenus* equalled 4 Roman miles, ignored the conflicting statement made by Pliny that the *schoenus* equalled 5 Roman miles, but then accepted the passage of Pliny's that stated that 1 *schoenus* equalled 40 of the *stadia* used by Eratosthenes. This, according to d'Anville, made Eratosthenes' *stade* 1/10 of a Roman mile or ≈148 m. While d'Anville's later

[47] J.B.B. d'Anville, 'Mémoire sur la mesure du schène égyptien, et du stade qui servant à le composer' *Mém. Acad. Inscript. et belles-lettres* 26 (1759), pp.82–91.

[48] d'Anville, 'Mémoire sur la mesure', pp.82–91.

[49] See: Engels, 'Length', p.301.

[50] Diod. Sic. 1.31.6.

[51] J.B.B. d'Anville, 'Écclaircissements géographiques sur l'ancinne Gaule' *Precedés d'un traité des mesures itinéaires des romains, et de la lieue gauloise* (Paris, chez la Veuve Estienne, 1741).

publication revised this work, it was this earlier idea over the size of the *stade* which found favour with some later scholars.

In 1929, Lehmann-Haupt also presented the idea that Eratosthenes' *stade* was around 148 m in length.[52] Lehmann-Haupt compared Pliny's *schoenus* to a Babylonian cubit of 49.5 cm. As a *stade* was known to be 400 cubits in size, this would make the *stade* 198 m in length (400 x 49.5 cm = 198 m). Lehmann-Haupt then multiplied this length by 30 (as per Pliny 5.11) to reach a size of the *schoenus* of 5,940 m. This value was then divided by 40 (as per Pliny's reference to the number of Eratosthenes' *stadia* per *schoenus* (12.53)) to reach a value for Eratosthenes' *stade* of 148.5 m. However, the size of the Babylonian cubit upon which Lehmann-Haupt's calculations are based is taken from a non-metrological relief housed in the Louvre, and the size of this unit has been variously measured between 49.59 cm and 50.625 cm.[53] Shcheglov therefore states that the size of Lehmann-Haupt's *stade* must fall within arrange of 148.77 m to 151.88 m.[54]

More recently, Fischer also followed the passage of Heron that stated that 1 *schoenus* equalled 40 *stadia* and 4 Roman miles, and that a *stade* was 1/10 of a Roman mile.[55] Based on a variability in estimates in the size of the Roman mile, Fischer then gave a range for Eratosthenes' *stade* of between 148.5 m and 158 m, which she said makes Eratosthenes' result (based on the 252,000 *stade* circumference found in the later ancient writers) too small by 6.75 per cent and 0.5 per cent, respectively.[56] Fischer argued that Eratosthenes could not have been using a *stade* that was 1/8 of a Roman mile (i.e. a *stade* of 185 m) because, as the Tropic was south of Syene, the use of this unit would have made Eratosthenes' result far too small and would give his results a margin of error of around 16.5 per cent (based on a circumference of 252,000 *stadia*).[57]

The main problem with these theories is that they all suggest the use of a unit of measure by Eratosthenes that did not exist in the ancient Greek world. Those offered by d'Anville and Lehmann-Haupt, for example, even if rounded up, only come to 150 m in length. This is far too short to be comparable to even the smallest known system of measurements—the Messenian Standard with its

[52] C.F.F. Lehmann-Haupt, 'Stadion (Metrologie)' in G. Wissowa and W. Kroll (eds.), *Paulys Real-Enccyclopädie* (Stuttgart, J.B. Metzlersche, 1929), pp.1961–1963.

[53] See: M.A. Powell, 'Maße und Gewichte' in A. Bramanti, E. Ebeling, and M.P. Streck (eds.), *Reallexikon der Assyriologie und Vorderasiatischen Archäologie VII* (Berlin, DeGruyter, 1990), pp.457–517.

[54] Shcheglov, 'Itinerary Stade', p.157.

[55] Fischer, 'Another Look', p.159.

[56] Fischer, 'Another Look', p.159.

[57] Fischer, 'Another Look', p.159.

stade of 160 m. Similar could be said for the lower-end estimate provided by Fischer, while Fischer's higher-end estimate of 158 m could be rounded up to the Messenian Standard.

A *stade* of the Messenian Standard was suggested as the size of the unit used by Eratosthenes by scholars in the nineteenth century. Girard (1830) and Letronne (1851), and Hultsch (1882) interpreted the passage of Pliny (12.53) which stated that there were 30 *stadia* to a *schoenus* as being a reference to Egyptian *stadia* 210 m in length—making 1 *schoenus* 6,300 m.[58] This value was then divided by 40 (as per Pliny 12.53) to make each *stade* 157.5 m. This is close to the size of the *stade* in the Messenian Standard (160 m).

Viedebantt (1915) analysed the same passage referring to the *schoenus* and Eratosthenes' *stade* through a comparison to the Roman foot. According to the metrological tables of Heron, 1 Roman mile equalled 5,400 Roman feet, which equalled 7.5 *stadia*.[59] Viedebantt set the Italian foot at 29.6 cm, which made the Roman mile 1,598.4 m—a value which is not found in any source.[60] This value was then divided by 7.5 to reach a *stade* length of 213.12 m—which is close to the Ptolemaic Standard (210 m)—and one *schoenus* of 6,393.6 m (213.12 x 30 = 6,393.6). This size for the *schoenus* was then divided by 40 (as per Pliny 12.53) to reach a size for Eratosthenes' *stade* of 159.84 m—which is again close to the Messenain Standard and the findings of Girard, Letronne and Hultsch. Viedebantt then supported this claim by stating that Alexandria had a stadium 159.8 m in length.[61] However, there is no stadium of such a size in Alexandria. The main venue in Alexandria was the Lageion, named after the father of the first Ptolemy, Lagos, and was first used for the Ptolemaieia festival in 279/278 BC.[62] The Lageion was a hippodrome, used for horse and chariot racing, although athletic contests could also be held there. Archaeological plans from the eighteenth century show that the Lageion had an overall length of around 560 m—which converts to around 3 x 191 m Olympic *stadia*.[63] Strabo refers to a 'so-called hippodrome' (ἱππόδρομος καλούμενος) at

---

[58] See: P.S. Girard, 'Sur la coudée septénnaire des anciens Égyptiens et les différents étalons qui en ont été retrouvés jusqu'a présent' *Mem. Acad. Sci. Inst. Fr.* 9 (1830), pp.591–608; J.A. Letronne, *Recherches critiques: historiques et géographiques sur les fragment d'Heron d'Alexandrie, ou de systéme métriques égyptien* (Paris, l'Imprimeire Nationale, 1851), pp.110–130; Hultsch, *metrologie*, pp.54, 60–63, 363–364.

[59] Hero. *Met.*

[60] Viedebantt, 'Eratosthenes', pp.215–216.

[61] Viedebantt, 'Eratosthenes', p.213.

[62] Hdn. *Gr.* 2.1–2; *SIG* Vol.I #390.

[63] C-L. Balzac, F-C. Cécile, and G-J-G. de Charbol de Volvic (eds.), *Description de l'Égypte, ou Recueil des observations et des recherches qui ont été faites en Égypte pendant l'expédition de l'armée française* (Paris, l'Imprimerie Royale, 1812), p.21, pl.#39; see also: J.S. McKenzie, S. Gibson, and A.T. Reyes, 'Reconstructing the Serapaeum in Alexandria from Archaeological Evidence' *JRS* (2004), p.103.

nearby Nicopolis, but no trace of this structure has ever been found.[64] There was a stadium for running races in Fayum, south-west of modern Cairo, but again no trace of the structure has been discovered.[65]

Diller also favoured the use of a *stade* around 160 m in length by Eratosthenes. Diller based his analysis on the 252,000 *stade* figure for the Earth's circumference that is found in the Roman period sources and stated that Pliny converted everything on the basis of 8 *stadia* to 1 Roman mile of 1,488 m.[66] Diller further suggested that Eratosthenes had to be using a *stade* less than 159.2 m, a figure which was obtained by dividing a circumference value of 40,120 km by 252,000 *stadia*, as this would make the conversion 9.3 *stadia* per Roman mile.[67] Diller then justifies this claim by following similar steps to those found in the works of earlier scholars: 12,000 cubits multiplied by the Egyptian standard of 52.5 cm equals a *schoenus* of 6,300 m which, when divided by 40 as per Pliny (12.53) makes Eratosthenes' *stade* 157.5 m.[68] One major flaw with Diller's hypothesis is that there is no evidence for the use of a unit less than 160 m in length in the ancient Greek world.

Dreyer stated that Eratosthenes must have used a *stade* less than 185 m (400 cubits of 0.462 m)—which he mislabelled as the Olympic Standard—or the 210 m *stade* (400 cubits of 0.525 m) of the Ptolemaic Standard.[69] Dreyer also accepted the positions of prior scholars by following Pliny (12.53) in that Eratosthenes set an Egyptian *schoenus* as 40 stadia, and a *schoenus* was 12,000 Ptolemaic cubits of 0.525 m. Thus, a *stade* for Eratosthenes was 157.5 m, and 252,000 x 157.5 equalled a circumference of the Earth of 39,690 km.[70]

Firsov used geography to come up with a similar value for Eratosthenes' *stade*. Firsov examined 81 different distance measurements that are given in the works of Strabo, but which are attributed to Eratosthenes. Firsov divided the straight-line distance between the locations given by the number of *stadia* that are listed to determine the lengths of the *stade* being used in each measurement. Firsov then averaged the lengths of all 81 *stadia* to arrive at a figure of 157.7 m.[71] There are a number of issues with Firsov's methodology.

[64] Str. *Geog.* 17.1.10; J.H. Humphrey, *Roman Circuses: Arenas for Chariot Racing* (Berkeley, University of California Press, 1986), p.509.
[65] J. McKenzie, *The Art and Architecture of Alexandria and Egypt 300BC–AD700* (New Haven, Yale University Press, 2007), p.48.
[66] Diller, 'Ancient Measurements', pp.7–8.
[67] Diller, 'Ancient Measurements', p.8.
[68] Diller, 'Ancient Measurements', p.8.
[69] Dreyer, *History of Astronomy*, p.176.
[70] Dreyer, *History of Astronomy*, p.175.
[71] L.V. Firsov, 'Eratosthenes's Calculation of the Earth's Circumference and the Length of the Hellenistic Stade' *VDI* 121 (1972), pp.154–175.

Firstly, many of the sites listed by Strabo have not been positively identified and so measuring the distance between some modern locations may not be accurate (see the following section 'Eratosthenes and the 180 m *Stade*'). Secondly, it is unlikely that Eratosthenes was working with linear, 'as the crow flies', measurements. Strabo, in a commentary on the critique of Eratosthenes' work made by Hipparchus, states that the measurements given by Eratosthenes are not linear distances, as Hipparchus says they were, as these would be shorter than on the ground measurements, and that Eratosthenes had taken straight lines only roughly (i.e. the most direct land route) 'as is proper to do in geography'.[72] Consequently, the distances between two locations will be larger than Firsov assumed based on a linear measurement. Lastly, as with many prior theories, there is no evidence for the use of a unit of 157.7 m in length in the ancient Greek world. It is possible that it should be the 160 m *stade* of the Messenian Standard. However, use of a Messenian *stade* would give Firsov's results a higher margin of error, as the conversions into modern equivalents would result in a greater distance between two points than his average.

Dutka also favoured the use of a *stade* slightly less than 160 m by Eratosthenes. However, rather than base his examinations of the 252,000 *stade* figure found in the later ancient sources as many prior scholars had done, Dutka accepted the 250,000 *stade* value given by Cleomedes and suggested that the 252,000 *stade* figure was mathematical convenience.[73] Dutka based his estimations on Eratosthenes using land survey data in Egypt, and an alteration of the observed shadow cast by the *gnomon* in Alexandria from 1/50 of the Earth's circumference to 1/48.[74] From such assumptions, Dutka stated that Eratosthenes' *stade* was 157.5 m in length—mirroring the theories proposed by Girard, Letronne, Hultsch, Diller, and Firsov. This value was also used by Pinotsis, who cites Hultsch, in his examination of the errors in Eratosthenes' calculations.[75]

MacLeod also seems to have accepted the 250,000 *stade* result given by Cleomedes. In his work on the Alexandria Library, MacLeod makes the general statement that Eratosthenes' result was out by around 8 km.[76] If this is based upon a comparison of the result to the current polar circumference of

---

[72] Str. *Geog.* 2.3.39.
[73] Dutka, 'Eratosthenes' Measurement', p.60.
[74] Dutka, 'Eratosthenes' Measurement', pp.56–57, 61.
[75] Pinotsis, 'Significance and Errors', p.58.
[76] MacLeod, 'Introduction', p.6.

the Earth of 40,007 km, then MacLeod must be assuming the use of a 160 m *stade* by Eratosthenes and accept Cleomedes' result of 250,000 (250,000 x 160 m = 40,000 km).

Most recently, Nicastro also supported the notion of a *stade* less than 160 m in length by Eratosthenes. Nicastro based his analysis on the 252,000 *stade* circumference of the Earth, and an adjusted distance between Alexandria and Syene to 5,040 *stadia*, and a *stade* of 157.5 m in length (based on the work by Firsov)—stating that this is the closest that one can get to the real distance, despite also acknowledging that a reference to what calculates as a 157.5 m *stadia* appears only in Pliny and in no other source.[77] Nicastro concluded that such parameters would make Eratosthenes' calculation quite accurate (252,000 x 157.5 = 39,690 km)—a result that Nicastro claimed to be able to make with 95 per cent confidence.[78] Nicastro additionally argued against the use of a larger unit by Eratosthenes (see following) by suggesting that the Delphic Standard *stade* of around 177 m—which he mislabels as the Attic Standard—would make Eratosthenes' result out by 15 per cent and so is unlikely to have been what Eratosthenes used.[79] In regard to larger *stadia* like the 185 m unit of the Attic Standard, Nicastro claimed that the chance of that being used by Eratosthenes was less than one in ten-thousand.[80] Clearly, there are many issues with Nicastro's hypothesis. By accepting the units and arguments of earlier examinations, with little original analysis, Nicastro has simply reiterated many of the issues found in these prior works—such as the resultant unit not existing in the ancient Greek world, the favouring of one value and/or one passage in an ancient text over the others with little justification for the exclusion of the rest. Furthermore, there is little reason to alter the reported distance from Alexandria to Syene to 5,040 *stadia* other than to make the mathematics of the calculation of the Earth's circumference work. As such, Nicastro's work, and many of those upon which it is based, can only be considered selective in their use of source material.

Similarly, Russo, citing Hultsch, and basing his conclusions on the 252,000 *stade* circumference found in the later sources, suggests that the size of Eratosthenes' *stade* was within the range of 155–160 m—giving his result a margin of error of −2.4 per cent to +0.8 per cent.[81] Russo additionally suggests that Eratosthenes' result was 252,000 *stadia* as it was a number that was divisible

[77] Nicastro, *Circumference*, pp.26–27, 119–123, 128.
[78] Nicastro, *Circumference*, pp.124, 128.
[79] Nicastro, *Circumference*, pp.26, 125; on p.123, Nicastro correctly labels the 185 m *stade* as Attic.
[80] Nicastro, *Circumference*, p.128.
[81] Russo, *Forgotten Revolution*, p.273; Russo later (*Forgotten Revolution*, p.276) gives Eratosthenes' result an average margin of error of less than 1 per cent.

by all numbers between one and ten.[82] This suggestion, in addition to ignoring ancient passages that provide a different value for the result, also suggests that the result was based upon the ideals of Platonic numerical perfection. Russo concludes that, due to this result, Eratosthenes must have created and entirely new *stade* that was equal to a 'convenient fraction of the meridian'.[83] Such a conclusion seems highly unlikely and there is no reference to Eratosthenes using a novel unit of measure in his calculations.

Gulbekian also used an altered distance between Alexandria and Syene of 5,040 *stadia* in his examination. Gulbekian examined the differences between the latitudes of Alexandria and Syene, a linear distance between the two locations of 840 km, and a mean radius of the Earth of 6,371 m to reach a conclusion that Eratosthenes' *stade* was 166.7 m in length.[84] This unit is too big to be part of the Messenian system, but too small to be part of the Delphic Standard. Gulbekian justified this finding via a comparison to the length of the Pharos Peninsula in Alexandria, which was known as the 'Heptastadion' (ἑπταστάδιον) or 'Seven Stadia'.[85] A nineteenth-century survey of the area stated that the length of the peninsula was 1,202 m.[86] According to Gulbekian, 1,202 m/166.7 m = 7.2 *stadia*—hence the name of the peninsula with the value rounded down. There are many issues with Gulbekian's reasoning. In addition to the errors found in some prior theories, such as the use of an adjusted value for the distance between Alexandria and Syene, basing his calculations on an unlikely linear distance, and suggesting the use of a *stade* by Eratosthenes that does not seem to have existed in the ancient Greek world, Gulbekian does not seem to have considered that the name for the peninsula that he uses as support for his conclusions could also be called the exact same thing if units of other sizes were used. For example, 1,202 m divided by the 177 m unit of the Delphic Standard equals 6.8 *stadia*, divided by the 180 m Pan-Hellenic Standard equals 6.7 *stadia*, and divided by the 185 m Attic standard equals 6.5 *stadia*. All of these could have been used in the measurement of the peninsula and then the values rounded up (as opposed to down as per Gulbekian) to give the area its name. As such, Gulbekian's use of sources to support his arguments can only be considered selective.

Rawlins argued for the use of a *stade* of the Attic Standard (185 m) in his examinations of Eratosthenes. He called this unit the 'standard Greek *stade*'

---

[82] Russo, *Forgotten Revolution*, p.277.

[83] Russo, *Forgotten Revolution*, p.277.

[84] Gulbekian, 'Stadion Unit', p.362.

[85] Gulbekian, 'Stadion Unit', p.362; for the Heptastadion, see: Str. *Geog.* 17.1.10.

[86] See: H. Kiepert, 'Zur Topographie des alten Alexandrien' *Zeitschr. d. Gesell. f. Erdkunde zu Berlin* 7 (1872), pp.337–349.

and compared 10 *stadia* to 1 nautical mile of 1,852 m in length.[87] Rawlins stated that this unit was the correct size, that all other notions for the use of a smaller unit by Eratosthenes were just 'folly', and that some scholars were unwilling to accept that Eratosthenes' calculation could be out by 17 per cent—which would be the result if the 250,000 *stade* value for the Earth's circumference found in Cleomedes is used (250,000 x 185 m = 46,250 km).[88]

Similarly, Engels suggested the use of a unit of 184.98 m in length by Eratosthenes, called suggestions for the use of smaller units of 148 m and 157 m 'entirely fictional', and declared that the arguments in support of these smaller *stadia* 'involve some tendentious mathematical calculations' and 'bogus reasoning'.[89] Engels debunks theories in favour of the 157.5 m *stade* by stating that they assume that Pliny correctly converted Eratosthenes' *stadia* into the Egyptian *schoenus*, but incorrectly in terms of the Roman mile, and that many of the early measurements are not accurate, nor were they based on linear distances.[90] Engels then refers to his own earlier work on the march rates of the Macedonian army to support the use of a 185 m *stade* by Eratosthenes.[91] However, the results of Engels' earlier work seem to be in error (see the following section 'Eratosthenes and the 180 m *Stade*').

Walkup also supported the idea of the use of the 185 m Attic *stade* by Eratosthenes stating: 'that 1 *stade* = 185 m ... is well established' and based her examination on the acceptance of the passage of Pliny which gives 1 *stade* equal to 1/8 of a Roman mile.[92] One Roman mile was about 5,000 Roman feet (which is just shorter than an English foot of 12 inches or 30.48 cm). Walkup took the Roman foot to be 11.65 inches (29.59 cm), a Roman mile to be 1,479 m, and 1/8 of a Roman mile gives a *stade* of 184.8 m.[93] Similar to both Walkup and Engels, Shcheglov also called the claims by earlier scholars in favour of the smaller size *stadia* 'unfounded' and critiqued the work found in many prior studies.[94] Shcheglov based his examination on the 252,000 *stade* circumference found in the later ancient texts to support the call for the use of a 185 m Attic *stade* by Eratosthenes.[95] Shipley claimed that Eratosthenes' result—which he took to be the 250,000 *stade* figure found in Cleomedes—resulted in a modern

[87] Rawlins, 'Nile Map', pp.211, 218.
[88] Rawlins, 'Too Big Earth', p.10; Rawlins, 'Nile Map', p.211; this argument was also echoed by Walkup, 'Mystery of the Stadia' *s.v. How Long is a Stade?*
[89] Engels, 'Length', pp.298–299.
[90] Engels, 'Length', pp.305–307.
[91] Engels, 'Length', p.309; Engels, *Logistics*, pp.157–158.
[92] Walkup, 'Mystery of the Stadia' *s.v. How Long is a Stade?*
[93] Walkup, 'Mystery of the Stadia' *s.v. How Long is a Stade?*
[94] Shcheglov, 'Itinerary Stade', pp.155–170.
[95] Shcheglov, 'Itinerary Stade', pp.154, 170.

equivalent of 46,000 km.[96] This is close to the conclusion presented by Rawlins and would make Eratosthenes' *stade* exactly 184 m—which, while close to the Attic standard (185 m), is a unit which is not supported by available literary and archaeological evidence.

All of these prior theories present arguments that, while mathematically solid, are still somewhat doubtful as many of them are selective in the source material used to substantiate their results, they mix measurements that are in different units, exclusively favour one of the values for the circumference of the Earth over the other, and/or contain parts that are clearly speculative. Many also result in the suggested use of a *stade* for which there is no proof of its existence in the ancient Greek world. Additionally, many of these theories do not engage with any source material that is directly attributable to Eratosthenes himself (or even attributed to Eratosthenes by later sources such as Pliny or Strabo). Many previous scholars have engaged with the passage from Pliny (12.53) which states that, for Eratosthenes, 8 *stadia* was equal to one Roman mile, although even this passage is disputed and altered to change the conversion rate to 10 *stadia* per Roman mile by some. What almost no prior scholar has engaged with is another ancient text which, similarly, directly states how many of the *stadia* used by Eratosthenes equalled one Roman mile.

This passage is found in an appendix attached to the tactical treatise of Aelian (the *Tactica*). Although written in the second century AD, Aelian asserts that 'you will find, contained within, Alexander of Macedon's manner of marshalling an army'.[97] Thus the work is an analysis of the units, formations and tactics used by the Macedonians in the fourth century BC. The appendix is found in a compendium of tactical works, containing Aelian's *Tactica*, called the *Codex Laurentianus 75.6* from the fourteenth century which is now housed in the Biblioteca Medicea Laurenziana in Florence. This work, also known as the *Sylloge*, is a copy of a Byzantine Era treatise on military tactics that is attributed to the Byzantine Emperor Leo VI (AD 912–959).[98]

A key passage from the appendix for understanding the work of Eratosthenes states: 'According to the accurate geographers, Eratosthenes and Strabo, the mile equals 8 1/3 *stadia*, but according to current custom it is 7 1/2 stadia.'[99] This passage directly states that Eratosthenes, and Strabo who

[96] Shipley, *Greek World*, p.362.
[97] Ael. Tact. *Tact*. Prae.
[98] A. Dain, *Sylloge Tacticorum quae elim 'inedita Leonis tactica' dicebantur* (Paris, Société d'édition 'Les Belles lettres', 1938); R. Vari, 'Die sog. Inedita Tactica Leonis' *BZ* 27 (1927), pp.241–270.
[99] Ael. Tact. *Tact*. App.3: τὸ μίλιον κατὰ μὲν Ἐπατοσθένην καὶ Στράβωνα τοὺς ἀκριβεῖς γεωγράφους ἔχει σταδίους η καὶ γ΄, κατὰ δὲ τὸ νῦν ἐπικρατοῦν ἔθος σταδίους ζ c΄.

regularly cites Eratosthenes in his work on geography, were working with *stadia* in the Pan-Hellenic Standard of 180 m in length (1 Roman mile of 1,480 m/8.3 ≈ 180 m).

Diller, one of the few of the prior examiners of Eratosthenes who engages with this passage, points out that, as far is known, Eratosthenes did not write a comparative examination of units of measure, whereas Strabo did, and suggests that the author of this passage must be citing Strabo and assigning an additional, fictitious, authorship to Eratosthenes, and that the author of the *Sylloge* is paraphrasing Strabo rather than citing direct quotes from Eratosthenes himself, as Strabo does not actually cite these values attributed to Eratosthenes anywhere in his work.[100]

However, in a similar passage, Strabo does state that the majority of geographers use 8 *stade* per Roman mile, except for Polybius, who uses 8 1/3.[101] This mirrors the information found in the passage of Pliny (12.53), with the addition of Polybius using a different conversion rate to most other ancient geographers. Dicks finds it surprising that, if Eratosthenes did use 8 1/3 *stadia* per mile as per the appendix in the *Tactica*, that Strabo did not mention as such considering that Eratosthenes was one of Strabo's main sources.[102] Engels, on the other hand, points out that Eratosthenes himself would be unlikely to refer to anything in Roman miles.[103] This in itself may have been enough for Strabo to omit any reference to Eratosthenes in a statement as to how many *stadia* per Roman mile were used by which geographers. Additionally, the omission of a main source was quite common in ancient writing. In the preface of his work on tactics for example, Aelian (whose work forms part of the *Sylloge*) refers to all of the prior works he had consulted as part of his research except for his main source—the earlier tactical treatise by Asclepiodotus.[104] As such, it cannot be ruled out that Eratosthenes did use 8 1/3 *stadia* to the Roman mile.

Such a conclusion would seem to conflict with the passage from Pliny (12.53) that is often cited by modern scholars which states that Eratosthenes used 8 *stadia* per mile. However, barring the possibility of a transmission/transcription error in the text of Pliny, the two texts can be reconciled. Pliny states that 40 *stadia* was equal to 5 Roman miles—making it 8 *stadia* per Roman mile. Alternatively, if the 40 *stadia* found in Pliny are divided by the 8

[100] Diller, 'Ascalon', pp.22–25.
[101] Str. *Geog.* 7.7.4.
[102] Dicks, *Fragments of Hipparchus*, pp.44–45.
[103] Engels, 'Length', p.308.
[104] For Aelian's sources, see: Ael. Tact. *Tact.* 1. See also Matthew's commentary on the text in the introduction to his translation of *The Tactics of Aelian* (pp.xiii–xiv).

1/3 figure found in the *Sylloge*, this results in 4.8 Roman miles, and Pliny may have simply rounded the number up to 5 Roman miles.

The consequence of this is that there are now two ancient texts that are the only ones which directly attribute a size for the *stade* used by Eratosthenes and compare it to another known unit of measure. Both of them also suggest that Eratosthenes was using units in the Pan-Hellenic Standard of 180 m in length in his work. This is a value for the size of the *stade* that has never been applied to an examination of Eratosthenes' calculation of the circumference of the Earth by any prior scholar, and has far-reaching implications for understanding the accuracy of his results. As will be shown in the following sections, the use of the Pan-Hellenic *stade* by Eratosthenes is also confirmed by other evidence that demonstrates that, as per the passage found in the appendix to Aelian's *Tactica*, both Eratosthenes and Strabo were working in units of 180 m.

5

The Re-Examination

5.1 Eratosthenes And The 180 m *Stade*

One thing that many of the prior examinations into Eratosthenes' calculation of the circumference of the Earth have in common is that they rarely deal with any form of evidence over the possible size of the *stade* that is directly attributable to Eratosthenes himself. Yet such evidence exists, in two different ancient sources, and in two different contexts, which demonstrate the size of the unit of measure that Eratosthenes used in his works. A review of this evidence, combined with confirmatory analysis using modern technologies such as satellite mapping and imagery, demonstrates that Eratosthenes used the 'Pan-Hellenic' *stade* of 180 m in length from the Classical Period of Greek history in his calculations of distances.

One piece of textual evidence that directly attributes a series of distances to Eratosthenes can be found in the geographical works of Strabo. Strabo recounts how Eratosthenes had recorded the distances between numerous sites in modern Iran and Afghanistan, in his own work on geography, based upon the figures taken by the surveyors (*bematists*) that had accompanied the army of Alexander the Great on their expedition to the east in the late fourth century bc.

The *bematists* were professional surveyors who were trained to walk with a regular pace in order to gauge distances.[1] Pliny refers to them as the 'surveyors of marches' (*itinerum eius mensores*).[2] Engels proposes that the accuracy of the measurements taken by the *bematists* suggests that they may have used

---

[1] Ath. *Diep.* 10.442b; Plin. (E) *HN* 6.61–62. In the nineteenth century, the British employed professional pace-counters, called Pundits, to undertake surveys of India and the Himalaya. The Pundits were trained, much like the Macedonian *bematists*, to walk with a regular pace—in this instance, 2,000 paces to the English mile—and each one hundred paces was recorded by removing one bead from set of Buddhist prayer beads that the Pundit would carry, and then placing that bead inside a prayer wheel. Some distances covered by the Pundits, such as the journey from Sikkim to Lhasa, covered 2,540 km or a staggering 3,160,000 paces (see: I. Cameron, *Mountains of the Gods* (London, Century, 1984), pp.119–125). Pace-counting is still used as a technique for navigation by modern soldiers and by people doing orienteering.

[2] Plin. (E) *HN* 6.61.

Eratosthenes and the Measurement of the Earth's Circumference (c.230bc). Christopher A. Matthew, Oxford University Press.
© Christopher A. Matthew (2023). DOI: 10.1093/oso/9780198874294.003.0005

some type of odometer, such as the one described by Heron of Alexandria.[3] Kegerreis, on the other hand, suggests that they just used 'advanced pace-measuring.'[4] Their collective name may derive from *bēma* (βῆμα)—the Greek word for 'pace'—which would indicate their methodology for measuring distances. This seems to be confirmed by the ancient grammarian Hesychius who states that the term *bematist* is Macedonian in origin and that they 'measured with their feet' (βηματίζει—τὸ τοῖς ποσὶ μετρεῖν. ἔστι δέ πως ἡ λέξις Μακεδονική).[5]

Kegerreis argues that because there is no record of any measurements taken by the *bematists* prior to the arrival of Alexander's army at the Caspian Gates in modern-day Iran, the unit of *bematists* could only have been formed partway through the campaign.[6] However, there is a potential flaw in Kegerreis' reasoning. Even if the unit was created midway through the campaign as he suggests, it would still take considerable time for a *bematist* to be trained to walk, and survey, using a regular pace. The average soldier could not have simply been 'promoted' or transferred to the unit and be expected to perform his duties with any level of accuracy. This suggests that, even if the unit itself had not existed prior to the middle of Alexander's campaign, individual *bematists*, would have been accompanying the expedition—possibly right from the very beginning. Kegerreis even suggests, somewhat contradictorily, that the origin of the term 'bematist' was first used in the Macedonian army 'shortly before or during the campaign.'[7] More so, these *bematists* would have learnt their craft in Greece prior to the campaign, which would suggest that teams of *bematistai* had been working for the different city-states of Greece for a long time. Kegerreis, suggests that anyone who was a 'day runner' (i.e. courier), such as Alexander's *bematist* Philonides, was also a surveyor.[8] Based on such reasoning, famous day runners from the fifth century BC, such as the Athenian Pheidippides, may have also acted as a *bematist*—even if the term itself may not have existed.[9] Additionally, Democritus mentions the so-called *arpedonaptai*

[3] Engels, *Logistics*, p.158; Hero. *Dioptr.* 39.
[4] C. Kegerreis, 'Setting a Royal Pace: Achaemenid Kingship and the Origin of Alexander the Great's *Bematistai*' *AHB* 31.1–2 (2017), p.40.
[5] Hsch. *Lex. s.v.* βηματίζει.
[6] Kegerreis, 'Royal Pace', pp.39, 43–53.
[7] Kegerreis, 'Royal Pace', p.41.
[8] An excellent examination of the *bematists* can be found in Y. Tzifopoulos, "Hemerodromoi' and Cretan 'Dromeis': Athletes or Military Personnel? The Case of the Cretan Philonides' *Nikephoros* 11 (1998), pp.137–170.
[9] Pheidippides is known for two great feats of endurance running (but not surveying). The first occurred in 490 BC during the Persian invasion of Greece. Pheidippides was sent on an urgent errand from Athens to request military aid from Sparta. According to the historian Herodotus (6.105–106), Pheidippides covered the 246 km distance from Athens to Sparta in only two days. This is the same

(ἀρπεδονάπται)—which translates as 'rope fasteners'—who seem to have used a cord of a set length which was stretched out between two points in order to measure distances.[10] This would suggest that measuring distance through pacing, and/or the use of a rope of a set length, may have had a long tradition in the ancient Greek world. Regardless, the army of Alexander the Great clearly contained several *bematists*, and they recorded many distances covered by the army during its long campaign.[11]

The methodology employed by the *bematists* to measure distances (i.e. pacing), and anthropological data on the ancient Greeks, indicate the unit of measure that these surveyors would have been using. An analysis of skeletal remains of ancient Greeks, dating from the sixth to the third centuries BC, place the average height of a male in this time at between 162 cm and 165 cm.[12] The average stride length for a male of this stature is around 68 cm (with an approximate range of 67.25 cm to 68.5 cm).[13] A 68 cm pace is approximately 1.5 times the 45 cm *cubit* of the Pan-Hellenic system of measurements that

distance (and route) over which the modern Spartathalon ultra-endurance race is run each year in which competitors have thirty-six hours to complete the course. The other feat that Pheidippides is known for is that, following the Greek victory over the Persians at the Battle of Marathon in 490 BC, Pheidippides ran the 42 km back to Athens to announce the victory (Luc. *Pro Lapsu* 3). The modern Olympic marathon race is run over this same distance in emulation of Pheidippides.

[10] Ps. Dem. *Alc.* Fr.68B 299.

[11] The names of several of Alexander's *bematists* have come down to us through the ancient sources: Philonides (who is also described as a personal courier for Alexander), Baeton, Diognetus, and Amyntas. For the activities of these individuals, and the *bematists* in general, see: Paus. 6.16.5; Ath. *Diep.* 2.74, 10.59 11.59, 11.102, 12.9, 12.39; Ael. *NA* 5.14, 17.17; Eus. *DE* 9.36; Hyg. *Ast. Po.* 2.30; Plin. (E) *HN* 2.181, 2.73, 6.44–45, 6.61–63, 6.69, 7.11, 7.20, 7.84; Str. *Geog.* 2.1.5–8, 2.1.23, 11.8.9, 11. 9.1, 15.1.28.

[12] Angel first reached this conclusion via the examination of around sixty-one skeletons in 1944 (see: J.L. Angel, 'A Racial Analysis of the Ancient Greeks: An Essay on the Use of Morphological Types' *Am J Phys Anthropol* 2 (1944), pp.331, 334; J.L. Angel, 'Skeletal Material from Attica' *Hesperia* 14 (1945), pp.284–285, 324). For a discussion of some of the potential issues with Angel's findings—such as the small sample size and the possibility that the skeletons were all from the upper class of ancient Greek society—see: L. Foxhall and H.A. Forbes, 'Σιτομετρεία: The Role of Grain as a Staple Food in Classical Antiquity' *Chiron* 12 (1982), p.47. Other scholars have suggested a height of around 170 cm, see: L. van Hook, 'On the Lacedaemonians Buried in the Kerameikos' *AJA* 36 (1932), pp.290–292; R. Gabriel and K. Metz, *From Sumer to Rome: The Military Capabilities of Ancient Armies* (Connecticut, Greenwood Press, 1991), p.71; W. Donlan and J. Thompson, 'The Charge at Marathon: Herodotus 6.112' *CJ* 71 (1976), p.341; V.D. Hanson, 'Hoplite Technology in Phalanx Battle' in V.D. Hanson (ed.), *Hoplites: The Classical Greek Battle Experience* (London, Routledge, 1991), pp.67–68; Matthew, *Storm of Spears*, p.8. An examination of approximately 1,000 skeletons from the region of Magna Graecia (the area of Greek control in southern Italy) also found the range in height for males to be between 162 cm and 165 cm—confirming the earlier findings by Angel (see: M. Henneberg and R.J. Henneberg, 'Biological Characteristics of the Population Based on Analysis of Skeletal Remains' in J.C. Carter (ed.), *The Chora at Metaponto: The Necropoleis Vol.II* (Austin, University of Texas Press, 1998), pp.509–514, 517–521.

[13] This can be determined by simply multiplying the height of an individual by 0.415 to determine their average stride length (see: R. Guest, O. Miguel-Hurtado, S. Stevenage, and S. Black, 'Exploring the Relationship Between Stride, Stature and Hand Size for Forensic Assessment' *J Forensic Leg Med* 52 (2017), p.47). Taking the mid-point in the range given by Henneberg and Henneberg (see n.12) as 163.5 cm, this then gives an average stride length of (163.5 x 0.415 =) ≈ 68 cm.

contains a 180 m *stade*. This seems to be confirmed by the information found in a metrological table that forms an appendix to some editions of the *Tactica* by the ancient writer Aelian, in which the conversion rate from cubits to paces is given as 1:1.6.[14] This equates to a pace for someone slightly taller than the average Greek of the fourth century, with an average height of around 172 cm. It is most likely that this larger ratio is an attribution to the stature of later Roman legionaries—from the time Aelian was writing. Vegetius, for example, outlines how one of the requirements for enlistment into Rome's legions was a minimum height of 172 cm.[15] A person of such height would have a pace of around 71 cm, and the ratio of this pace to the size of the earlier Greek cubit would be 1:1.6. Thus, taking the ratio to be 1:1.5 as per the anthropological data for the Hellenistic Greeks, the procedure for the recording of a distance measurement by a *bematist* would likely take the following steps:

1) The distance is initially measured in the number of paces taken to traverse the distance (e.g. 300,000 paces).
2) The number of paces is then divided by 1.5 to find the number of *cubits* (e.g. 300,000 paces/1.5 = 200,000 *cubits*).
3) The number of *cubits* is then divided by 400 (as per *Suda s.v.* Πλέθρον) to find the number of *stadia* traversed (e.g. 200,000 *cubits*/400 = 500 *stadia* of 180 m each).

Being professional measurers, a high level of accuracy would be expected in the distances recorded by the *bematists*. The US Army, for example, considers a 3–5 per cent margin of error acceptable for pace-counting navigation over flat terrain.[16] Engels, in his examination of the records of the *bematists*, gave the figures found in Strabo an average margin of error of 2.3 per cent (median 1.4 per cent) compared to the actual distances.[17] However, there are a number of issues with Engels' determinations (see following). Regardless, by understanding the methodology and unit used by the *bematists*, it would be expected that, when their measurements, as recounted by Strabo, are converted using a 180 m unit of measure, the results would have a similar margin of error of around 3 per cent (or better) when the recorded distances are compared to actual distances.

---

[14] See: Diller, 'Julian of Ascalon', p.23.
[15] Veg. *Mil.* 1.5.
[16] For example, see: US Department of the Army, *Army Field Manual 3-25.26: Map Reading and Land Navigation Handbook* (Eugene, Doublebit Press, 2019), pp.5–8; M. Guadognoli, G. Fober, and P. Terry, 'Accuracy of Pace Count as a Distance Estimation Procedure' *Mil. Psychol.* 2–3 (1990), pp.183–191.
[17] Engels, *Logistics*, p.157.

In relation to the distances recorded by Alexander's *bematists*, and then recounted by Eratosthenes, Strabo states:

> Eratosthenes gives the distances as follows: From . . . Alexandreia in the country of the Arians, . . . to the city of Bactra, also called Zariaspa, three thousand eight hundred and seventy [stadia]; . . . He also gives the distance from the Caspian Gates to . . . Hecatompylos, they say one thousand nine hundred and sixty stadia; to Alexandreia in the country of the Arians, four thousand five hundred and thirty [stadia]; then to Prophthasia in Drangê, one thousand six hundred [stadia]; . . . then to the city Arachoti, four thousand one hundred and twenty [stadia]; then to Ortospana, . . . two thousand [stadia] . . .'[18]

By comparing the figures given by Eratosthenes, through Strabo, to the distances between known modern locations using basic satellite mapping software such as Google Earth and Google Maps, it can be seen that Eratosthenes was citing distances that had been measured using a *stade* 180 m in length and that these measurements had an average margin of error of <1 per cent.

5.1.1 Caspian Gates To Hecatompylos

Geographically speaking, in terms of following the path of the locations given by Strabo from west to east, the distance from the Caspian Gates to Hecatompylos (1,960 *stadia*) comes first. However, the determination of this distance is somewhat problematic, as the locations of both of these sites have not been positively identified.

Even in antiquity, the location of the Caspian Gates was not certain. Pliny, for example, notes that there were a number of mountain passes in the land of the 'Caspian People' (*gens Caspia*) that could be referred to as the 'Gates', and recommends that the best way to determine their location would be to refer to the distances recorded by Alexander (which is what Eratosthenes did).[19] Unfortunately, the main surviving narratives of Alexander's campaign that we have, those of Arrian and Curtius, do not provide a lot of detail that helps narrow down the location of the Gates. Arrian described the Gates as a significant boundary between the eastern and western parts of the Persian Empire and that they mark the 'limit of cultivated land'.[20] Arrian also notes, in a different part of his narrative, that the Gates were about one day's march from

---

[18] Str. *Geog.* 11.8.9.
[19] Plin. (E) *HN* 6.15.40.
[20] Arr. *Anab.* 3.21.1, 5.25.5.

Rhagae—the suburb of Shahr-el-Rey in the south-eastern parts of present-day Tehran.[21] Strabo, citing a lost work by Apollodorus, states that it was even further eastward—500 *stadia* from the city.[22] Consequently, most modern scholars have placed the Gates somewhere to the east of Tehran.

The corpus of modern works attempting to determine the location of the Gates is vast and offers two possible directions that Alexander's army may have taken eastwards from the vicinity of Tehran. Alexander's army would have most likely traversed the easiest terrain, which has not changed significantly over the last 2,300 years, and many modern roads would follow the same, or a very similar, route. One route follows a north-easterly path through the modern town of Pardis before turning south towards Semnan and then turning north-easterly again. The second route follows a south-easterly path through Garmsār. Modern roadways follow both of these routes (see Fig. 67 following).

Modern estimations for the location of both the Gates and Hecatompylos lie on both of these routes. Van Donzel and Schmidt, for example, locate the Gates at Shahr-el-Rey in eastern Tehran.[23] Jackson, basing his conclusion on the passages of Arrian cited above, places the Gates at Sar-i-Darreh, 82 km east of Shahr-el-Rey, and close to Garmsār, on the southern route.[24] This location is also suggested by Heckel and Trittle.[25] Sar-i-Darreh is one of the two 'most favoured' sites according to Kegerreis, with the other being the Hableh Rud Gorge further to the east on the southern route.[26] Fuller states that the Gates were approximately 71 km east of Shahr-el-Rey, and that Alexander's army camped near the modern city of Aradan on the day after passing through the Gates.[27] This would place the Gates to the east of Garmsār on the southern route. Bosworth, on the other hand, states that the Gates were north-east of Tehran and was made up of a 'complex of defiles'.[28]

---

[21] Arr. *Anab.* 3.20.2.

[22] Str. *Geog.* 11.9.1.

[23] E.J. Van Donzel and A.B. Schmidt, *Gog and Magog in Early Christian and Islamic Sources: Sallam's Quest for Alexander's Wall* (Leiden, Brill, 2010), p.11.

[24] A.V.W. Jackson, *From Constantinople to the Home of Omar Khayyam* (New York, Macmillan, 1911), pp.127–137.

[25] W. Heckel and L. Trittle, *Alexander the Great: A New History* (Oxford, Wiley, 2011), p.41.

[26] Kegerreis, 'Royal Pace', p.55; see also: A. Anderson, 'Alexander at the Caspian Gates' *TAPA* 59 (1928), pp.130–163; J. Hansman, 'The Problems of Qūmis' *J.R. Asiat. Soc.* 100.3–4 (1968), pp.111–139; J.F. Standish, 'The Caspian Gates' *G&R* 17 (1970), pp.17–24; A.B. Bosworth, *A Historical Commentary on Arrian's History of Alexander Vol.I* (Oxford, Clarendon Press, 1980), pp.333–341; R. Stoneman, 'Romantic Ethnography: Central Asia and India in the Alexander Romance' *AncW* 25.1 (1994), pp.99–100.

[27] J.F.C. Fuller, *The Generalship of Alexander the Great* (Hertfordshire, Wordsworth, 1998), p.113.

[28] A.B. Bosworth, *Conquest and Empire: The Reign of Alexander the Great* (Cambridge, Cambridge University Press, 2008), p.94.

Locating the site of ancient Hecatompylos is somewhat easier, but still shrouded in scholarly controversy. The ancient writer Curtius simply calls it 'a famous city'.[29] The site was first connected with the modern Iranian town of Damghan by Houtum-Schindler in 1887.[30] Hansman excavated the site of Shar-i-Qūmis, 40 km south-west of Damghan, in the 1960s and 1970s and, based upon finds of period relative pottery and his interpretation of the distances recorded by Alexander's *bematists*, concluded that this was the site of Hecatompylos.[31] Bosworth, citing Hansman, also took Qūmis to be the location of Hecatompylos.[32] Kegerreis, however, questions Hansman's methodology on a number of key points: (a) Hansman declared a 'scholarly consensus' for Sar-i-Darreh being the location for the Caspian Gates—despite a clear lack of such a consensus; (b) Hansman's acceptance of the *bematists'* distance of 1,960 *stadia*, when applied to the actual distance between Sar-i-Darreh and Shar-i-Qūmis, results in an uncommon size for the *stade* of 163 m—which Hansman then uses to reach an equivalent modern distance that matches the distance between these two sites given in another ancient source—Pliny.[33] Hansman's *stade* is not far removed from the Messenian *stade* of 160 m. Kegerreis, on the other hand, suggests that Strabo's distance is actually given in Attic *stadia* of 185 m, which would make the distance 144 miles (231 km). Kegerreis also points out that Hansman has ignored the fact that the distance given in Strabo is longer than that given by Pliny—which he calculates would be equal to 225 miles (362 km)—a figure also given by Engels in his work on the logistics of the Macedonian Army—who also uses an 'Attic' *stade* of 185 m in his calculations.[34] Based on this, Kegerreis argues that Qūmis cannot be Hecatompylos and favours the more likely location of Damghan. Woods also located Hecatompylos at Damghan.[35]

Locating ancient Hecatompylos at Damghan allows the location of the Caspian gates (or at least one set of the Caspian Gates) to be determined by working backwards using the value given by Strabo. Using Shahr-e-Rey and Damghan as the most westerly and easterly possible sites for the

---

[29] Curt. 6.2.15; see also: Diod. Sic. 17.75.1.

[30] A. Houtum-Schindler, 'Notes on Some Antiquities Found in a Mound Near Damghan' *J.R. Asiat. Soc,* 9 (1877), pp.425–427.

[31] Hansman, 'Qūmis', pp.111–139; this is the location used by Engels (*Logistics*, p.85) in his work on the logistics of the Macedonian army.

[32] Bosworth, *Conquest and Empire*, p.96.

[33] Kegerrris, 'Royal Pace', p.56; Pliny (*HN*, 6.17.44) gives the distance between the Caspian Gates and Hecatompylos as 133 Roman miles or around 122 modern miles (196 km). Hansman's calculations, with the small *stade*, equal 128.5 miles (206 km).

[34] Kegerreis, 'Royal Pace', p.56; Engels, *Logistics*, pp.157–158.

[35] M. Woods, *In the Footsteps of Alexander the Great* (London, BBC Worldwide, 2001), p.127.

Caspian Gates and Hecatompylos, respectively, and plotting these locations into a modern satellite mapping program such as Google Earth Pro, shows the distances between both of these two points via both the northern and southern routes which then include all of the other sites that have been suggested by scholars attempting to identify these landmarks (Fig. 67):

Fig. 67 The routes between the Caspian Gates (Shahr-e-Rey) and Hecatompylos (Damghan).[a]

[a] Image created using Google Earth Pro.

As can be seen in Fig. 67, the distance along the northern route—from Shahr-e-Rey to Pardis, Damavand, Firuzkuh, south-east to Semnan and then north-east through Qūmis to Damghan—is 351 km. The southern route—from Shahr-e-Rey to Garmsār, Aradan, Semnan, Qūmis, and Damghan—is 332 km. Modern mapping software packages such as Google Maps and Google Earth take any undulation or incline in the terrain into consideration when they determine distance. This means that their 'on-the-ground' distance estimates can be considered highly accurate. This was not possible with the linear distances obtained from topographical maps that many scholars have relied upon in the past—except through very painstaking calculations based on the contour lines on the map. The values obtained from Google Earth can then be used as a basis of comparison for conversions of the distance given in Strabo (1,960 *stadia*) using the various sizes for the *stade* that have been offered in prior studies on both the location of these sites and on the work of Eratosthenes (Table 6):

Adopting the US Army's ±3 per cent as the best-case margin of error, the conversions using all but the 177 m *stade* and the 180 m *stade* along the northern route between Shahr-el-Rey and Damghan can be discounted. Thus, a

Table 6 The conversion of distances for the different routes between the Caspian Gates and Hecatompylos, by known sizes of the *stade*, with margins of error.

Stade size	Conversion (of 1,960 stadia)	Margin of error North Route (351 km)	Margin of error South Route (332 km)
160 m (Messenian)	314 km	−11.8%	−5.7%
177 m (Delphic)	347 km	−1.2%	+4.3%
180 m (Pan-Hellenic)	353 km	+0.6%	+5.9%
185 m (Attic)	363 km	+3.3%	+8.5%
191 m (Olympic)	374 km	+6.1%	+11.2%
210 m (Ptolemaic)	412 km	+14.8%	+19.4%

route linking two of the proposed sites for the Caspian Gates and Hecatompylos (Shahr-e-Rey and Damghan, respectively) can be shown to be within a margin of error of +0.6 per cent when the distance of 1,960 *stadia* given in Strabo is converted into a modern equivalent using the Pan-Hellenic *stade* of 180 m in length and following the route suggested by Bosworth. This is the smallest margin of error for any of the conversions.

What may contribute to this margin of error is that the ancient narratives do not specifically state where in each of these locations the measurements were taken. For example, was the distance between two cities measured from gate to gate? Or from marketplace to marketplace? Or from the centre of one site to the centre of the other? The modern mapping software used in this examination measures distances between selected points, but these points may not fully align with the positions the *bematists* measured from and to. Depending upon the size of the settlement or location at either end of these measured distances, the overall length of a route between them may be out by several hundred metres—which, over a distance of around 350 km, for example, could equate to a variation of around 0.1 per cent. Even if this potential margin of error is added to the conversions detailed above in Table 6, it is only the conversion using the Pan-Hellenic Standard that results in a value with a margin of <1 per cent. This not only suggests that, contrary to several long-running scholarly debates, Shahr-e-Rey and Damghan are the locations of the Caspian Gates and Hecatompylos respectively that are mentioned in the works of Strabo, but also that the distance between them is likely given in units incorporating the Pan-Hellenic 180 m *stade* (contra estimates made by both Hansman and Kegerreis and others).

The accuracy in the conversion to a modern unit of measure using a 180 m *stade* additionally suggests that the distance had not been recorded using one of the other units of measure that are attributed to Eratosthenes' calculation of the circumference of the Earth by previous scholars. Additionally, it shows that the two locations given in the work of Strabo have to be referring to both Shahr-e-Rey and Damghan as the calculated distances between any of the other proposed sites results in a higher margin of error (Table 7):

As can be seen from Table 7, the locations for the Caspian Gates and Hecatompylos cannot be Sar-i-Darreh and Qūmis respectively as the conversion into any of the known *stadia* results in large margins of error. Similarly, the locations cannot be Sar-i-Darreh and Damghan as, although they are lower, the margins of error are all still very high. Furthermore, the locations can also not be Shahr-e-Rey and Qūmis as, although this is a greater distance which results in a lower margin of error compared to the other conversions, the margins are all still high which makes these two locations unlikely.

Converting the distance between Shahr-e-Rey and Damghan into a modern equivalent using proposed *stadia* from the Hellenistic Period of a larger size—191 m and 210 m—makes the distance between the Gates and Hecatompylos quite large (374 km and 412 km respectively). This would require the site of

Table 7 The conversion of distances between alternative sites for the Caspian Gates and Hecatompylos by known sizes of the *stade*, with margins of error.

Stade size	Conversion (of 1,960 *stadia*)	Margin of Error Sar-i-Darreh to Qūmis (204 km)	Margin of Error Sar-i-Darreh to Damghan (250 km)	Margin of Error Shahr-e-Rey to Qūmis (286 km)
160 m (Messenian)	314 km	35.0%	20.4%	8.9%
177 m (Delphic)	347 km	41.2%	28.0%	17.6%
180 m (Pan-Hellenic)	353 km	42.2%	29.2%	19.0%
185 m (Attic)	363 km	43.8%	31.1%	21.2%
191 m (Olympic)	374 km	45.5%	33.2%	23.5%
210 m (Ptolemaic)	412 km	50.5%	39.3%	30.6%

the Caspian Gates to be further west than Shahr-e-Rey, or Hecatompylos further east than Damghan, or both. As such, it is only through the use of the Pan-Hellenic 180 m *stade* that the distance stated in Strabo and any of the proposed sites for these locations closely align. This shows that while Pliny describes the presence of a number of passes in the region that could be referred to as the Gates, and while some of the sites on the southern route proposed by some modern scholars may be the locations of these other Gates, in regard to the route taken through the Gates by Alexander's army as recounted by Eratosthenes and then Strabo, they took the more northerly of the two options and that the locations of the Caspian Gates and Hecatompylos are Shahr-e-Rey and Damghan.

Furthermore, it is unlikely that Strabo has converted Eratosthenes' work into units more contemporary with his own time as the distances would not then correlate with a conversion to a known standard from the Classical and Hellenistic Periods. This then leaves that the distances are just as Alexander's surveyors recorded them, and just how they were reported by Eratosthenes. The important result of this is that Eratosthenes seems to have been using measurements in the Pan-Hellenic standard for his discussion of the distance between the Caspian gates and Hecatompylos.

5.1.2 Hecatompylos To Alexandria Aerion

The next distance, in geographical order from west to east, as reported in Strabo is that from Hecatompylos to Alexandria Aerion. The latter has been identified as modern Herat in north-western Afghanistan.[36] Hammond alternatively suggests that the site was near Artacoana—which, in turn, is near modern Herat.[37] Strabo gives the distance between these two locations as 4,350 *stadia*. The modern route from Damghan to Herat—which follows the path taken by Alexander's army as described in the ancient narratives— via Shahroud, Sabzevar, Neyshabur, Farīmān, Torbat-e-Jam, Taybad, and Ghourian, is just over 815 km (Fig. 68).

Converting the distance given in Strabo (4,530 *stadia*) again confirms the unit of measure that is being used (Table 8):

As can be seen from Table 8, most of the conversions, except for those using the 177 m and 180 m *stade*, fall outside the US Army's 3 per cent margin of error—with the conversion using the 185 m *stade* close to the cut-off

[36] H.H. Wilson, *Ariana Antiqua* (London, East India Co., 1841), p.151; V. Tscherikower, 'Die hellenistischen Städgrundüngenvon Alexander dem Grossem bis auf die Römerzeit' *Philologus Suppl.* 19.1 (1927), pp.1–216; Engels, *Logistics*, p.85.

[37] N. Hammond, *The Genius of Alexander the Great* (London, Duckworth, 1997), p.131.

Fig. 68 The route between Hecatompylos (Damghan) and Alexandria Aerion (Herat).[a]

[a] Image created using Google Earth Pro.

(2.8 per cent). When converted using a 180 m *stade*, the distance between Hecatompylos and Alexandria Aerion given in Strabo also calculates as 815 km. This represents a margin of error of 0.0 per cent in relation to the figure stated in Strabo and the modern distance—but assumes that the route plotted into the mapping software (815.25 km) and the distance used in the conversion calculations (815 km) differs by a quarter of a kilometre due to the start and end points used by Alexander's *bematists* being unknown. Regardless, the conversion using the 180 m *stade* is far more accurate than the estimates found in prior studies. Engels, for example, estimated the modern distance to be 855 km, which gave a margin of error between his converted distance (838 km—based on a 185 m *stade*) and his modern distance of −2.0 per cent.[38]

Table 8 The conversion of the distance between Hecatompylos and Alexandria Aerion by known sizes of the *stade*, with margins of error.

Stade size	Conversion (of 4,530 *stadia*)	Margin of Error (vs 815 km)
160 m (Messenian)	725 km	−12.4%
177 m (Delphic)	802 km	−1.6%
180 m (Pan-Hellenic)	815 km	0.0%
185 m (Attic)	838 km	2.8%
191 m (Olympic)	865 km	5.8%
210 m (Ptolemaic)	951 km	14.3%

[38] Engels, *Logistics*, p.157.

Kegerreis similarly suggested that the modern distance was 836 km.[39] While these values would conform to the use of a 185 m *stade*, any alignment is only due to an incorrect value for the actual modern distance between the locations. Consequently, the conversion of the stated distance between Hecatompylos and Alexandria Aerion also indicates that the values reported by Eratosthenes, and cited by Strabo, are given in *stadia* 180 m in size.

5.1.3 Alexandria Aerion To Prophthasia

As with the locations for the Caspian Gates and Hecatompylos, there has been scholarly debate over the location of Prophthasia. Droysen stated that he was not able to determine a location for the city.[40] Berve suggested it was the site of modern Farrah in southern Afghanistan.[41] Tarn suggested that 'Prophthasia' was merely a nickname given to the city of Alexandria in Sestian which he identified as Zaranj further south in Afghanistan.[42] Engels offered that it was not easy to identify the location, but settled on the site of Juwain, near Farrah, in his analysis.[43] Fraser argued against Tarn, and supported Berve, in suggesting that Prophthasia is Farrah.[44] Bosworth similarly located the site at Farrah and referred to it as the Darangian capital.[45] Hammond suggests that Alexander encamped at Phrada (Farrah), but then founded a new city in Darangiana, which he called Prophthasia.[46] Cohen simply states that 'the exact location of Prophthasia is not yet known.'[47]

However, contrary to Hammond, in the ancient sources, both Plutarch and Stephanus state that Alexander the Great renamed the city of Phrada to Prophthasia (which means 'anticipation').[48] This renaming was most likely done as a conspiracy against Alexander was uncovered at the site.[49] Strabo states that the distance between Alexandria Aerion and Prophthasia is 1,600 *stadia*. Using

[39] Kegerreis, 'Royal Pace', p.54.
[40] See: J.G. Droysen, *Histoire de l'hellénisme* (Paris, E. Leroux, 1883), pp.674–676, 692.
[41] H. Berve, *Das Alexanderreich auf Prosopographischer Grundlage Vol.I* (München, Beck, 1926), p.293.
[42] Tarn, *Alexander Vol.II*, p.233.
[43] Engels, *Logistics*, p.158.
[44] P.M. Frazer, *Cities of Alexander the Great* (Oxford, Oxford University Press, 1996), pp.139, 249.
[45] Bosworth, *Conquest and Empire*, pp.100, 101, 104.
[46] Hammond, *Genius of Alexander*, pp.132–133.
[47] G.M. Cohen, *The Hellenistic Settlements in the East from Armenia and Mesopotamia to Bactria and India* (Berkeley, University of California Press, 2013), p.284.
[48] Plut. *Mor.* 328F; Steph. Byz. *s.v. Φράδα.*
[49] Arr. *Anab.* 3.26–27; Curt. 6.7.1–7.2.38; Diod. Sic. 17.79–80; Plut. *Alex.* 48–49.

Farrah as the consensus location for Prophthasia, modern satellite mapping software shows that the distance between modern Herat and Farrah through Adraskan and Khaki Safed is 287 km (Fig. 69).

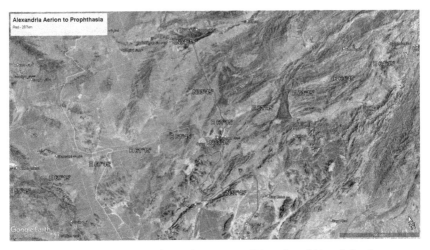

Fig. 69 The route between Alexandria Aerion (Herat) and Prophthasia (Farrah).[a]

[a] Image created using Google Earth Pro.

Converting the stated distance (1,600 *stadia*) into modern equivalents again demonstrates a correlation to the use of the Pan-Hellenic *stade* of 180 m in length (Table 9):

The conversion of the distance between Alexandria Aerion and Prophthasia—two relatively well identified locations—is much simpler than with the calculations of distance between disputed sites like the Caspian

Table 9 The conversion of the distance between Alexandria Aerion and Prophthasia by known sizes of the *stade*, with margins of error.

Stade size	Conversion (of 1,600 *stadia*)	Margin of Error (vs 287 km)
160 m (Messenian)	256 km	−12.1%
177 m (Delphic)	283 km	−1.3%
180 m (Pan-Hellenic)	288 km	0.3%
185 m (Attic)	296 km	3.0%
191 m (Olympic)	306 km	6.1%
210 m (Ptolemaic)	336 km	14.6%

Gates and Hecatompylos. Combined with an expected margin of error of 3 per cent, the data for the conversions between all of these sites strongly favours the use of *stadia* of either 177 m or 180 m by Eratosthenes and Strabo, and would seem to discount the use of any larger or smaller units.

Furthermore, the use of satellite mapping to obtain a more accurate distance between these two sites highlights some of the errors found within prior examinations. For example, a margin of error in the ancient measurements between Alexandria Aerion and Prophthasia of only 0.3 per cent, when a 180 m unit of measure is used in the conversion, is far more accurate than the work of Engels—who estimated the modern distance to be 304 km, which resulted in a margin of error between his converted distance (257 km based on the use of a 185 m *stade* and setting the location at Juwain) and his modern distance of −10.5 per cent.[50] Kegerreis similarly suggested the modern distance was 296 km based on a 185 m *stade* and a location at Farrah for Prophthasia.[51] Furthermore, the distance from Herat to Zaranj (as per Tarn) is 549 km, which would give the conversion a margin of error of 91 per cent. Such a huge margin of error would immediately discount the site of Zaranj as a possible location for ancient Prophthasia. Based upon this analysis of the source material, it can be stated with a high level of probability that modern Farrah is ancient Prophthasia. The analysis of the source material also further supports the conclusion that Eratosthenes' figures are based on a *stade* 180 m in length.

5.1.4 Prophthasia To Arachoti Polis

There have also been several possible locations for the site of Arachoti Polis proposed by modern scholars. Boyce and Grenet, for example, have suggested that it is the city of Kandahar—383 km east of Farrah.[52] Engels placed Arachoti Polis at the site of modern Qalat-el-Gilzay to the north-east of Kandahar on the main road to Kabul and 511 km distant from Farrah.[53] Kegerreis called Engels' conclusion 'educated guesswork'.[54] The distance given in Strabo does not initially aid this contention. The stated distance given is 4,120 *stadia*, which equates to a distance of around 742 km when converted by the *stade* of 180 m

---

[50] Engels, *Logistics*, p.157.
[51] Kegerreis, 'Royal Pace', p.54.
[52] M. Boyce and F. Grenet, *A History of Zoroastrianism Vol.III: Zoroastrianism under Macedonian and Roman Rule* (Leiden, Brill, 1991), p.128.
[53] Engels, *Logistics*, pp.157–158.
[54] Kegerreis, 'Royal Pace', p.56.

that has the smallest margin of error in the other distance measurements.[55] As Kegerreis points out, this would suggest a location much further east than either Kandahar or Qalat-el-Gilzay if the starting point of the calculation was Farrah.[56] However, the figure given in Strabo can be reconciled with a positive identification of the site with Qalat-el-Gilzay as per Engels through an understanding that what this passage is referring to is not actually the distance between Prophthasia and Arachoti Polis.

As noted, the distance from modern Farrah (Prophthasia) to Qalat-el-Gilzay is 511 km (Fig. 70).

Fig. 70 The route between Prophthasia (Farrah) and Arachoti Polis (Qalat-el-Gilzay).[a]

[a] Image created using Google Earth Pro.

If this is added to the distance from the previously listed distance from Alexandria Aerion (Herat) to Prophthasia (Farrah) of 287 km, this comes to a total distance of 797 km (Fig. 71).

This equates to a distance of 4,428 *stadia* when converted using a 180 m *stade*—a difference to the distance given by Strabo of 308 *stadia* (or around 55 km). This seems to highlight a transcription error in the text of Strabo. Up to this point, the text had been ascribing distances between two locations, and then having the finishing point for one distance act as the starting point for the next distance in the list (e.g. Caspian Gates to Hecatompylos; Hecatompylos to Alexandria Aerion; Alexandria Aerion to Prophthasia). However, the

---

[55] Engels (*Logistics*, p.157) calculated this distance as 845 km, while Kegerreis ('Royal Pace', p.54) stated that the distance was 835 km.

[56] Kegerreis, 'Royal Pace', p.56.

Fig. 71 The route between Alexandria Aerion (Herat) and Arachoti Polis (Qalat-el-Gilzay).[a]

[a] Image created using Google Earth Pro.

next distance in the list, rather than following the format set out for the other measurements and being the distance from Prophthasia to Arachoti Polis (2,839 *stadia* based on the location of Qalat-e-Gilzay), is really the distance from Alexandria Aerion (Herat) to Arachoti Polis with a possible margin of error of around 7 per cent. As the text seems to have been corrupted in regard to the locations, a corruption of the figures themselves also cannot be discounted. It is possible that the value given as 4,120 *stadia* (͵δρκ) should have been 4,420 *stadia* (͵δυκ) which would make the distance out by only 8 *stadia*.

The identification, and subsequent correction, of such corruptions in the text then also correlates with the use of a 180 m *stade* in the record of this distance (Table 10):

Table 10 The conversion of the distance between Alexandria Aerion and Arachoti Polis by known sizes of the *stade*, with margins of error.

Stade size	Conversion (of 4,420 *stadia*)	Margin of Error (vs 797 km)
160 m (Messenian)	707 km	−12.7%
177 m (Delphic)	782 km	−1.9%
180 m (Pan-Hellenic)	796 km	−0.2%
185 m (Attic)	818 km	2.5%
191 m (Olympic)	844 km	5.6%
210 m (Ptolemaic)	928 km	14.1%

As can be seen, a proposed value of 4,420 *stadia* converts to approximately 796 km using a 180 m *stade*—different from the modern distance by only 1 km and giving the value a margin of error on only −0.2 per cent. Again, if a worst-case margin of error of 3 per cent is assumed, then the measurements have to have been given using either 177 m, 180 m, or 185 m units—with the conversion using the 180 m Pan-Hellenic *stade* again having the smallest margin of error. Additionally, the site of Arachoti Polis is unlikely to be that of Kandahar as Boyce and Grenet suggest, as this would equate to a total distance from Herat of 653 km, or around 3,628 *stadia*, which would result in a greater margin of error of around 12 per cent. This suggests that a corruption of both the place names and the values within the text of Strabo is more than likely.

5.1.5 Arachoti Polis To Hortospana

Hortospana has been identified as the city of Kabul.[57] Strabo lists the distance from Arachoti Polis to Hortospana as 2,000 *stadia*. This helps identify the site of Arachoti Polis with present Qalat-e-Gilzay (see previous). The route from Kabul to Qalat-e-Gilzay is 366 km (Fig. 72).

Fig. 72 The route between Arachoti Polis (Qalat-el-Gilzay) and Hortospana (Kabul).[a]

[a] Image created using Google Earth Pro.

---

[57] K. Fischer, 'Zur Lage von Kandhar an Landverbindungen zwischen Iran und Indien' *Bonner Jahrb.* 167 (1967), pp.129–252.

Table 11 The conversion of the distance between Arachoti Polis and Hortospana by known sizes of the *stade*, with margins of error.

Stade size	Conversion (of 2,000 *stadia*)	Margin of Error (vs 366 km)
160 m (Messenian)	320 km	−14.4%
177 m (Delphic)	354 km	−3.4%
180 m (Pan-Hellenic)	360 km	−1.7%
185 m (Attic)	370 km	1.1%
191 m (Olympic)	382 km	4.2%
210 m (Ptolemaic)	420 km	12.9%

Converting the stated distance (2,000 *stadia*) further correlates with the use of the Pan-Hellenic *stade* of 180 m in the writings of Eratosthenes and Strabo (Table 11):

This shows that, when a *stade* 180 m in length is used, the stated and actual distances disagree by only 6 km—a margin of error of around −1.7 per cent. Engels estimated the modern distance to be 372 km, and the ancient distance to be 370 km, which gave him a margin of error of −0.5 per cent based on the use of a 185 m *stade*. However, Engel's modern estimate is different from the actual distance by 6 km, and due to the relatively small distance between the two locations, this affected the margins in his determination.[58] Kegerreis similarly suggested 370 km for the modern distance.[59] While the conversion of the stated distance into a modern equivalent has a smaller margin of error when a 185 m *stade* is used of 1.1 per cent, as the distance between the two locations is so small (366 km), the difference between the margins of error for the conversions using the 180 m and 185 m *stadia* (2.8 per cent) represents a distance of only 10.25 km. This variance could be attributable to only a slight deviation between Alexander's route and the modern road, and/or the use of slightly different start and end points in the ancient and modern measurements. The higher margin of error for the conversion using the Delphic Standard unit of 177 m (−3.4 per cent) falls outside of the best-case margin. This would suggest that this unit was not used in the determination of these distances that are reported by Eratosthenes/Strabo—despite the close correlation between recorded and converted distances between other sites when using this unit. If

[58] Engels, *Logistics*, p.157.
[59] Kegerreis, 'Royal Pace', p.54.

it is assumed that Eratosthenes used the same unit in all of his reported measurements (and the *bematists* would have undoubtedly used the same unit), this narrows the possibilities for which unit was being used and further indicates that Alexander's *bematists*, or Eratosthenes, or both, were using a *stade* of 180 m in length in their determinations of distances.

5.1.6 Alexandria Aerion To Bactra-Zariaspa

One of the other distances listed by Strabo is not part of this continuing trail—that from Alexandria Aerion (Herat) to Bactra-Zariaspa (Balkh). This is another determination that is simplified by the fact that both locations have been positively identified. Arrian calls Zariaspa the capital of the region of Bactria.[60] The distance given by Strabo is 3,870 *stadia*. The modern distance by road between Herat and Balkh, via Kala Nau, Bala Murghab, Maimana, and Andkhui, is 708 km (Fig. 73).

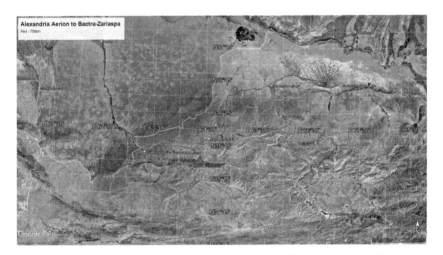

Fig. 73 The route between Alexandria Aerion (Herat) and Bactra-Zariapsa (Balkh).[a]

[a] Image created using Google Earth Pro.

Yet again, a conversion of the stated distance into modern equivalents using *stadia* of different lengths indicates a use of the Pan-Hellenic Standard or the Attic Standard in the recording of these measurements (Table 12):

---

[60] Arr. *Anab.* 4.2.

Table 12 The conversion of the distance between Alexandria Aerion and Bactra-Zariaspa by known sizes of the *stade*, with margins of error.

Stade size	Conversion (of 3,870 *stadia*)	Margin of Error (vs 708 km)
160 m (Messenian)	619 km	−14.3%
177 m (Delphic)	685 km	−3.4%
180 m (Pan-Hellenic)	697 km	−1.6%
185 m (Attic)	716 km	1.1%
191 m (Olympic)	739 km	4.2%
210 m (Ptolemaic)	813 km	12.9%

As with the route from Arachoti Polis to Hortospana (see previous), the lower margins of error suggest the use of a 180 m *stade* or the 185 m *stade* in the determination of this distance. Engels, using the larger 185 m *stade* as a basis, offered that the ancient measurement converted to a distance of 716 km, and that the actual distance was 705 km (giving him a margin of error of 1.5 per cent).[61] However, Engels' margin differs from that given here because his estimation of the modern distance being slightly too low. Also, similar to the conversion of the distance from Arachoti Polis to Hortospana, the distance between the two sites is 708 km—which makes the difference between the margins of error for the 180 m and 185 m *stade* conversions (2.7 per cent) equal to about 19 km. This again could be attributable to a small deviation between Alexander's route and that of the modern road and/or slight differences in the start and end points of the measurement.

When *stadia* of 177 m, 180 m, and 185 m are considered, the conversion of the distances recorded by Eratosthenes, and then later recounted by Strabo, are accurate for the distances between known, or now confirmed, locations with a margin of error of around ±2.0 per cent (Table 13).

This level of accuracy for recordings made by Alexander's *bematists* is better than had been suggested by Engels, and is far better than the acceptable margins for pace-counting used by the US Army. As can be seen, the most accurate conversions are those done using the Pan-Hellenic *stade* of 180 m in length, which has an average margin of error of around −0.4 per cent. The conversions using both the smaller 177 m *stade* of the Delphic Standard, and the larger 185 m *stade* of the Attic Standard, both sit within 2 per cent (on average) on either side of the Pan-Hellenic conversions. While this does place all of these conversions within the best-case margin of error of 3 per cent, the level of

[61] Engels, *Logistics*, p,157.

Table 13 Eratosthenes' distances, as cited in Strabo, converted to modern distances using a 177 m, 180 m, and 185 m *stade*, compared to actual distances, and with margins of error.

Ancient Names	Distance (Str.Geog.1.8.9)	Distance (A) (GoogleEarth)	177 m Stade		180 m Stade		185 m Stade	
			Converted Distance (S)	Margin (S-A/S)x100	Converted Distance (S)	Margin (S-A/S)x100	Converted Distance (S)	Margin (S-A/S)x100
Caspian Gates to Hecatompylos	1,960 *stadia*	351 km	347 km	−1.2%	353 km	+0.6%	363 km	+3.3%
Hecatompylos to Alexandria Aerion	4,530 *stadia*	815 km	802 km	−1.6%	815 km	0.0%	838 km	+2.8%
Alexandria Aerion to Prophthasia	1,600 *stadia*	287 km	283 km	−1.3%	288 km	+0.3%	296 km	+3.0%
Alexandria Aerion to Arachoti Polis(corrected)	4,420 *stadia*	797 km	782 km	−1.9%	796 km	−0.2%	818 km	+2.5%
Arachoti Polis to Hortospana	2,000 *stadia*	366 km	354 km	−3.4%	360 km	−1.7%	370 km	+1.1%
Alexandria Aerion to Bactra-Zariaspa	3,870 *stadia*	708 km	685 km	−3.4%	697 km	−1.6%	716 km	+1.1%
				Avg ≈ −2.0%		Avg ≈ −0.4%		Avg ≈ +2.3%

accuracy of the Pan-Hellenic conversions suggests that this was the unit used by Alexander's *bematists*, recorded by Eratosthenes, and recounted by Strabo.

The small variations in the distances recorded by the *bematists*—either to the actual distances and/or when comparing the conversions using different *stadia*—can be attributed to several factors:

- possible rounding of the numbers by Alexander's *bematists*, Eratosthenes, and/or Strabo;
- slight differences in the routes measured by Alexander's *bematists* and through modern satellite imagery and mapping software;
- measurements to slightly different start and end points by Alexander's *bematists* and through modern satellite imagery and mapping software;
- some of the distances are quite short and a small variation in reported and actual distances creates a larger margin of error; and
- the differences between the size of the units being considered is only small (3 m between the Delphic and Pan-Hellenic Standards, and 5 m between the Pan-Hellenic and the Attic). Over short distances, the difference between converted measurements will be small and this, in combination with other factors listed above, would result in close margins of error.

However, these factors, either individually or in combination, do not have a significant impact on the accuracy of the measurements. Nicastro says it does not matter how accurate Eratosthenes was with the figures cited in Strabo so long as he overestimated as much as he underestimated.[62] We have no way of knowing whether Eratosthenes converted the distances recorded by Alexander's *bematists* into 180 m units, or if the *bematists* themselves used this unit—but it seems likely that this was the unit the *bematists* used based upon the anthropological evidence. Regardless, Nicastro is somewhat correct in this matter. Statistically, if the two outriders in the dataset for the Pan-Hellenic *stade* (+0.6 per cent and –1.7 per cent) are removed, the mean margin of error is still only –0.4 per cent. When the outriders for the conversions using the *stadia* of other sizes are similarly removed, the mean margins of error for those conversions are: 160 m = –12.9 per cent; 177 m = –2.1 per cent; 185 m = +2.4 per cent; 191 m = +5.4 per cent; 210 m = +14.0 per cent. This also demonstrates that Eratosthenes was most likely using a 180 m unit for all of these distances.

---

[62] Nicastro, *Circumference*, p.126.

Nicastro, based upon the errors resultant from the use of a 160 m *stade* in his conversions, also suggests that fixing accurate measurements between locations over long distances was problematic until the invention of modern surveying techniques.[63] Clearly, Alexander's surveyors were capable of measuring distances with a level of accuracy that was much higher than Nicastro's statement would imply. The inaccuracy assumed by Nicastro has come from the use of a *stade* of the incorrect size for his analysis. The real level of accuracy, taken with nothing more than the measured step of a professional surveyor, demonstrates not only the precision that Alexander's *bematists* could accomplish, but also that Eratosthenes was using a *stade* of 180 m in the recording of this information (regardless of whether the *bematists* themselves used this unit of measure or not). If Eratosthenes is simply recounting the values recorded by Alexander's surveyors, then it can be stated, with a high degree of confidence, that the '*bematists*' stade' equalled the unit 180 m in length of the Pan-Hellenic Standard. There is also one other measurement of distance, attributed to Eratosthenes but connected to a work independent of the recording of geographical distance in Iran and Afghanistan, that confirms his use of a *stade* of this size.

5.1.7 Alexandria To Syene

The other recorded distance that is found in an ancient text, and that is directly attributable to work by Eratosthenes, is the figure for the distance between Alexandria and Syene that he used in his calculation of the circumference of the Earth. Cleomedes states that this value was 5,000 *stadia*.[64] Similarly, Strabo states that the city of Meroe, situated on the equator, was 10,000 *stadia* from Alexandria, and therefore Syene, being close to the Tropic of Cancer, and halfway between the two locations, was 5,000 *stadia* from Alexandria.[65] Both the accuracy of this given figure and where it was derived from have been, much like most of the modern work on Eratosthenes, mired in scholarly debate.

In his account of Eratosthenes' experiment and calculations, Cleomedes does not elaborate where the value for the distance to Syene came from—other than implying that was a piece of established information that Eratosthenes

---

[63] Nicastro, *Circumference*, p.26.
[64] Cleom. *De motu* 1.10.
[65] Str. *Geog.* 2.5.7.

was working with.[66] The later writer Martianus Capella (fourth century AD) says that Eratosthenes got the distance from Syene and Meroe, presumably for his work on geography and/or his map, from Ptolemy's *bematists* (*per mensores regios Ptolemaei*).[67] This account of where Eratosthenes obtained some of his values from is found in no other ancient source. Despite such a reference, scholars have proposed numerous theories over the years as to where Eratosthenes may have obtained the 5,000 *stadia* figure for the distance between Alexandria and Syene.

One theory is that Eratosthenes had previously obtained, or determined, the distance for the composition of his map of the *oikoumene*. Walkup, for example, states that 'there is little doubt that Eratosthenes got this figure directly from his earlier map of the known world'.[68] Walkup goes on to say that we may never know how Eratosthenes got this value for his map, but is doubtful that he measured it himself.[69] Goldstein goes as far as to suggest that Eratosthenes made no actual measurements at all, but relied on estimates and approximations for distances and angles, and that Cleomedes (or his sources) misunderstood what Eratosthenes did when they recounted his experiment.[70] Such a conclusion seems unlikely. Not only is there little reason to question Cleomedes' reporting but, despite the simplicity of what Eratosthenes is reported to have done, the accuracy of his results would suggest that he did undertake some observations, measurements, and/or calculations. Indeed, if Cleomedes' account is accurate, then Eratosthenes followed many of the hall-marks of modern scientific practice where a hypothesis was formulated, an experiment was devised to test the hypothesis, observations were made, and then results and conclusions were reached.

Shipley, conversely, suggests there was no such thing as 'science' in the ancient Greek world, at least not in terms of the modern definition of it being 'an objective branch of knowledge, based on systematic observation, experiment, and tests and aimed at understanding the material world', in part because there was no formal qualification in scientific investigation in ancient Greece.[71] However, Eratosthenes' experiment to calculate the circumference of the Earth would seem to comply with all of the criteria that Shipley lists. Russo outlines how '[t]o build a scientific theory, then, it is not enough to be able to deduce one statement from another; one must choose appropriately

---

[66] Cleom. *De motu* 1.10.
[67] Mart. Cap. *Phil.* 6.598.
[68] Walkup, 'Mystery of the Stadia' *s.v. Distance from Alexandria to Syene*.
[69] Walkup, 'Mystery of the Stadia' *s.v. Distance from Alexandria to Syene*.
[70] Goldstein, 'Eratosthenes', pp.411–416.
[71] Shipley, *Greek World*, pp.326–327.

the premises and terms of discourse. Also essential was the use of elements other than verbal argumentation, drawn from observation and from technical activities . . . .'[72] Again, the elements of Eratosthenes' experiment, such as the calculated distance between Alexandria and Syene, would seem to conform with these principles.

Schwarz and Russo, following the reference in Capella, suggest that Eratosthenes obtained the distance value from *bematists*.[73] Cuomo suggests that the distance had at least been recorded by surveyors.[74] It has also been suggested, by Fraser and Nicastro, that Eratosthenes either hired these *bematists*, or was given the use of them by one of the ruling Ptolemies (either Ptolemy III or Ptolemy IV).[75] Nicastro suggests that Eratosthenes needed 'as the crow flies' distances, and that any hired *bematists* would have required surveyors to go with them to ensure they were walking in a straight line.[76] There are several things wrong with the claims of Fraser and Nicastro. Firstly, it seems unlikely that Eratosthenes would have been working with an 'as the crow flies' linear measurement for the distance between Alexandria and Syene. Indeed, the evidence indicates that the 5,000 *stade* value is based upon a route which, for the most part, follows the course of the river Nile (see following). As such, no additional personnel would be required to ensure that a measurement was being taken in a straight line over hundreds of kilometres.

Furthermore, while the account of Capella states that Eratosthenes obtained the data from the Royal *Bematists*, this does not necessarily mean that Eratosthenes used the *bematists* himself to determine the distance from Syene to Meroe (or from Alexandria to Syene). It is just as probable that records that had already been taken by the *bematists* were housed in the collection of the Library where Eratosthenes was working. Both Herodotus and Strabo describe how the size of the arable land of Egypt was surveyed each year following the annual flood of the Nile in order to establish the levels of taxation for the next twelve months.[77] These measurements, which would have been taken by

[72] Russo, *Forgotten Revolution*, p.173; For a discussion on the debate over whether the ancient Greeks understood the concepts behind scientific methodology or not, see Russo, *Forgotten Revolution*, pp.194–196.

[73] K.P. Schwarz, 'Zur Erdmessung des Eratosthenes' *AVN* 82 (1975), p.8; Russo, *Forgotten Revolution*, p.275.

[74] S. Cuomo, *Ancient Mathematics* (London, Routledge, 2005), p.85.

[75] See: Fraser, *The Cities of Alexander*, pp.80–81; Nicastro, *Circumference*, p.111.

[76] Nicastro, *Circumference*, p.111.

[77] Hdt. 2.109: 'What is more, this king, so they said, divided the country among all the Egyptians by giving each [of them] an equal square parcel of land, and made this the source of his revenue, initiating the payment of an annual tax' [κατανεῖμαι δὲ τὴν χώρην Αἰγυπτίοισι ἅπασι τοῦτον ἔλεγον τὸν βασιλέα, κλῆρον ἴσον ἑκάστῳ τετράγωνον διδόντα, καὶ ἀπὸ τούτου τὰς προσόδους ποιήσασθαι, ἐπιτάξαντα ἀποφορὴν ἐπιτελέειν κατ᾽ ἐνιαυτόν]. Similarly: Str. *Geog.* 17.1.3: 'In Egypt, the Nile passes in a straight

the Royal *Bematists*, could have been used to determine a distance along the course of the Nile with only a slight variation (see following). Eratosthenes had employed such a method in the past. Strabo details how, in his work on *Geography*, Eratosthenes had relied on registered measurements that had been kept by the Persian Court to determine the distance from the Euphrates River to the Caspian Gates.[78]

Strangely, despite suggesting the employment of both *bematists* and survey-ors, Nicastro follows the description found in Strabo to claim that Eratosthenes may have calculated the linear distance 'geometrically' by laboriously going over the tax maps which divided all of the arable (i.e. taxable) land along the river into measured parcels.[79] Nicastro also states that, due to the topo-graphical obstacles of taking such a measurement (i.e. traversing the Nile delta, patches of desert, and crossing the river) the result could only be a very good approximation at best.[80] It is somewhat unclear what Nicastro is implying with this statement. The annual surveying of arable land (if that was the basis for Eratosthenes' figure) would not require the river to be crossed, nor to have the surveyors venture into the desert. If, on the other hand, Nicastro is suggest-ing the possibility of Eratosthenes hiring *bematists* to measure the distance from Alexandria to Syene, then they may have been required to traverse a small section of desert at least (see following). If the annual tax records were kept in the Library in Alexandria, the use of predetermined measurements, recorded by *bematists*, may be the foundation for Capella's reference to where Eratosthenes obtained his value for the Alexandria-Syene distance.

This has not prevented other scholars from forwarding their own hypothe-ses on where the 5,000 *stade* figure came from. Kline, for example, in an

---

line from the little cataract above Syene and Elephantine, at the boundary of Egypt and Ethiopia, to the sea. The country was divided into *nomes* [provinces], which were subdivided into sections [...] There was need of this accurate and minute division because of the continuous confusion of the boundaries caused by the Nile at the time of its increases [flooding], since the Nile takes away and adds soil, and changes conformations of land, and in general hides from view the signs by which one's own land is distinguished from that of another. Of necessity, therefore, the lands must be re-measured again and again. And here it was, they say, that the science of land-measuring originated.'

[78] Str. *Geog.* 2.1.24: 'He divides the whole into portions, as he found registered measurements recorded as follows: he started back at the Euphrates crossing near Thapsacus and from there to the Tigris, where Alexander crossed, he records 2400 stadia; from here to several locations in order, through Gaugamela and the Lycus and Arbela and Ecbatana, where Darius fled from Gaugamela to the Caspian Gates, he calculates the 10,000 stadia' [κατὰ μέρος δὲ διαιρῶν, ὡς ἀναγεγραμμένην εὗρε τὴν μέτρησιν οὕτω τίθησιν, ἔμπαλιν τὴν ἀρχὴν ἀπὸ τοῦ Εὐφράτου ποιησάμενος καὶ τῆς κατὰ Θάψακον διαβάσεως αὐτοῦ. μέχρι μὲν δὴ τοῦ Τίγριδος, ὅπου Ἀλέξανδρος διέβη, σταδίους δισχιλίους καὶ τετρακοσίους γράφει· ἐντεῦθεν δ᾽ ἐπὶ τοὺς ἑξῆς τύπους διὰ Γαυγαμήλων καὶ τοῦ Λύκου καὶ Ἀρβήλων καὶ Ἐκβατάνων, ᾗ Δαρεῖος ἐκ τῶν Γαυγαμήλων ἔφυγε μέχρι Κασπίων πυλῶν, τοὺς μυρίους ἐκπληροῖ].

[79] Nicastro, *Circumference*, p.111; Nicastro also concedes that we have no way of knowing how Eratosthenes reached his figure.

[80] Nicastro, *Circumference*, p.111.

unreferenced passage, suggests the distance was measured by knowing that camel trains, which could travel 100 *stadia* per day, take 50 days to travel to Syene.[81] Goldstein and Russo suggest that the distance was based on the time taken to walk or sail between the locations multiplied by the average distance travelled per day, and that the 5,000 *stade* figure may have been a traditionally accepted value in Eratosthenes' time.[82] Dueck similarly suggests that the distance was based upon the movement of camel caravans, and that any resultant value would only be an estimation, as it would be dependent upon such things as the weather and the means of transport.[83] Russo, who assumes that Eratosthenes had undertaken multiple observations to determine the location of the Tropic, despite no ancient source stating this in regards to the calculation of the circumference of the Earth, states that the accuracy of these observations would be pointless if the distance from Alexandria to Syene was then only approximated based on travel times between the two sites. Dutka additionally dismisses the idea that the distance was based upon travel times as camel caravans did not become common in Egypt until the Christian Era—several centuries after the time of Eratosthenes.[84]

Dutka further dismisses the idea that the 5,000 *stade* figure came from either Eratosthenes' map or from surveyors as, he claims, no source mentions these.[85] This would clearly ignore the statement made by Capella. Rather, Dutka suggests that Eratosthenes may have got the figure from travellers' reports and official records kept in the Library.[86] Dutka does not elaborate on what he thinks these official records are, but some of them may have been the tax surveys carried out by the *bematists* that are mentioned by other scholars. The use of texts housed within the Library would also correlate with the statement of Cleomedes, which does not actually mention Eratosthenes undertaking his own measurements. Walkup suggests that Eratosthenes may have used the annual land surveys and that, as it is a round figure, it may have been the traditionally accepted value for the distance—a value that had been established long before his time.[87] Rawlins offers that Eratosthenes thought he was using land measurements but in fact was using measurements drawn from astronomical data.[88] This would also imply that Eratosthenes was drawing upon

[81] Kline, *Mathematical Thought*, p.161.
[82] Goldstein, 'Eratosthenes', pp.411–412; Russo, *Forgotten Revolution*, p.273.
[83] Dueck, *Geography*, p.70.
[84] Dutka, 'Eratosthenes' Measurement', p.58.
[85] Dutka, 'Eratosthenes' Measurement', pp.60–62.
[86] Dutka, 'Eratosthenes' Measurement', p.62.
[87] Walkup, 'Mystery of the Stadia' *s.v. Distance from Alexandria to Syene*.
[88] Rawlins, 'Geodesy', pp.211–219, especially 215 n.15.

some other text. Indeed, there would have been several texts in the Library's collection that would have mentioned distances to Syene. As early as the fifth century BC, the historian Herodotus gave the distance from the sea to Elephantine (near Syene) as 7,920 *stadia*.[89] Herodotus had travelled in Egypt as part of the research for his work, but his text is notorious for its use of exaggerated and incorrect numbers and cannot be totally relied upon.[90] Interestingly, however, this distance is very close to distance from Halicarnassus in south-western Turkey (Herodotus' home town), across the eastern end of the Mediterranean to Alexandria, along the western edge of the Nile delta to Memphis (Cairo) and then up the river to Syene (a distance of around 1,700 km) when converted into *stadia* using the Ptolemaic Standard unit 210 m in length (8,095 *stadia*). It is possible that what Herodotus is describing is the distance he travelled from his home, or some other starting location in south-western Turkey, to the furthest point south that he got to in his travels. Regardless, Herodotus' account of distances in Egypt, and other works with a higher level of accuracy, may have been the sources that Eratosthenes drew upon for his distance to Syene.

In the end, how or where the 5,000 *stade* figure for the distance from Alexandria to Syene was obtained is something of a moot point as it has no bearing on the determination of the accuracy of Eratosthenes' calculations other than further confirming the use of a 180 m unit of measure as it appears that, for his calculation of the circumference of the Earth, Eratosthenes was using a latitudinal distance of around 4,500 *stadia* between these two sites. This, however, has not prevented modern scholars from engaging with the 5,000 *stade* figure found in the passage of Cleomedes.

Several scholars have suggested that the 5,000 *stade* figure found in Cleomedes' account is rounded.[91] Similarly, Fischer has stated that 'some modern writers correctly stress that Eratosthenes' round number of 5,000 *stadia* for the Alexandria-Syene distance was only an approximation'.[92] This doubt over

---

[89] Hdt. 2.9: 'the distance from the sea inland to Thebes is 6,120 *stadia*. It is a further 1,800 *stadia* from Thebes to Elephantine' [ὅσον δέ τι ἀπὸ θαλάσσης ἐς μεσόγαιαν μέχρι Θηβέων ἐστί, σημανέω: στάδιοι γὰρ εἰσὶ εἴκοσι καὶ ἑκατὸν καὶ ἑξακισχίλιοι. τὸ δὲ ἀπὸ Θηβέων ἐς Ἐλεφαντίνην καλεομένην πόλιν στάδιοι χίλιοι καὶ ὀκτακόσιοι εἰσί].

[90] For some examples of commentaries on Herodotus' numbers, see: W.W. How and J. Wells, *A Commentary on Herodotus Vol.II* (Oxford, Clarendon Press, 1912), p.212; C. Hignett, *Xerxes' Invasion of Greece* (Oxford, Clarendon Press, 1963), p.39; R.M. Macan, *Herodotus: The 7th, 8th and 9th Books Vol.I* (New York, Arno Press, 1973), p.lxxxii; A.B. Lloyd, *Herodotus Book II: Commentary 1–98* (Leiden, Brill, 1976), pp.41–45; J.F. Lazenby, *The Defence of Greece 490–479BC* (Warminster, Aris and Phillips, 1993), 90.

[91] For example, see: Evans, *Ancient Astronomy*, p.64; Dreyer, *Thales to Kepler*, p.175 Rawlins, 'Geodesy', p.261.

[92] Fischer, 'Another Look', p.153.

the accuracy of the figures found in Cleomedes and Strabo has led to many theories that examine Eratosthenes' calculations, which alter the value given in these ancient texts in order to reach a conclusion. Viedebantt, for example, basing his analysis on the 252,000 *stade* figure for the Earth's circumference found in the later ancient texts (as opposed to the 250,000 *stade* figure given in Cleomedes), and favouring the use of a *stade* around 160 m in length by Eratosthenes, stated that there must be an error in the reported 5,000 *stade* figure found in Cleomedes and suggested that the 'real' figure should be 5,250 *stadia*.[93] The basis of this assumption is that 5,250 x 160 m = 840 km—which is the linear, 'as the crow flies', distance between Alexandria and Syene that Viedebantt used in his examination.[94] Viedebantt also bases his analysis on a passage found in Strabo, which states that the distance from the Cataract on the Nile south of Syene to the Sea is 5,300 *stadia*.[95] Both Dreyer and Nicastro also base some of their conclusions on this passage, despite the fact that in another part of the text, Strabo specifically states that the distance from Syene to Alexandria is 5,000 *stadia*.[96] The issue with the later passage found in Strabo is that is does not specifically designate Alexandria and the end point for the measurement, merely the sea. As such, it could be the measured distance to anywhere within the Nile Delta. Additionally, the Cataracts of the Nile are south of Syene. Much of the river south of modern Aswan was inundated with the construction of the Aswan High Dam in the 1960s. However, there is a set of narrows in Lake Nasser—which was formed with the building of the dam. These narrows are located approximately 54 km south of modern Aswan— a distance of 300 *stadia* when converted using a 180 m *stade*. Strabo refers to the 'little cataract' which was south of Syene and on the border of Egypt and Ethiopia.[97] This would suggest a location over one hundred kilometres south of Syene. However, another cataract, not as far south of Syene, may be the Cataract that Strabo is referring to in his later passage, which provides a distance of 5,300 *stadia* to the sea.

Alternatively, Shcheglov offers that the 5,300 *stade* distance from Syene to the Sea found at Strabo 17.1.2 considers the bends in the river, whereas the 5,000 *stade* value at Strabo 2.5.7 is a rounded number to reflect the latitudinal difference between the two sites.[98] Parts of this conclusion have merit. Plotting

[93] Viedebantt, 'Eratosthenes', pp.211, 216.
[94] Viedebantt, 'Eratosthenes', p.215.
[95] Viedebantt, 'Eratosthenes', p.212; Str. *Geog.* 17.1.2.
[96] Dreyer, *Thales to Kepler*, p.175; Nicastro, *Circumference*, p.84; Str. *Geog.* 2.5.7.
[97] Str. *Geog.* 17.1.2.
[98] Shcheglov, 'Itinerary Stade', p.171.

a course that follows the river between Alexandria and Syene in Google Earth gives a distance of approximately 977 km (Fig. 74).

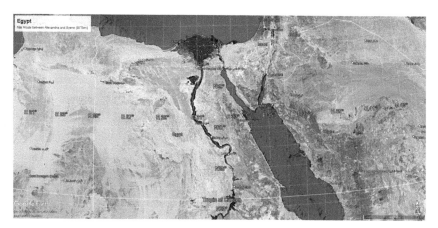

Fig. 74 The route between Alexandria and Syene (Aswan), following the course of the Nile.[a]

[a]Image created using Google Earth Pro.

A total of 5,300 *stadia*, converted using the 180 m unit that was employed for the distances between locations in Iran and Afghanistan, equates to 954 km—a margin of error of 2 per cent. However, despite this low margin in the conversion, this does not account for Strabo specifically stating that the 5,300 *stade* distance was measured from a point south of Syene. As such, the two figures given in Strabo (5,000 *stadia* and 5,300 *stadia*) appear to be giving the distances between different locations, or different routes, in the same geographical area. Unfortunately, Shcheglov then uses an 'as the crow flies' distance of 843.6 km from Alexandria to Syene, determined using Google Maps, to justify the use of a 157.5 m *stade* by Eratosthenes when it is unlikely that he was actually basing his calculations on a record of 'on the ground' linear distance.[99]

Fischer suggests that Eratosthenes revised his distance from 5,000 *stadia* to 5,040 *stadia* so that a result of 252,000 *stadia* would divide evenly into 4,200 *stadia* per *hexacontade*.[100] Nicastro also states that Eratosthenes' figure was 5,040 *stadia*.[101] Gulbekian suggests that it is the 250,000 *stade* figure for the circumference that is rounded, that it should be 252,000 *stadia* as per the later texts, and therefore the distance to Syene is 252,000 x 1/50 = 5,040 *stadia*.[102] As

[99] Shcheglov, 'Itinerary Stade', p.172.
[100] Fischer, 'Another Look', p.154.
[101] Nicastro, *Circumference*, pp.27, 117.
[102] Gulbekian, 'Origin and Value', pp.361–362.

previously noted, the value of 5,040 *stadia* may be an adjustment of the figure to match the ideals of Platonic numerical perfection, rather than the actual distance. Rawlins begins his examination with a value of 5,000 *stadia*, which he states is rounded, but then increases it to 5,050 *stadia* based upon adjusting the observed angle cast by the *gnomon* that is reported in Cleomedes (1/50 or 7°1") to 7°12'30" due to the error that he calculated was caused by the umbra of the shadow itself.[103] Fischer, however, goes even further in her revisions of the figure given in Cleomedes' text. Firstly, she suggests that the figure of 5,040 *stadia* is too short by 163 *stadia* because Eratosthenes would have known that Syene was north of the Tropic. Fischer then suggests that this additional figure was rounded up to 168 *stadia*. Then she further suggests that, because Syene is on the edge of the 'sub-solar zone' during the Solstice, an additional 150 *stadia* (half of the 300 *stade* wide 'sub-solar zone') needs to be added to that figure. This results in an additional 318 *stadia* added to the starting value of 5,040 *stadia*, giving a total of 5,358 *stadia*. Fischer made these adjustments to the figures so that the total adjustment to the circumference would equate to 15,900 *stadia*—which she states is the alteration to Eratosthenes' circumference made by Hipparchus (for a further discussion of this alteration by Hipparchus, see section 5.2, 'The 180 m Stade and the Circumference of the Earth').[104]

The problem with all of these analyses is that there is no ancient text that mentions Eratosthenes altering the figures that he was basing his calculations on, and no ancient text refers directly to any value other than 5,000 *stadia* as the distance from Alexandria to Syene. Such reworkings of a value that is specifically stated in two ancient sources can only be considered mathematical expediency. Additionally, there is no evidence for the use of a 160 m *stade* by Eratosthenes (or commonly in the ancient Greek world in general), and it is unlikely that Eratosthenes was basing his determinations on a linear distance. Indeed, there is no need to over complicate things, or to try to discern what Eratosthenes 'really meant', as many scholars have attempted to do in the past. There is also no reason to consider the figures given in Cleomedes as anything other than accurate (at least from Cleomedes' perspective). Rather, it is far more prudent to examine the data as it has been presented to us via the ancient texts and determine how Eratosthenes arrived at his conclusions. This will then allow for the accuracy of his observations, methodologies, and calculations to be determined. By accepting the distance from Alexandria to Syene as it is stated in the oldest surviving record of the calculations that we

[103] Rawlins, 'Geodesy', pp.259–261, 264.
[104] Fischer, 'Another Look', pp.155–156.

have—Cleomedes' value of 5,000 *stadia*—it is shown that, even in the recording of this distance, the unit of measure used was the Pan-Hellenic *stade* 180 m in length.

Cleomedes states that one of the pieces of assumed knowledge that Eratosthenes was working from was that Alexandria and Syene sit on the same longitudinal meridian.[105] However, Eratosthenes must have been aware that there was a significant bend in the river to the east, approximately 700 km to the south of Alexandria. The reason why he would have known this is that the important city of Thebes (modern Luxor) is situated on this bend, and Luxor had been a prominent Egyptian city for at least 1,900 years prior to the time of Eratosthenes. The result of this is that, even if Eratosthenes had been using land-survey data as some scholars suggest, he cannot have included any area that had surveyed land along the bend in the Nile in order to calculate a distance, as straight as possible, that went from Alexandria to Syene. If, on the other hand, Eratosthenes did use *bematists* to measure the distance, they must have been instructed to follow the river as closely as possible, but to cut across the 55 km stretch of desert that negated the bend in the river. Regardless of the method, the distance between Alexandria and Syene must have been based upon a route that began in Alexandria, followed the western edge of the Nile Delta as far as Memphis (modern Cairo), then followed the west bank of the river down to modern Nagga Hamadi where the bend in the river began, cut across the desert to the vicinity of modern Tod to pick up the river again, and then followed it south to Syene. This is the opposite of Nicastro's suggestion that traversing desert and crossing the Nile would make the resultant measurement anything more than an approximation. Rather, staying confined to the west bank of the river, and cutting across the desert, resulted in the most accurate figure for the distance to Syene that could be obtained without calculating an actual linear measurement. Plotting the course outlined above into Google Earth Pro results in a distance of just over 907 km (Fig. 75).

As with the distance measurements for locations in Iran and Afghanistan that are given in Strabo, this distance can then be used to compare the figure given in *stadia* by Cleomedes, with conversions based upon all of the units that have been proposed to have been used by Eratosthenes (Table 14):

As can be seen, as with many of the other distance measurements attributed to Eratosthenes, the stated value of 5,000 *stadia* converts into a modern equivalent, based on a 180 m *stade*, with a margin of error of less than 1 per cent. Conversion into the Messenian and Delphic standards result in a value that

---

[105] Cleom. *De motu* 1.10.

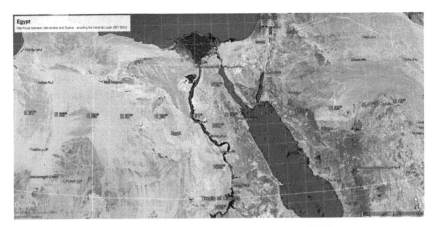

Fig. 75 The route between Alexandria and Syene (Aswan), avoiding the bend in the Nile at Luxor.[a]

[a] Image created using Google Earth Pro.

Table 14 The conversion of the distance between Alexandria and Syene by known sizes of the *stade*, with margins of error.

Stade size	Conversion (of 5,000 *stadia*)	Margin of Error (vs 907 km)
160 m (Messenian)	800 km	−13.4%
177 m (Delphic)	885 km	−2.5%
180 m (Pan-Hellenic)	900 km	−0.8%
185 m (Attic)	925 km	+1.9%
191 m (Olympic)	955 km	+5.0%
210 m (Ptolemaic)	1,050 km	+13.6%

is too short for the distance between Alexandria and Syene and can be easily dismissed. Conversions into the Attic standard results in a distance of 925 km. This is too long for the more direct route used here, but is also too short (by 52 km or ≈5.6 per cent) if the bend in the river is considered (977 km). This indicates that the Attic standard was not used in these measurements either. Conversion using the larger Olympic standard (955 km) results in a value that is either too long for the direct route (907 km), or too short for the full course or the river (977 km), while the use of the Ptolemaic standard results in a distance that is too long (1,050 km) even for the full length of the river.

Combining the percentage differences from Table 14 with the data from the other conversions of distance examined earlier indicates that Eratosthenes used the Pan-Hellenic *stade* of 180 m in length in his geographical works, and so would have done similarly in his calculation of the circumference of the

Earth. Furthermore, as the figures given for the distances between locations in Iran and Afghanistan, and the stated distance between Alexandria and Syene, are found in different sources (Strabo and Cleomedes respectively), it cannot be the case that a later author has converted Eratosthenes' work into something more contemporary. Thus, the only conclusion left is that these figures are just as Eratosthenes (or Strabo and Cleomedes) recorded them.

It also does not seem likely that the figure for the distance from Alexandria to Syene was a liner, 'as the crow flies', measurement as Viedebantt, Nicastro, and Shcheglov used in their examinations of Eratosthenes' work. The linear distance from Alexandria to Syene is approximately 840 km (Fig. 76).

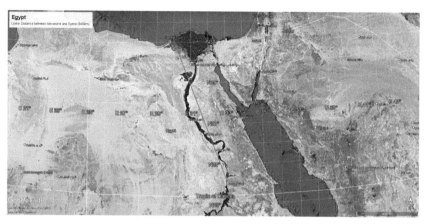

Fig. 76 The linear distance between Alexandria and Syene (Aswan).[a]
[a] Image created using Google Earth Pro.

Basing the 5,000 *stade* distance given in Cleomedes as equal to this linear distance, and converting the 5,000 *stadia* using all of the known sizes for the *stade*, yields the following results (Table 15):

As can be seen, all of the conversions, based on the linear distance between Alexandria and Syene are larger than some of the conversions of the 'on the ground' route which, mostly, follows the west bank of the river (see Fig. 75 and Table 14). The smallest margin of error is for the conversion using the Messenian Standard with a 160 m *stade*. However, this conversion is still out by a margin of −5 per cent. The conversion using the Delphic Standard, with a 177 m *stade*, is out by a similar amount (+5.1 per cent) but the conversion results in a larger distance, compared to a lesser distance when using the Messenian Standard. This explains why scholars such as Viedebantt, Nicastro, and Shcheglov have suggested the use of a *stade* of around 158 m in length by Eratosthenes, in conjunction with a 5,300 *stade* linear distance between

Table 15 The conversion of the linear distance between Alexandria and Syene by known sizes of the *stade*, with margins of error.

Stade size	Conversion (of 5,000 *stadia*)	Margin of Error (vs 840 km)
160 m (Messenian)	800 km	−5.0%
177 m (Delphic)	885 km	+5.1%
180 m (Pan-Hellenic)	900 km	+6.7%
185 m (Attic)	925 km	+9.2%
191 m (Olympic)	955 km	+12.0%
210 m (Ptolemaic)	1,050 km	+20.0%

Alexandria and Syene (as per their interpretation of the passage found in Strabo) as this would then allow the converted distance (5,300 x 158 = 837 km) to closely match with the actual linear distance and their ideas of how accurate Eratosthenes' calculations should be. The problem with such conclusions is that, firstly, there is no evidence for the use of a *stade* 158 m in length in the ancient Greek world, and so even if the 160 m *stade* had been used by Eratosthenes, it would subsequently need to be concluded that his results were not as accurate as they should be. Secondly, it is unlikely that Eratosthenes was basing his calculation on a linear distance between Syene and Alexandria as this would have required complex and painstaking calculations based on the survey data for the Nile, but there is no reference to him doing that in the surviving accounts of his experiment and calculations. Lastly, as has been shown, the 5,300 *stade* distance given by Strabo is to the small cataract south of Syene and is not to Syene itself. As such, it becomes clear that a linear distance was not being used by Eratosthenes in his calculation of the circumference of the Earth as some scholars have suggested.

Berger suggested that the ancients could not accurately determine the size of the Earth because they could not measure terrestrial distances accurately.[106] However, a review of the figures given in both Strabo and Cleomedes for the distances between different locations in Iran, Afghanistan, and Egypt, via a comparison to modern satellite imagery and mapping, demonstrates that the ancient *bematists* and other surveyors were capable of taking measurements over long distances, and over both flat and mountainous terrain, with a very high level of accuracy. Furthermore, as will be shown, and of utmost importance for the understanding of his calculation of the circumference of the Earth, this review additionally demonstrates that Eratosthenes' work was

---

[106] E.H. Berger, *Geschichte der wissenschaftlichen Erdkunde der Griechen* (Leipzig, Von Veit & Co., 1903), pp.267, 410, 591.

based on the use of a *stade* 180 m length and a latitudinal distance between Alexandria and Syene of around 4,500 *stadia*, and not the 5,000 *stade* value found in the account of Cleomedes.

5.2 The 180 m *Stade* And The Circumference Of The Earth

The use of a 180 m *stade* by Eratosthenes throws an interesting light on his calculations of the circumference of the Earth. Basing the calculation on the 250,000 *stadia* figure for the circumference that is reported by Cleomedes, the result is an answer with a large margin of error:

$$250{,}000 \times 180 \text{ m} = 45{,}000 \text{ km} \qquad \text{[eq. 5.2.1]}$$

This differs from the current value for the polar circumference of the Earth (40,007 km) by around 11 per cent.[107] How could Eratosthenes have got the figure so wrong?

Eratosthenes' experiment was sublime in its simplicity—merely analysing the angle of a shadow cast in Alexandria and using Euclidean geometry to apply that angle to the overall arc length distance between Alexandria and Syene to calculate the full circumference of a circle. The only major sources of uncertainty in the experiment come from Eratosthenes' ability to determine correctly the distance from Alexandria to Syene, and his ability to calculate accurately the angle cast by the *gnomon* in Alexandria. As such, and barring a major error in the reading of the shadow and its associated angle, a high level of accuracy, to within a few per cent, would be expected. This then suggests that what should be considered is whether there is something wrong with the reported account of Eratosthenes' experiment.

Interestingly, Pliny the Elder states that, in the second century BC, Hipparchus, who was famous for his critique of Eratosthenes' work, 'adjusted' Eratosthenes' result (given as 252,000 *stadia*) by adding 'not much less than 26,000 *stadia*'.[108] This was most likely done in his polemic work *Against Eratosthenes*, which has not survived, but which is mentioned in the

---

[107] For the current polar circumference of the Earth, see: Defense Mapping Agency, *Supplement to Department of Defense World Geodetic System 1984 Technical Report*, p.3–46.

[108] Plin. (E) *HN* 2.112: 'The above is all that I consider worth relating about the length and the breadth of the Earth. But Eratosthenes, a man who was particularly well skilled in all the more subtle areas of learning, and in this above everything else, and a person whom I consider to be accepted by everyone, has stated that the whole of the circumference to be 252,000 stadia, which, according to a Roman estimate, equals 31,500 miles. The attempt is bold, but it is supported by such subtle arguments

Suda.[109] Strabo also states that Hipparchus adopted a figure of 252,000 *stadia* and that he corrected Eratosthenes.[110]

How then, is the apparent 'adjustment' by Hipparchus to be seen? This is an important passage for understanding the accuracy of Eratosthenes' determination of the circumference of the Earth, but is another problematic passage to interpret. Unfortunately, it is also a passage that has been often neglected by modern scholars. Some scholars do not even refer to it in their examinations of Eratosthenes' calculations.[111] Other scholars make only passing reference to it. Carmen and Evans, for example, cite the passage in a footnote, stating that 26,000 *stadia* was added to the total, but then do not refer to it again in the rest of their examination.[112] Engels similarly refers to Hipparchus' addition without any further analysis.[113]

Other scholars have attempted to work out what was meant by Hipparchus' adjustment and why it may have been made. In the eighteenth century, d'Anville suggested that Hipparchus added 25,000 *stadia* (as opposed to the 26,000 *stadia* mentioned by Pliny) because Eratosthenes had underestimated the size of the distance by 10 per cent. This was based on d'Anville's assumption that Eratosthenes was using a *stade* of 149 m in length for his calculations.[114] Not only is there no evidence for Eratosthenes using a *stade* of this size, but there is also no evidence for its use throughout the ancient Greek world. Berger suggests that Pliny, or his source, may have confused the 38,000 *stade* north-south extent of the known world (taken from Eratosthenes' work on geography and/or his map), and the apparent 63,000 *stadia* per quadrant of a globe with a circumference of 252,000 *stadia* (63,000–38,000 = 25,000), or that it may be a mean value between Eratosthenes' result and the 300,000 *stade* figure given by Archimedes.[115]

---

that we cannot refuse our acceptance. Hipparchus, whom we must admire, both for the ability with which he counters Eratosthenes, as well as for his diligence in everything else, has added to the above number not much less than 26,000 stadia.' (*de longitudine ac latitudine haec sunt, quae digna memoratu putem. universum autem circuitum eratosthenes, inomnium quidem litterarum subtilitate, set in hac utique praeter ceteros solers, quem cunctis probari video, cclii milium stadiorum prodidit, quae mensura romana conputatione efficit trecentiens quindeciens centena miliapassuum: inprobum ausum, verum ita subtili argumentatione conprehensum, ut pudeat non credere. hipparchus, etin coarguendo eo et in reliqua omni diligentia mirus, adicit stadiorum paulo minus xxvi M.*)

[109] *Suda s.v.* Ἵππαρχος; see also: Str. *Geog.* 1.1.12, 2.1.38–41.
[110] Str. *Geog.* 2.1.38, 2.5.34.
[111] For example, see the works by Viedebantt, Diller; Rawlins, Dutka, Bowen, Walkup, Nicastro, and Shcheglov that have been cited previously.
[112] Carmen and Evans, 'Two Earths', p.3 n.8.
[113] Engels, 'Length', pp.302–303.
[114] d'Anville, 'Mémoire sur la mesure', pp.82–91; J.B.B. d'Anville, 'Discussion de la mesure de la terre par Ératosthène' *Mém. Acad. Inscript. et belles-lettres* 26 (1759), pp.92–100.
[115] E.H. Berger, *Die geographischen Fragmente des Eratosthenes* (Teubner, Leipzig, 1880), p.130.

Hultsch suggests that the text is corrupt and that the addition was only 2,600 *stadia*.[116] Similarly Fisher, somewhat confusingly, suggests that Hipparchus accepted Eratosthenes' result, and then suggests that Hipparchus calculated a refinement, possibly in an attempt to determine a better angle for the ecliptic as, Fischer states, Hipparchus would have been aware that Syene was north of the Tropic, but decided that it was either not certain enough, or significant enough, or both, to apply it.[117] Not only does such a premise go against the text of Pliny, which states that Hipparchus adjusted Eratosthenes' result, but it is uncertain how an alteration of around 10 per cent cannot be considered significant. Fischer additionally suggests that the value of 26,000 may be a transcription error where an extra X has been added to the Roman numerals for 16,000 (XVI M)—a claim for which there is no proof other than making the data fit with her ideas for the size of the *stade* that Eratosthenes was using.[118] Dicks calls the passage of Pliny an 'egregious blunder' and dismisses it based on the fact that an addition of 26,000 *stadia* to Eratosthenes' result (again taken as 252,000), for a total of 278,000 *stadia*, would then make the passage of Pliny conflict with that of Strabo which refers to the use of a 252,000 *stade* total by Hipparchus.[119] It can only be assumed that such omissions in analysis, as are found in the works of some modern scholars, are to make any subsequent examination into Eratosthenes ring true with other sources which give a result of either 250,000 or 252,000 *stadia* without really accounting for what the 'adjustment' actually was.

However, there is another way in which the 'adjustment' of Hipparchus can be seen. Hipparchus may have added around 26,000 *stadia* to the result of Eratosthenes' calculations in order to reach the figure of 250,000 *stadia* given by Cleomedes. It is unlikely that an adjustment was made to reach the 252,000 figure found in the Roman-period sources as this figure seems to be an alteration intended to make the results fit with the ideals of Platonic numerical perfection. Such an interpretation would mean that Eratosthenes' result was actually in the order of 224,000 *stadia*. Pliny states that the figure given for the adjustment is only an approximation (i.e. 'not much less'), which suggests that the adjustment was slightly smaller than 26,000 *stadia*. If a value of 25,900 *stadia* is taken as the size of Hipparchus' adjustment, and this is

[116] Hultsch, *metrologie*, pp.63–64.
[117] Fischer, 'Another Look', pp.154, 156.
[118] Fischer, 'Another Look', p.156.
[119] D.R. Dicks, *The Geographical Fragments of Hipparchus* (London, Athlone Press, 1960), p.153.

subtracted from the value given by Cleomedes of 250,000 stadia, this results in a value for Eratosthenes' calculation of the circumference of the Earth of 224,100 *stadia*. Such an interpretation has considerable consequences for the examination of the accuracy of Eratosthenes' calculations and the determination of the size of the *stade* that he was using in his calculations. Not only would this negate the need for any adjustment to the results of Eratosthenes' calculations to accommodate such things as a division into units of exact degrees—as some scholars have done despite the calculations being made in a time when such divisions were not in use—but would also potentially set the 'adjusted' work of Hipparchus as the initial source for the later passages, including Cleomedes' report of the experiment that gives a figure of 250,000 *stadia*.

The accuracy of a true result of Eratosthenes' calculation of the circumference of the Earth of around 224,100 *stadia* (conforming with the report of Cleomedes post Hipparchus' adjustment), when compared to the actual polar circumference of the Earth of 40,007 km, is confirmed when these distances are converted into modern distances using the *stadia* of different sizes that were either used in the ancient Greek world and/or by previous scholars in their examinations of Eratosthenes' work (Table 16).

As can be seen from Table 16, a conversion of a 224,100 *stade* circumference by a 180 m *stade* results in a distance in modern units of 40,338 km. This result has a margin of error of 0.8 per cent when compared to the modern value of the polar circumference of the Earth. If adjusted by—0.13 per cent to account for the errors in Eratosthenes' methodology, then the circumference equates to 40,286 km and the margin of error reduces to 0.7 per cent. If Eratosthenes' result was almost exactly 224,000—which would assume the

Table 16 The conversion of a 224,100 *stade* circumference of the Earth by known sizes of the *stade*, with margins of error.

Stade size	Conversion (224,100 *stadia*)	Margin of Error (vs 40,007 km)
160 m (Messenian)	35,856 km	−10.4%
177 m (Delphic)	39,666 km	−0.9%
180 m (Pan-Hellenic)	40,338 km	0.8%
185 m (Attic)	41,459 km	3.6%
191 m (Olympic)	42,803 km	7.0%
210 m (Ptolemaic)	47,061 km	17.6%

Hipparchus' adjustment was considerably 'not much less' than 26,000 *stadia*— and is then further reduced by −0.13 per cent to account for the errors, then the circumference equates to 40,266 km and the margin of error reduces further to 0.6 per cent. Newton claims that a level of accuracy of around 0.5 per cent was beyond the capability of measurements in Eratosthenes' time.[120] However, as has been shown, measurements taken by the *bematists*, for example, were regularly made to a very high degree of accuracy with an average margin of error of 0.4 per cent. A result with a margin of error of <1 per cent also falls within the parameters of the margins of error for the measurements found in Cleomedes and Strabo for distances in Central Asia and Egypt, which are attributed to Eratosthenes, when they are converted into a modern equivalent using the same unit.

Conversion of the possible value for the circumference using the Messenian, Attic, Olympic or Ptolemaic standards (160 m, 185 m, 191 m, and 210 m, respectively) results in a margin of error that is much higher—either positively or negatively—and none of these systems of measure seem to have been used. Interestingly, conversion of a 224,100 *stade* circumference using the Delphic standard of 177 m yields a result that has a similar margin of error to the conversion using the Pan-Hellenic (albeit in the opposite direction). However, the use of this unit of measure does not correlate with the evidence for Eratosthenes' use of the Pan-Hellenic standard for the distances in Iran, Afghanistan, and Egypt. Additionally, the size of the units in both the Delphic and Pan-Hellenic systems are very similar (177 m and 180 m), and this would account for the similarities in the margins of error for the conversion of the 224,100 *stade* distance.

If this is the case, the closer correlation of the conversion of a 224,100 *stade* circumference, using a 180 m *stade*, to the real value of the Earth's polar circumference, makes a figure in the vicinity of 224,000 *stadia* the more likely value of Eratosthenes' true result as it aligns with the level of accuracy that would be expected from the simplicity of the experiment. However, Pliny does state that Hipparchus accepted a value of 252,000 *stadia* as the result of Eratosthenes' work, which would then make the true total 226,100 *stadia* prior to the adjustment. However, this would then not account for why Cleomedes, in the earliest record of the calculations that we have, reports that the result was 250,000 *stadia*. It seems that Pliny's reporting of a result of 252,000 may be the later conversion of the value into something that better conformed with the ideals of Platonic numerical perfection which was a common element in the later reporting of Eratosthenes' results.

---

[120] Newton, 'Sources of Eratosthenes' Measurement', p.379.

Taking 224,100 *stadia* as Eratosthenes' result, the number of times that the reported 5,000 *stade* distance from Alexandria to Syene fits into a circumference of 224,100 *stadia* can be determined using the simple formula: circumference/distance:

$$224,100\ stadia/5,000\ stadia = 44.8 \qquad \text{[eq. 5.2.2]}$$

If the angle created by a shadow cast by a *gnomon* on a sundial in Alexandria on the Summer Solstice equated to a distance between Alexandria and Syene of 1/44.8 of the circumference of the Earth, then the actual angle that would have been created by that shadow would be:

$$\frac{\text{distance} \times 360°}{1/44.8 \times 36° = 8.0°} \qquad \text{[eq. 5.2.3]}$$

Another way to calculate the angle is by using the formula:[121]

$$\text{angle}/360° = \text{distance }(stadia)\ /\text{circumference }(stadia) \qquad \text{[eq. 5.2.4]}$$

Thus:

$$\text{angle}/360° = 5000/224,100$$
$$\text{angle}/360° = 0.0223$$
$$\text{angle} \quad = 360° \times 0.0223$$
$$= 8.0°$$

However, the figure given in Cleomedes (1/50th of the circumference), which is mostly likely based upon the distance between the base of the *gnomon* and the intersection of the noon and Summer Solstice lines on a sundial calibrated for Alexandria, equates to 7.2° (see section 2.5). While there is only a 0.8° difference between these values, and while this may be within the margins of error for ancient instrumentation, the variance suggests that there may be something wrong in the values reported by Cleomedes.[122]

---

[121] This formula was used by Nicastro in some of his calculations (see: *Circumference*, p.27).

[122] Plato (*Resp.* 530b) states that astronomy should be treated in the same way as geometry, which sets out problems for solution. Hon ('Is There a Concept of Experimental Error in Greek Astronomy?' *Br. J. Hist. Sci.* 22.2 (1989), p.131) suggests that, based on this ideal, observational results in ancient Greek astronomy were therefore used to corroborate theories, rather than test them. Hon further suggests (pp.129, 131, 146–147) that in ancient Greek astronomy, 'an awareness of error is at best only implicit or indeed lacking altogether' and that 'in a science where empirical results are used more often for the purpose of illustrating and supporting theories rather than testing them, one would not expect a clear grasp of the concept of experimental error'. Aaboe and Price ('Qualitative Measurements in Antiquity: The Derivation of Accurate Parameters from Crude but Crucial Observations' in A. Koyré (ed.), *L'aventure de la Science, Mélanges* (Paris, Hermann, 1964), pp.3–4) point out that the purpose of many instruments used in ancient observations was to serve convenience rather than to provide

Strabo states that Eratosthenes had determined the distance from Alexandria to Rhodes as being 3,750 *stadia* using the differences in the equinoctial lines inscribed on sundials calibrated for each location, presumably for his work on geography and/or his map.[123] The latitudinal difference between these locations (5.2°) is around 5/6 of a full *hexacontade* equal to 6°. This would suggest that, for Eratosthenes, each *hexacontade* on a sundial equalled:

$$(3{,}750/5.2) \times 6 \approx 4{,}327 \; stadia \qquad\qquad \text{[eq. 5.2.5]}$$

However, this does not seem to be correct.

The latitudinal distance (β) between Alexandria (31.2°N) and Rhodes (36.4°N) can be calculated as:

Difference in latitude : $36.4° - 31.2° = 5.2° = 5.2 \times (\pi/180) = 0.0907$ radians.
Average Earth radius : 6,371km
$\beta = 0.0907 \times 6{,}371 \approx 578$km

$$\text{[eq. 5.2.6]}$$

If this distance is divided by the reported distance of 3,750 *stadia*, this would make the unit of measure used by Eratosthenes 154 m (578,000 m/3,750 $\approx$ 154 m). A unit of this size is not detailed in the literary and/or archaeological evidence for the Greek world. It is possible that it should be in units of the Messenian Standard, with its 160 m *stade*, which would make the distance 600 km and give the calculation a margin of error of just over 3.5 per cent. This high margin of error, compared to margins for Eratosthenes' other calculations and distances, makes the use of this unit unlikely and suggests that there is something else at issue with the reported value.

---

precision, and that the measurement itself depended more on the choice of what was to be observed rather than the choice of instrumentation. If the results of the observation could be matched with a mathematical calculation, then the agreement between theory and observation was perfect. Even if this is true, that does not mean that the observations themselves could not be highly accurate. Rawlins ('Geodesy', p.263), for example, suggests that the accuracy of Eratosthenes' measurements may have been to within less than one arc-minute. Aaboe and Price ('Qualitative Measurements', p.9) additionally suggest that, at the Solstice, the Sun is at its maximum declination, but that declination changes by only 6' over the next five days. As such, the exact time of a Solstice could not have been determined with any accuracy better than within a three- to four-day window, which would have a small effect on the length of any shadow observed. Lewis ('Greek and Roman Surveying and Surveying Instruments' in R.J.A. Talbert (ed.), *Ancient Perspectives: Maps and their Place in Mesopotamia, Egypt, Greece and Rome* (Chicago, University of Chicago Press, 2012), pp.136–137), on the other hand, suggests that, based upon experiments conducted with reconstructions of ancient instruments such as the *dipotra*, which could be used to survey horizontal and vertical angles, the use of such devices may have been accurate to within 0.007 per cent.

[123] Str. *Geog.* 2.5.4; Neugebauer (*Exact Sciences*, p.653) suggests that the figure of 3,750 *stadia* was the result of a rough estimate.

If it is assumed that Eratosthenes was working in the same 180 m *stade* of the Pan-Hellenic Standard that he uses in his other calculations, this would make the distance equal to 675 km (3,750 x 180 m = 675 km) which is too great for the latitudinal distance between Alexandria and Rhodes. Interestingly, the converted distance is equal to the exact latitude of the city of Halicarnassus (37.0°N), located to the north of Rhodes. Whether the 3,750 *stadia* value is an incorrect attribution of the location by Eratosthenes, his source, or a later copyist, is the result of an error in the reading of the sundial in Rhodes, or is the result of a sundial calibrated for Halicarnassus being located in Rhodes, is unknown.

Additionally, *skaphe*-type sundials were accurate when used in any location within a range of around 55 km north or south of the place they were calibrated for (see section 2.5). The latitude of Halicarnassus (37.0°N) is approximately 60 km north of Rhodes (36.4°N). As such, a sundial made for use in Halicarnassus would be reasonably accurate if used in Rhodes—although there would be some small errors if, for example, the details of the lines of the *arachne* were determined mathematically. Interestingly, 55 km north of Rhodes sits the latitude of the island of Kos. This was where, at least according to Vitruvius (9.6.2), Berossus had established a school of astrological enquiry not long before the time of Eratosthenes, and who is credited as one of the inventors of the hemispherical *skaphe*. A sundial calibrated for Kos (36.9°N) would be accurate in both Rhodes and Halicarnassus, and this may have been the instrument on which Eratosthenes was basing his calculations. Consequently, it cannot be ruled out that Eratosthenes was basing these calculations on a sundial that was not in its totally correct position. This would then account for why the value that Eratosthenes gives for the distance between Alexandria and Rhodes is not correct. However, what is important for the understanding of the accuracy of Eratosthenes' calculations is that the redesignation of this measurement from Rhodes to Halicarnassus or Kos has important consequences for his determination of the distance between Alexandria and Syene.

The difference between the latitudes of Alexandria (30.96°N as per Eratosthenes) and Halicarnassus (37.0°N) or Kos (36.9°N) is almost exactly 6° or one full *hexacontade*. Based upon these values, for Eratosthenes, one *hexacontade* was equal to 3,750 *stadia*.[124] The distance between the base of the *gnomon* and

---

[124] There is another way of determining this distance quite accurately using two sundials that were calibrated for the two locations. In the sixteenth century, German instrument maker Georg Hartmann outlined how there is a point on the bowl of the hemispherical *skaphe* that would act as the pivot point when a set of dividers were used to inscribe the equinoctial and Solstice lines on the 'face' of the sundial.

the intersection of the noon and Summer Solstice lines on the sundial that Eratosthenes would have examined in Alexandria—representing the length of the *gnomon's* shadow at midday on the Solstice and the fractional distance between Alexandria and Syene—was 1.2 *hexacontade* (see section 2.5). For a sundial calibrated for Syene, assumed to be on the Tropic, the Solstice line would run right under the *gnomon*—resulting in a fractional difference of zero for that instrument. The difference in latitude for the two locations would therefore be indicated by the fractional distance for the sundial in Alexandria (1.2 *hexacontade*) minus the fractional distance for Syene (zero), or a total of 1.2 *hexacontade*. Consequently, based upon one *hexacontade* equalling 3,750 *stadia*, the distance between Alexandria and Syene, was 3,750 *stadia* x 1.2 = 4,500 *stadia*. This is very close to the latitudinal distance between these locations of 4,450 *stadia* determined using the Pan-Hellenic *stade* of 180 m with a margin of error of around 1 per cent (see following). This not only demonstrates Eratosthenes' use of the 180 m *stade* in his calculations, but also confirms that the distance between Alexandria and Rhodes given by Strabo is actually to Halicarnassus or Kos.

The question then remains as to how Eratosthenes may have reached a value of 224,100 *stadia* for the circumference of the Earth. Cleomedes presents us with a series of values that work mathematically (50 x 5,000 = 250,000), but these same values cannot be used to reach 224,100. However, by using the basic formula of angle x distance = circumference, as per Cleomedes, and solving for a latitudinal distance between Alexandria and Syene of 4,500 *stadia*, based on the 180 m *stade*, results in the following:

$$\text{angle} = 4,500/224,100 = 1/50 \qquad \text{[eq. 5.2.7]}$$

This is the exact figure reported by Cleomedes—and a difference of only 50 *stadia* (−1.12 per cent) to the actual latitudinal distance of 4,450 *stadia*. Such

---

If the midday axis, running from one side of the upper edge of the bowl, through the *gnomon*, to the other edge is divided into degrees—with the upper rim (representing the horizon) set at 0° and the base of the *gnomon* set at 90°—then the number of degrees from the 0° horizon to this point is equal to the latitude of the location for which the sundial is calibrated (see: G. Hartmann, *Practika*, P.27r–P.28r). However, Eratosthenes divided circles into sixty segments (*hexacontade*), each of which would equal 6° (Str. *Geog.* 2.5.7). Thus, the point on the sundial for the latitude of Alexandria (31.2°N) would have been approximately five and one-sixth *hexacontade* down from the upper rim of the sundial. For a sundial calibrated for the latitude of Halicarnassus (37.0°N), the spot would have been almost exactly six and one-sixth *hexacontade* from the rim—making a difference between the two values of one *hexacontade*. We are told that Eratosthenes, using sundials, determined that the distance between Alexandria and, incorrectly, Rhodes was 3,750 *stadia* (Str. *Geog.* 2.5.4; Plin. *HN* 5.132). As such, for Eratosthenes, each *hexacontade* on the sundial must have represented 3,750 *stadia*. This could then be used to determine the latitudinal distance between any two locations, as long as the positions of the pivot points for each of the sundials were measured.

a method would have given Eratosthenes a close latitudinal distance between the two locations and a method of calculating the circumference of the Earth.

A latitudinal distance of around 4,500 *stadia* between Alexandria and Syene, and the use of the 180 m *stade* by Eratosthenes, can also be demonstrated mathematically (Fig. 77):

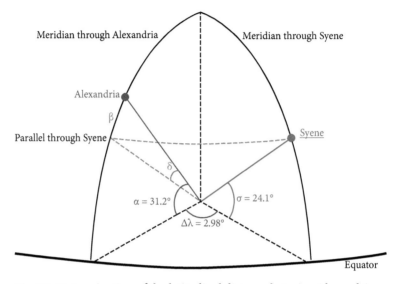

Fig. 77 Determination of the latitudinal distance between Alexandria and Syene (with the angles and curvature exaggerated).

The difference between the latitudes of Alexandria (30.96°N for Eratosthenes) and Syene (24.1°N) (δ) ≈ 7.2° or 7.2 x (π/180) = 0.1257 radians. The latitudinal difference between Alexandria and Syene in kilometres (β) = δ x R (where R is the mean radius of the Earth: 6,371 km):[125]

$$\beta = 0.1257 \times 6{,}371\text{km} \approx 801\text{km} \qquad \text{[eq. 5.2.8]}$$

This distance is equal to 4,450 *stadia* when converted using the *stade* of 180 m in length from the Pan-Hellenic system of measurements (801,000 m/180 m = 4,450). When converted using the 191 m *stade* of the Olympic system, the distance converts to approximately 4,194 *stadia* (801,000 m/191 m ≈ 4,194)—a difference of 252 *stadia* or around 6 per cent. Using the 160 m *stade* of the Messenian standard, the smallest size for the *stade*

---

[125] For the mean radius of the Earth, see: Defense Mapping Agency, *Supplement to World Geodetic System 1984*, pp.3–46.

used in the ancient Greek world, the distance converts to around 5,006 *stadia* (801,000 m/160 m ≈ 5,006).

Mapping software such as Google Earth shows that the linear distance between Alexandria and Syene is 840 km (see Fig. 75). Cleomedes states that Eratosthenes based his work on a distance between the two locations of 5,000 *stadia*. The latitudinal distance between Alexandria and Syene, in *stadia*, is short of this value even when converted using either the Pan-Hellenic or Olympic Standards, and is slightly larger when converted using the Messenian Standard. The value could have still been rounded to the 5,000 *stade* figure attributed to Eratosthenes, but it is a fairly substantial rounding of the figure depending upon which unit is used in the conversion. However, a difference between the figure given by Cleomedes and the latitudinal distance between the two sites is critical to understanding the accuracy of Eratosthenes' work and how it was determined.

Eratosthenes would have had access to a sundial calibrated for Alexandria— which was presumably in the grounds of the Library—and so would have been able to easily observe the location of the Summer Solstice line inscribed upon it. All he then needed to do was compare the position of the midday point along that line, to the same point on a sundial calibrated for, and located in, Syene—which was assumed to be on the Tropic. However, Eratosthenes did not actually have to see this sundial. As he was working on the assumed knowledge that the Sun was directly overhead in Syene at midday on the Solstice, the location of the midday point on the Solstice line for Syene would have been directly beneath the *gnomon*—as no shadows are cast. As such, it seems likely that Eratosthenes calculated the circumference of the Earth using a latitudinal distance between Alexandria and Syene of around 4,500 *stadia* that had probably been determined using earlier observations of the elevation of the Sun in the two locations in order to create sundials, and that this information was later used by Eratosthenes, in conjunction with possible confirmatory observations in Alexandria, in the calculation of the circumference of the Earth.

Later, Hipparchus added around 26,000 *stadia* to Eratosthenes' result to make it the 250,000 *stade* value found in Cleomedes (Pliny's attribution of the adjustment being made to a result of 252,000 then comes later). However, Eratosthenes' original calculation would no longer work as, for example, 50 multiplied by a distance of approximately 4,500 *stadia* does not equal 250,000 unless it is significantly rounded up: 50 x 4,500 = 225,000. Consequently, the distance to Syene was altered to 5,000 *stadia*. This alteration was probably based upon the use by Hipparchus of the larger distance to Syene of 5,000 *stadia* based upon the overland route between the two locations which avoided the bend in the Nile at Luxor measured using a 180 m *stade*. This distance had possibly been measured by Eratosthenes for his earlier work on geography and/or his map, but the two sites were incorrectly assumed to be on the same meridian. The result of this is that, for those who did not undertake confirmatory examinations, the overland distance to Syene would have been incorrectly assumed to have been the latitudinal distance as well. If this is so, then the later attribution by Martianus Capella for the use of *bematists* by Eratosthenes to determine this distance may possibly belong to his geographical research instead of to his calculation of the Earth's circumference. Alternatively, Hipparchus may have used surveyors to measure the route from Alexandria to Syene, found that the result was larger than what Eratosthenes had used, and then made the 'adjustment' in order to correct what he deemed to be an error in Eratosthenes' work. Regardless, Hipparchus, in redoing Eratosthenes' work, may have based his own calculations on the geographical distance (possibly given in Eratosthenes' geography), rather than the latitudinal distance, and found that the math did not work without alteration. The result was then recalculated and 'adjusted' by around 26,000 *stadia*. These adjusted values of Hipparchus were then recounted by Cleomedes, but they were attributed directly to Eratosthenes.

Consequently, a result of around 224,100 *stadia* for Eratosthenes' calculation of the circumference of the Earth allows for a timeline of the reporting of the results to be constructed:

1. Third century BC: Eratosthenes calculates the circumference of the Earth as ≈224,100 *stadia* based upon the use of a 180 m *stade* and a latitudinal distance between Alexandria and Syene of around 4,500 *stadia* determined using the lines inscribed on sundials for both locations. This converts into a calculated circumference of 40,338 km—a margin of error compared to the actual polar circumference of <1 per cent.

2. Second century BC: A generation later, Hipparchus adjusts the result to 250,000 *stadia* by adding ≈26,000 *stadia*. It seems likely that Hipparchus used the overland distance to Syene in his calculations, rather than the latitudinal distance, and found that the math did not work without alteration. An account of this adjustment then survives in the later work of Pliny (first century AD) who bases his account on a result of 252,000 *stadia* (see following).

3. First century AD: Hipparchus' figure of 250,000 *stadia* for the Earth's circumference, and the 5,000 *stade* value for the distance from Alexandria to Syene, is then reported by Cleomedes who attributes the numbers directly to Eratosthenes.

4. First to second century AD: Whether there is a later transcription/transmission error, or whether the value was altered to suit mathematical ideals of Platonic numerical perfection, Roman-period sources then give the result as 252,000 *stadia*.

5. Fourth century AD: Later sources, such as Marcian of Heraclea, report the figure as 259,200 *stadia*—either due to a transmission error (as per Letronne), or as this value divides evenly into degrees, minutes, and seconds (as per Gosselin).

However, it is the two values found in the earlier sources (the 250,000 *stadia* reported in Cleomedes, and the 252,000 *stadia* reported by the Roman period writers) that have been used in every modern examination of Eratosthenes' work (until now). This has led to the creation of numerous theories that have attempted to examine the size of the unit of measure that Eratosthenes was using, and subsequently, gauge the accuracy of his results. Most scholars agree that Eratosthenes' work should result in a relatively accurate value for the Earth's circumference. Unfortunately, by working backwards from a point which includes the incorrect value for his result, this has resulted in many scholars suggesting that Eratosthenes was using a *stade* of a size that does not correlate with other sources of evidence that are attributed to him, and for which there is no evidence of its use in the ancient Greek world. In many cases, it seems that these uncommon sizes for the *stade* are presented out of mathematical convenience, as they then allow a scholar to present a thesis that matches with their preconceived idea of how accurate they believe Eratosthenes' work actually was. This has led, in some cases, to pieces of evidence being ignored or glossed over, for questions to be raised over the values presented in Cleomedes in regard to the angle of the shadow that Eratosthenes observed in Alexandria and/or the stated distance from Alexandria to Syene, and for alternative values to be presented in order for a particular model to fit within a pre-determined margin of error for Eratosthenes' work. A consequence of this method that has been used in the past is that it does not present a hypothesis that can correlate with all of the available evidence, and therefore their conclusions cannot be considered complete.

It is only by working forwards rather than retrospectively, looking at the account of the experiment as presented by Cleomedes, and comparing that to all of the available evidence—literary, archaeological, scientific, mathematic, etc.—that a holistic understanding of Eratosthenes' calculations can be made. The analysis and use of independent sources of evidence for the unit of measure that Eratosthenes was working in—such as his values for distance measurements between locations in Iran and Afghanistan—further indicates not only the size of the unit of measure that he used in his calculation of the circumference of the Earth, but also that he was basing his calculation on a latitudinal distance, determined using sundials, between Alexandria and Syene. Not only does all of this evidence show that Eratosthenes was working in units of 180 m in length (the Pan-Hellenic Standard), but it also shows that the results of his calculations were highly accurate.

6

The Circumference

6.1 The Accuracy Of Eratosthenes' Circumference

Eratosthenes' measurement of the circumference of the Earth satisfied many of the requirements of research undertaken by a Chief Librarian in Alexandria. While it did not come up with something new—there had been prior attempts to determine the size the world—Eratosthenes developed a new, and improved, methodology for doing so which allowed him to present a different answer to what had been an ongoing question. This not only added to that debate, but also contributed to the expanding knowledge base of Hellenic scientific enquiry and tradition and allowed his patrons, the Ptolemies, to claim a position of intellectual superiority within the conflicted Hellenistic world. This would have been accomplished via the wide distribution of his results. Certainly, the fact that scholars such as Cleomedes were referring to Eratosthenes' work hundreds of years later testifies to the success of that dissemination. And yet, despite the simplicity of creating an experiment that could be easily repeated by others, the fickle nature of state-sponsored research in the Hellenistic world meant that Eratosthenes' work could not just be accepted, but needed to be improved upon and refined. This, in turn, has had a great impact on the written accounts of Eratosthenes' work that have come down to us over the centuries.

Cleomedes' account of the attempt by Eratosthenes to calculate the circumference of the Earth, for example, has been the cause of a scholarly debate that has been running for more than two millennia. Cleomedes presents a series of values which seem to be straightforward: an angle created by the shadow cast by the *gnomon* of a sundial in Alexandria equal to 1/50 of the circumference of a circle; an arc-length distance between Alexandria and Syene of 5,000 *stadia*, and a resultant circumference of the world of 250,000 *stadia*. The debate over this result has raged, in part, due to the inability to convert correctly the size of this unit of measure into a modern equivalent. However, the accomplished research objectives of this examination—to determine the size of the *stade* that Eratosthenes used, and thereby determine the accuracy of his calculation of

Eratosthenes and the Measurement of the Earth's Circumference (c.230bc). Christopher A. Matthew, Oxford University Press.
© Christopher A. Matthew (2023). DOI: 10.1093/oso/9780198874294.003.0006

the Earth's circumference—can now put Eratosthenes' calculations into their correct context within the history of astronomy.

As the foundation of his experiment, Eratosthenes used an instrument which was a comparatively recent invention in terms of ancient Greek timekeeping—the hemispherical *skaphe* (see section 2.4). Using this sundial, and several pieces of assumed knowledge, Eratosthenes devised a method for calculating the circumference of the Earth that followed modern scientific principles in being controlled, measurable, and repeatable. Despite the simplicity of the experiment and calculations, there were inherent errors in many of the pieces of assumed knowledge from which Eratosthenes was working. However, Eratosthenes was fortunate enough to be working in a location where many of these errors cancelled each other out, and the cumulative effect of these errors would have been an alteration of the angle created by the *gnomon*'s shadow of only −0.009731°, which would have been imperceivable to naked-eye observations and therefore would have had a minimal impact on the results (see section 3.2). Using the difference between the positions of the Summer Solstice lines on sundials calibrated for Alexandria and Syene, and the latitudinal distance between the two sites, Eratosthenes was able to present the results of a calculation of the circumference of the Earth in a unit of measure commonly used in his time—the *stade*. The modern debate over this result has predominantly raged via modern scholars attempting to determine the size of this unit of measure, and thereby the accuracy of Eratosthenes' work.

Through an analysis of various forms of evidence, it is clear that there were several systems of measurement in use across the Classical and Hellenistic periods of ancient Greece's history which included a *stade*, but these units were all of different sizes. This is confirmed via an examination of literary sources, such as ancient military manuals, which describe the dimensions of common infantry formations; an analysis of the metrological reliefs in the epigraphic record which graphically depict the component parts of many of these systems of measurement; and via a comparison of archaeological sites, predominantly the stadiums of ancient Greece (the size of which the *stade* is named after) using satellite mapping software such as Google Earth to determine the length of the 'stadium unit'.

From this examination, it becomes clear that there were six different systems of ancient Greek measurement, all incorporating *stadia* of different lengths: the Messenian Standard (160 m), the Delphic Standard (177 m), the Pan-Hellenic Standard (180 m), the Attic Standard (185 m), the Olympic

Standard (191 m), and the Ptolemaic Standard (210 m) (see section 4.1). However, the confirmation of the existence of these standards does not aid in establishing which *stade* Eratosthenes used in his calculations which, by default, sheds no light on determining how accurate his estimation for the circumference of the Earth actually was.

The contention over this topic has not been aided by passages from later writers from antiquity, who ascribe different values for the results of Eratosthenes' calculations (most commonly 252,000 *stadia*) or, in passages tangential to the topic, provide seemingly confused conversions of the size of the *stade* into other ancient units of measure such as the Egyptian *schoenus* or the Roman mile. This perplexing corpus of evidence has led to many theories being forwarded by modern scholars who have attempted to work out the size of Eratosthenes' *stade*. Unfortunately, all of the analyses on this topic that have come before are limited in their conclusions. This is due to the failure of these scholars to formulate a model that incorporates all of the available evidence. Many examinations are based on only one of the offered values for the result of Eratosthenes' calculations (mostly the 252,000 *stadia* figure found in the later sources), provide no justification for the acceptance of one figure over another, and/or engage in 'mathematical expediency' by altering some of the figures that have been provided in the ancient texts to accommodate a preconceived idea of how accurate the results should be. The result of such methods is that many previous attempts to ascertain the unit of measure that Eratosthenes used present conclusions that cannot be considered valid as they suggest the use of a unit of measure that did not exist in the ancient Greek world, or fail to incorporate evidence that would contradict the results of their analyses. This makes all prior studies on this subject incomplete, inconclusive, or simply incorrect (see section 4.2).

However, through a critical examination of all of the available evidence, it is now clear that Eratosthenes was using the Pan-Hellenic *stade* of 180 m in length in his calculations. For example, using novel technologies such as satellite mapping to analyse the distances between locations in Iran and Afghanistan, measured by the *bematists* accompanying the army of Alexander the Great, recounted by Eratosthenes, and then cited by Strabo, clearly shows that Eratosthenes was reporting distances in units of 180 m in length. This was accomplished by employing techniques that have not been applied to a study of Eratosthenes by previous scholars—such as comparing the stated distances that are given in the ancient texts to 'on the ground' measurements obtained via the use of mapping software like Google Earth (see section 5.1). The use of a unit of this size also correlates with the 5,000 *stade* distance between Alexandria and Syene (following the Nile, but avoiding the bend in the river at Luxor) which is given by Cleomedes as one of the elements of Eratosthenes' calculations—although this value seems to have come from a 'correction' made by Hipparchus and not Eratosthenes himself (see section 5.2). Consequently, not only does the attribution of a 180 m *stade* to the work of Eratosthenes correlate with the archaeological and geographical evidence, it also matches with the extant literary accounts. However, a unit of this size had not been applied to an examination of Eratosthenes' work until now.

Additionally, the evidence indicates that Eratosthenes used a latitudinal distance between Alexandria and Syene of around 4,500 *stadia*—which was determinable through the use of sundials—in his calculations, and not the 5,000 *stade* figure reported by Cleomedes. This would make the calculation of the Earth's circumference more accurate than using a linear, 'as the crow flies', distance between Alexandria and Syene, as some scholars have suggested, or an 'on the ground' measurement which seems to have been used later by Hipparchus. Calculations and re-creative experiments further show that the angle created by the shadow cast in Alexandria that was observed by Eratosthenes would have been equal to 1/50 of the circumference of a circle—just as is reported by Cleomedes. Furthermore, this examination of the evidence demonstrates that Eratosthenes' result was ≈224,100 *stadia* and not the 250,000 *stadia* reported by Cleomedes (see section 5.2). Converting the 224,100 *stade* value using the 180 m *stade* identified in this research gives a result for Eratosthenes' calculations of 40,338 km, which differs from the polar circumference of the Earth (40,007 km) by only 0.8 per cent. All of this demonstrates that Eratosthenes' calculation of the circumference of the Earth was, as is expected from the experiment's simplicity, highly accurate. Consequently Eratosthenes' work can be put in its rightful place as one of the first accurate observations of solar characteristics, and Eratosthenes himself can be seen as one of the first people in the history of science to correctly understand the nature of our world.

6.2 Eratosthenes' Legacy

Strangely, despite the accuracy of Eratosthenes' result, later scholars in Alexandria, like Hipparchus, made their own attempts to determine the circumference of the Earth using different methodologies and reached different results.

Posidonius (135–51 BC), for example, reached a result of either 240,000 or 180,000 *stadia* for the circumference of the Earth. Cleomedes reports that Posidonius assumed that the cities of Alexandria and Rhodes were situated on the same longitudinal meridian.[1] This bears many similarities to the work that Eratosthenes has previously done for his work on geography and to construct his map of the world. Cleomedes also reports that, for Posidonius, the distance between Alexandria and Rhodes 'seems to be' 5,000 *stadia* (και ίο το διάστημα το μεταξύ των πόλεων πεντακισ χ ι λ ίων σταδίων είναι δοκεί).[2] This again bears similarities with the previous work by Eratosthenes, but in this instance it mirrors the distance that Hipparchus later used for the distance between Alexandria and Syene in his replication of Eratosthenes' calculations. However, the term δοκεί in Cleomedes' text—meaning 'seems to be' or 'appears to be'—suggests that Cleomedes may have had doubts over the figure.

Posidonius' methodology differed from that of Eratosthenes in that, rather than using solar observations when the Sun was at its zenith as Eratosthenes had done, Posidonius used the differences in the observed angular height of the star Canopus in two locations. Cleomedes states that in Posidonius' calculations, Canopus was just above the horizon in Rhodes while, at the same time, the star was at a height equal to one-quarter of a zodiacal sign (i.e. 1/48 of a circle or 7.5°) in Alexandria.[3] Similar to how the difference in the angle of the shadow cast by the *gnomon* of a sundial was used by Eratosthenes to determine the circumference of the Earth, Posidonius used the observed difference in the altitude of Canopus to deduce the same thing. Thus, for Posidonius, the circumference of the Earth was the angle (1/48) multiplied by the distance between the two locations (5,000 *stadia*) to reach (48 x 5,000 =) 240,000 *stadia*.[4] As if to confirm his doubts over the distance between Alexandria and Rhodes that Posidonius had used, Cleomedes closes his account by stating that 'this assumes that from Rhodes to Alexandria is 5,000 *stadia*; but if it is not, it is in the same ratio as the [actual] distance.'[5] Adding to the confusion over Posidonius' result, Strabo in his examination of Posidonius' refutation of Aristotle's sizes for the different zones of the world (e.g. torrid, temperate), states that 'of the more recent measurements of the Earth, the one that makes the Earth

[1] Cleom. *De motu* 1.10.
[2] Cleom. *De motu* 1.10.
[3] Cleom. *De motu* 1.10.
[4] Cleom. *De motu* 1.10.
[5] Cleom. *De motu* 1.10: εάν ώσιν οι από Ρόδου εις Αλεξάνδρειαν πεντακισχίλιοι εί δε μη, πρὸς λόγον τού διαστήματος.

smallest in circumference [is] that of Posidonius, who estimates its circumference at about 180,000 *stadia*.[6] Fischer calls this value a misunderstanding by Strabo.[7] Diller, alternatively, suggests that the two figures given for Posidonius' results (240,000 *stadia* and 180,000 *stadia*) represent a 4:3 conversion of the result from one standard unit of measure to another.[8] However, there is no evidence for such a conclusion and Fischer states that attempts to reconcile the two figures given for Posidonius' result are 'simple classroom examples with arbitrary numbers, and not a real determination'.[9] In this smaller result, the length of arc between Alexandria and Rhodes would be approximately 3,750 *stadia*. Interestingly, this is the exact figure that is attributed to Eratosthenes for the distance between Alexandria and Rhodes that he had apparently calculated using sundials.[10]

The main issue with Posidonius' result—regardless of which figure it was—lies in his methodology. Fischer calls Posidonius' method crude compared to that used by Eratosthenes and suggests that this was why Posidonius' work is only briefly covered in geodesy texts.[11] The main problem with Posidonius' work is that due to the refraction of light by the Earth's atmosphere, a star such as Canopus, which was said to have been viewed when it was just on the horizon in Rhodes by Posidonius, would not have really been above the horizon at all, but would have actually been below an observer's line of sight by more than 0.5 degrees (see section 3.2.7).[12] Additionally, close to the horizon, observed angles are further influenced by disturbances in the atmosphere, which can alter any observed angle by a further 0.3° or more.[13] Both of these factors can result in an overestimate of calculated results by 20 per cent. This would mean that mean that Posidonius' observed angle should have really been closer to 7.2°, the same result that Eratosthenes has reached, depending upon how much atmospheric disturbance was encountered near the horizon.[14]

Additionally, the distance between Alexandria and Rhodes that Posidonius based his calculations on is incorrect. Strabo states that Eratosthenes had measured the distance as 3,750 *stadia* using sundials.[15] As has been

---

[6] Str. *Geog.* 2.2.2.
[7] Fischer, 'Another Look', p.160.
[8] Diller, 'Ancient Measurements', p.9.
[9] Fischer, 'Another Look', p.160.
[10] Str. *Geog.* 2.5.4; Plin. *HN* 5.132.
[11] Fischer, 'Another Look', p.160.
[12] Meeus, *Astronomical Algorithms*, p.105.
[13] B.E. Schaefer and W. Liller, 'Refraction Near the Horizon' *Publ. Astron. Soc. Pac.* 102 (1990), pp.796–805.
[14] Nicastro (*Circumference*, p.150) states that the angle was really 5.5°.
[15] Str. *Geog.* 2.5.4; Plin. *HN* 5.132.

shown (section 5.2), this value was really the distance between Alexandria and the latitude Halicarnassus or Kos. The actual latitudinal arc length between Alexandria and Rhodes is 5.25° which equates to 578 km or around 3,211 *stadia* if converted using the Pan-Hellenic *stade* of 180 m. Interestingly, if Posidonius' angle is adjusted to 6° (1/60 of a circle or one *hexacontade*) to account for atmospheric refraction, and the distance taken as 3,750 *stadia* as per Eratosthenes' reported measurements, this would have then made Posidonius' circumference (60 x 3,750 =) 225,000 *stadia*—which is not far removed from Eratosthenes' initial result of 224,000 *stadia*. If, on the other hand, Posidonius had determined the correct angle based upon the arc length of 5.25°, or approximately 1/69 of a circle, and had used the correct latitudinal distance between Alexandria and Rhodes of 3,211 *stadia*, this would have resulted in a circumference of the Earth of (69 x 3,211 =) 221,559 *stadia*—which is also not too far removed from Eratosthenes' initial result. Fischer additionally points out that the angle Posidonius uses (7.5°) is too large by about 25 per cent compared to the actual arc length angle of 5.25°, and that the distance Posidonius uses (5,000 *stadia*) is similarly out compared to the actual distance of 3,750 *stadia* by about 25 per cent—so that the two errors effectively cancel each other out to reach a 'plausible' result of 240,000 *stadia*.[16] However, Fischer fails to consider that even the value of 3,750 *stadia* for the distance between Rhodes and Alexandria is incorrect. Nicastro, on the other hand, suggests that Posidonius later 'accepted' the smaller value of 180,000 *stadia* based on a corrected distance of 3,750 *stadia* and an adjusted angle of 5.25°.[17] It is uncertain how Nicastro reached such a conclusion as 5.25° (or 1/69 of a circle), multiplied by 3,750 *stadia*, equals 258,750 *stadia* which does not correlate with either of the figures provided in the ancient texts. Such a calculation would only work if Posidonius had retained his initial angle of 7.5° (1/48 of a circle), but then multiplied it by a revised distance figure of 3,750 *stadia*, to reach a result of 180,000 *stadia*. This may be what Nicastro is suggesting and, if this did occur, it would also account for the two different values assigned to Posidonius' result.

A revision of Posidonius' initial result seems to be confirmed as the value of 180,000 *stadia* was later used by Ptolemy in his work on geography instead of the 240,000 *stadia* value that was initially connected with Posidonius, or Hipparchus' adjusted 250,000 *stadia* value that is attributed to Eratosthenes by

[16] Fischer, 'Another Look', p.160.
[17] Nicastro, Circumference, p.150; Nicastro states that Posidonius' recalculation is recounted in a passage by Strabo (*Geog.* 2.2.2), but the text of the passage does not mention a recalculation nor an acceptance by Posidonius of one value over another.

Cleomedes.[18] Ptolemy begins his discussion with a summary of methodologies that had been employed by his predecessors to determine the size of the Earth—methods which are very similar to the methodology of Eratosthenes, although Ptolemy does not name him directly.[19] Ptolemy then outlines how he constructed new instruments—a *meteoroskopeion*, which appears to be a form of armillary sphere with nine rings, and a plinth-based quadrant—for examining the angle of the obliquity of the ecliptic via observation of the angular height of the North Celestial Pole from different locations.[20]

The first method for calculating the circumference of the Earth that Ptolemy then outlines involves measuring the angle between two targets (which are presumably stars) that are measured over the zenith point in two locations at the same time.[21] Similar to the work of Eratosthenes, Ptolemy concludes that if lines were extended downwards from these points, they would intersect at the centre of the Earth. The angle formed by that intersection would then equal a part of the circumference of the world; which could then be used to calculate the full circumference if the distance between the two locations was measured or known, and was then applied to the ratio of the angle compared to the number of degrees in a full circle.[22]

Ptolemy follows on to state that this method works if the two locations are located on the same meridian.[23] However, if they were not, then an observation of the angular height of the North Celestial Pole (most likely targeted on the star Polaris) could be used to calculate the Earth's circumference. The angle of the Pole could be compared to a measurement of the angle of where the 'greatest [latitudinal] circle drawn through the line of distance intersects with the meridian circle at the zenith point.'[24] The distance between the locations, multiplied by the ratio of the angle, then equates to the circumference of the Earth.[25] Thus, Ptolemy's method seems to be an improvement

---

[18] Ptol. *Geog.* 1.7, 7.5. Interestingly, in the *Almagest* (9.2), Ptolemy refers to Hipparchus as a 'great lover of the truth', which would suggest that Ptolemy had a lot of confidence in Hipparchus' findings. Why Ptolemy then chose to accept Posidonius' value for the circumference of the Earth over that of Hipparchus is far from certain, other than it seeming to match with the results of his own ideas.

[19] Ptol. *Geog.* 1.3.

[20] Ptol. *Geog.* 1.3; Proclus (*Hyp.* 6.3) states that the armillary sphere should have a diameter of not less than half a cubit (*ca.* 23 cm) and that the scale on the instrument should be in gradations of 0.1°. Pappus (*in Ptol. Alma.* 1.6), on the other hand, suggests that Ptolemy's armillary sphere had a diameter of one cubit (*ca.* 45 cm). It is possible that Proclus has confused radius with diameter. For a discussion of Ptolemy's instruments, see: J.P. Britton, *Models and Precision: The Quality of Ptolemy's Observations and Parameters* (New York, Garland, 1992), pp.4–11.

[21] Ptol. *Geog.* 1.3.

[22] Ptol. *Geog.* 1.3.

[23] Ptol. *Geog.* 1.3.

[24] Ptol. *Geog.* 1.3.

[25] Ptol. *Geog.* 1.3.

on the angular elevation method employed by Posidonius. The main point of improvement is that, by using a zenith point and one other celestial target at a high elevation, any observations made would be subject to significantly less atmospheric refraction. This should have made Ptolemy's method much more accurate.

In a probable reference to the work of Ptolemy, Simplicius (fifth–sixth century AD) states that the 'ancients' used a *dioptra* to observe two stars that were separated by one degree of arc.[26] The locations where these two stars were at the zenith at the same time (i.e. one star overhead in one location, and the second star overhead in the other) were identified and the distance between them was measured using a *hodometer*.[27] Earlier in *Geography*, Ptolemy does state that the distances between two locations could be determined using the observations of stars, and he clearly outlined a method for using stars which simultaneously culminated over two sites.[28] However, what Ptolemy does not specifically state is that the two target stars had to be separated by only 1°, such detail is only found in the account of Simplicius. Interestingly, Simplicius reports that the distance measured between the two observing sites was 500 *stadia*—the same number per degree of latitude as used by Ptolemy—and so the circumference of the Earth had to be (360 x 500 =) 180,000 *stadia*; which is also the circumference value favoured by Ptolemy.[29] As such, Simplicius seems to be referring to the method of angular observation employed by Ptolemy.

Ptolemy may have been additionally aided in his observations through the use of the *dioptra* which is referred to by Simplicius. First invented in the third century BC, the *dioptra* was initially designed as a surveying tool.[30] It consisted of a flat circular disk mounted on a thin column or set of legs. The outer edge of the disk was divided into the segments of a circle—presumably in *hexacontades* of 6° each when the instrument was first invented, and then into 360° from the first century BC onwards. The disk was mounted so that it was parallel to the ground and the 0° mark pointed to the north. On top of the disk was set an observing bar, just shorter that the diameter of the disk, and with an upright at either end punctuated with a hole through the middle. This bar was connected to the baseplate with some form of rivet, through both the mid-point of the length of the bar and the centre of the disk. This allowed the bar to be rotated.

---

[26] Simpl. *in Cael.* 548.27-549.10.
[27] Simpl. *in Cael.* 548.27-549.10.
[28] Ptol. *Geog.* 1.2, 1.3.
[29] Simpl. *in Cael.* 548.27-549.10.
[30] Hero. *Dioptr.* 3; Archim. *Psam.* 4.

One end of the bar presumably also possessed some form of pointer, which would point to the degree markings on the outer edge of the disk.[31]

In order for a surveyor to take a reading of an angle from one location to another, the *dioptra* would be set up at one site, with the observing bar and 0° mark pointing north. Then the bar would be rotated until the second location could be seen through the two holes in the uprights. When this was accomplished, the pointer on the end of the bar would indicate the angle (Fig. 78).

Fig. 78 Measuring lateral angles using a replica *dioptra* in a horizontal orientation.[a]
Image courtesy of Michael Lewis.

Hipparchus seems to have refined the design of the *dioptra* with some form of hinged mounting so that the baseplate could be rotated 90° into a vertical orientation. In this way, the *dioptra* could then be used in exactly the same way to measure angles in the vertical plane—including the angular elevation of stars and their angular separation (Fig. 79).[32]

According to Shipley, Hipparchus may have also used this instrument to produce the first star catalogue, calculate the precession of the equinoxes, and design the concepts of latitude and longitude.[33] However, in Eratosthenes' time, Euclid referred to using the *dioptra* to observe the angular elevation of

---

[31] For an examination of the use of the *dioptra* in ancient Greek surveying, including tests made using a replica instrument, see: Lewis, 'Greek and Roman Surveying', pp.130–142.

[32] Image courtesy of Michael Lewis.

[33] Ptol. *Alm.* 5.14; Procl. *Hyp.* 4.87.

Fig. 79 Measuring elevation angles using a replica *dioptra* in a vertical orientation.[a]
Image courtesy of Michael Lewis.

the constellation Cancer.[34] This would suggest that an alteration to the design of the *dioptra* to allow it to be used for vertical observations had been made prior to the time of Hipparchus, but also does not discount the possibility that Hipparchus had further refined the design of the instrument.

Regardless of whether Ptolemy used a *dioptra* or not, there are a number of issues with the zenith-angle methodology. Tupikova, for example, in her examination of the method outlined by Simplicius, which has the stars only 1° apart, calls such a small separation in the zenith angles 'a serious handicap'.[35] There were a number of characteristics that the stars in these close pairs would have had to have met in order for them to be useful to such a method:

1) Both stars in the pair had to have been known in antiquity. This can be confirmed by finding references to them in texts such as Ptolemy's *Almagest*. However, Ptolemy refers to the work of his predecessors— most likely Hipparchus—as well, and so the stars had to have been known, and culminated, in both the time of Hipparchus and the time of Ptolemy.

2) The pair of stars would need to have the same, or similar, right ascensions (α) so that they culminated at roughly the same time, while their

---

[34] Image courtesy of Michael Lewis.
[35] Shipley, *Greek World*, p.351.

declinations (δ), which are equal to each locations' latitude (φ), had to be φ and φ±1°, respectively.

3) Both stars in the pair had to be of a magnitude that was visible to the naked eye, and had to be bright enough to be able to see through the small observation holes in a *dioptra* at night.

4) At least one of the stars had to culminate over the place where Ptolemy was working—Alexandria in Egypt.

5) The zenith point at both locations needed to be determined so that the corresponding arc could be measured on the ground. This would suggest that both measurements were taken in, or close to, known settled areas or other landmarks.

Very few pairs of stars met these criteria in antiquity. Tupikova, in her 2014 study, and later with Geus in 2017, outlined only four pairs of stars listed in Ptolemy's *Almagest* that met some of these criteria (Table 17):[36]

Yet even these pairs do not meet all of the necessary criteria. ν and τ Andromeda, for example, did culminate over Alexandria, but with apparent magnitudes of around 4, they may have been too faint.[37] The pairings of ν Lyr/θ Her and ν Crb/o Crb culminated over mainland Greece.[38] In 2014, Tupikova suggested that the best pairing was that of ν and ξ UMa.[39] However, while ν UMa and ξ UMa are the brightest of the pairs, and are part of a constellation that was well known in ancient times, one of the stars (ν UMa) culminates over

Table 17 Possible pairs of stars used in Ptolemy's 'zenith-angle' method.

Star	Magnitude	Almagest #	RA	Dec
ν UMa	3	31	141.40°	42.41°
ξ UMa	3	32	141.50°	41.45°
ν Crb	>4	97	208.57°	38.47°
o Crb	5	98	208.56°	37.37°
ν Lyr	4	153	271.13°	37.30°
θ Her	4	154	271.02°	36.29°
ν And	4	352	358.19°	31.16°
τ And	4	353	358.33°	34.14°

[36] Euc. *Phae.* 1.

[37] I. Tupikova, 'Ptolemy's Circumference of the Earth' (Berlin, Max Planck Institute for the History of Science, 2014), p.2; see also: K. Geus and I. Tupikova, 'Astronomy and Geography: Some Unexplored Connections in Ptolemy' in F. Pontani (ed.) *Certissima Signa: A Venice Conference on Greek and Latin Astronomical Texts* (Venezia, Edizioni Ca' Fascari, 2017), p.66.

[38] See: Tupikova, 'Ptolemy's Circumference', p.2; Geus and Tupikova, 'Astronomy and Geography', pp.66–67.

[39] Tupikova, 'Ptolemy's Circumference', p.2.

the city of Lysimachia in NW Turkey rather than Alexandria in Egypt, and the other star culminates over no known settlement. Additionally, while ν UMa culminated over Lysimachia in the time of Ptolemy, it did not do so during the time of Hipparchus, and so cannot be one of the stars that Ptolemy refers to when he talks of the work of his predecessors.[40] Thus, none of the proposed pairs of stars meets the criteria for use in Ptolemy's zenith-angle method.

Geus and Tupikova suggest that Ptolemy's zenith-angle method was actually nothing more than a theoretical 'thought experiment' and that, in reality, only a single star was used.[41] Indeed, any pair of stars could have been used, and a separation by more than 1° would have provided greater precision so long as they culminated at the same time for the observers in the two different locations, and the distance between those locations was then subsequently measured. However, it seems that stars separated by a single degree fit nicely with the idea of single degrees of latitude that were a central part of Ptolemy's model. If, on the other hand, only a single star was used, the most likely candidate offered by Geus and Tupikova would have been β Gem (Pollux) in the constellation Gemini. The reason for this is that Pollux is one of the brightest stars in the sky (magnitude 1.16), and with a right ascension of 86.10°, and a declination of 30.03°, the star would have culminated almost exactly 1° south of Alexandria in the time of Hipparchus and Ptolemy.[42] However, there is an issue with this conclusion. Along the meridian that runs through Alexandria, there is nothing 1° south of the city but bare, open desert. Furthermore, the first part of a journey due south of Alexandria would require the crossing of a large lake, and the traversing of the reed beds of the western Nile Delta. As such, it would have been impossible for the distance between Alexandria and a southerly location to have been measured using a *hodometer* as is outlined by Simplicius. Interestingly, the non-existence of a southerly location was the reason why Geus and Tupikova dismissed the pairing of ν UMa/ξ UMa from the list of potential candidate pairs for the zenith-angle method. Yet, curiously, they have not applied the same criterion to β Gem.[43]

The lack of a suitable southerly location for observations of β Gem again suggests that Ptolemy's zenith-angle method may have been a 'thought experiment'. If Ptolemy was using an instrument like the *dioptra* or his plinth-quadrant to measure stellar elevations, the angle of any star from the zenith could have been measured. Only one actual star is really required, and the zenith point, determined with the *dioptra* or quadrant, could just be taken as an 'assumed star'. This would further suggest that the zenith-angle method was

---

[40] Tupikova, 'Ptolemy's Circumference', p.2.
[41] Tupikova, 'Ptolemy's Circumference', p.2.
[42] Geus and Tupikova, 'Astronomy and Geography', p.66.
[43] Tupikova, 'Ptolemy's Circumference', p.2; Geus and Tupikova, 'Astronomy and Geography', p.66.

more theoretical than actual. This seems to be confirmed by the terminology used in the two accounts of the method. In Ptolemy's *Almagest* and *Geography*, he refers to 'zenith points' (σημεῖα). However, several hundred years later, Simplicius refers to 'zenith stars' (ἀστέρες). It seems likely that Simplicius has confused a theoretical model with an actual procedure.

Ptolemy was certainly aware of the work that had been done before him and this appears to have strongly influenced his *Geography*. In his discussion of the circumference of the Earth and the amount of 500 *stadia* per degree, Ptolemy states that his result is 'a measurement which is proved by distances that are known and certain.'[44] Unfortunately, Ptolemy does not always outline who the sources were which proved these measurements, or provided measured and certain distance figures. Eratosthenes had outlined the use of different observed angles in his calculation of the Earth's circumference, and had provided many details of distances between locations in his work on geography, and Ptolemy would have known of both of these texts. The work of Hipparchus from the second century BC also seems to have had a strong influence on Ptolemy's geographical works. In *Geography*, for example, Ptolemy directly refers to Hipparchus' work to determine the angular elevation of the North Celestial Pole from various locations in order to determine latitude.[45] A century later, like Eratosthenes, Posidonius had used the differences in observed angles to determine a circumference of the Earth of 180,000 *stadia*, which equates to 500 *stadia* per degree of latitude—the exact same figures that are quoted by Ptolemy. In the first century AD, Heron of Alexandria had used the *dioptra* to determine the differences in the elevation angles of a lunar eclipse that was simultaneously observed in Alexandria and Rome, and distances determined using a *hodometer*, to calculate the circumference of the world.[46] In his *Geography*, Ptolemy also refers to eclipse observations being used to determine latitude, although he does not cite Heron as his source for this.[47] Ptolemy was also certainly aware of the (now lost) *Geography* of Marinus

[44] Tupikova, 'Ptolemy's Circumference', p.2; Geus and Tupikova, 'Astronomy and Geography', p.67.

[45] It is also interesting to note that, despite the issues that they outline in Ptolemy's method, Geus and Tupikova ('Astronomy and Geography', p.64) still state that the zenith method for determining the circumference of the Earth described by Ptolemy was superior to that of Posidonius and Eratosthenes as it could be employed easily at any time of year. It is uncertain how such a claim can be made when no stars can be found which fit the requirements of the zenith-angle method, the method would only work in clear weather, and when the stars were visible—and so could not have been undertaken at any time of year. Comparatively, the method employed by Eratosthenes, could have been undertaken at any time, on any day, and in any weather.

[46] Ptol. *Geog.* 1.11.

[47] Ptol. *Geog.* 1.4; For discussions of the influence of Hipparchus on Ptolemy, see: L. Russo, 'The Astronomy of Hipparchus and His Time: A Study based on Pre-Ptolemaic Sources' *Vist. Astron.* 38.2

of Tyre from the first century AD as he cites him several times in his text.[48] It is interesting to note just how many aspects of these previous works are found in the models laid out by Ptolemy and Simplicius. Thus, Ptolemy seems to have taken the principles of angular measurement set down by people like Eratosthenes, used instrumentation similar to Heron to measure those angles, and taken Posidonius' values for the circumference of the world and the number of *stadia* per degree of latitude, and merged them into one, all encompassing, hybrid theoretical method for calculating the size of the Earth.

It further seems that Ptolemy may have just accepted a 180,000 *stade* circumference for the Earth from the work of Posidonius without any comparison to other texts, or experimentation to try to confirm the value. For Eratosthenes, for example, at least as it is reported by Cleomedes, there was around 695 *stadia* per degree of latitude (360 x 695 ≈ 250,000) or 700 *stadia* based upon the 252,000 *stade* circumference found in later sources (360 x 700 = 252,000)— a figure that Strabo says was adopted by Hipparchus, the person who had 'adjusted' Eratosthenes' result and whose work had an influence on Ptolemy.[49] For Eratosthenes' 224,000 *stade* initial result, there were approximately 622 *stadia* per degree of latitude. This poses the question, raised by Russo: if both Marinus and Ptolemy knew of Eratosthenes' result, and knew that Posidonius had reached a different value (regardless of which 'result' for Eratosthenes they had access to), and they had access to the same information about Syene, why did they not try to repeat Eratosthenes' experiment?[50] If they had done so, they would have clearly seen that some of the values they were working with—such as the circumferences given by Posidonius and Hipparchus/Cleomedes—were incorrect.

The answer may lie in one of the underlying principles of Hellenistic science: to come up with something new, or to answer a long-standing question. Simply to repeat previous work was not enough. Thus, people like Posidonius, Heron, and Ptolemy attempted to come up with new, and potentially improved, methods and models to examine what others had already done

---

(1994), pp.207–248; D. Shcheglov, 'Hipparchus' Table of Climata and Ptolemy's Geography' *Orb. Terr.* 9 (2007), pp.159–192.

[48] Hero. *Dioptr.* 35; Heron also states that he has taken the value for the circumference as 252,000 *stadia*—a figure which he attributes to Eratosthenes. See also: N. Sidoli, 'Heron's *Dioptra* 35 and Analemma Methods: An Astronomical Determination of the Distance Between Two Cities' *Centaurus* 47 (2005), pp.236–258. Sidoli claims (p.237) that Heron's work may also be a summary of some earlier work. Rome ('Le probème de la distance entre deux villes dans la dioptra de Héron' *Ann. Soc. Sci. Bruxelles, Ser. A* 42 (1923), p.249) suggests that Heron's methodology was based upon an earlier work of Hipparchus.

[49] Ptol. *Geog.* 1.4.

[50] Ptol. *Geog.* 1.6–1.21.

beforehand. However, these attempts at improving on prior work seem to have not always been successful. This further suggests that Ptolemy had simply accepted Posidonius' figures and worked them into his theoretical model. In doing so, this meant that his method for the determination of the size of the Earth was, in part, just a recounting of Posidonius' earlier work but with slightly different (and from Ptolemy's perspective, improved) methodology. Importantly, and unfortunately, by simply accepting Posidonius' values, Ptolemy was also accepting all of the errors inherent in that work—including it being based on an incorrect distance between Alexandria and Rhodes. As a result, not only was Ptolemy's circumference of the Earth too small, but also the number of *stadia* per degree of latitude was additionally incorrect. It is clear that not everyone agreed with Ptolemy. Plutarch, for example, writing around the same time as Ptolemy, cites a figure of 40,000 *stadia* for the radius of the Earth.[51] This figure correlates with the circumference of 250,000 *stadia* that was refined from Eratosthenes' initial result by Hipparchus, and which is found in the works of Cleomedes. Plutarch's stated circumference additionally shows that even several hundred years after Eratosthenes' experiment, the scholarly debate over the size of the Earth was far from resolved.

The debate over the size of the circumference of the Earth was not aided by the fact that both worlds—the 250,000 *stade* globe reported by Cleomedes, and the 180,000 *stade* world of Ptolemy—implied that there was a vast amount of water in the outer ocean which was assumed to surround the known world (the *oikumene*) of Europe, North Africa, and the Near East. Eratosthenes, in his work on geography, had suggested that the known world was elongated in shape—with a north–south width of 38,000 *stadia* and an east–west breadth of 73,800 *stadia*—and was potentially surrounded by more than 170,000 *stadia* of water.[52] If the 250,000 *stadia* circumference of Hipparchus/Cleomedes is taken as correct, for example, this would mean that there was (250,000–73,800 =) 176,200 *stadia* of water or more in the outer ocean. For Ptolemy's smaller world, there was still 90,000 *stadia* of outer ocean.[53] In the second century BC, Crates of Mallus had attempted to account for this vast amount of ocean by suggesting the existence of four continents, one in each quadrant of the globe: the *oikumene*, the *perioeci* ('those who dwell adjacent'), the *antioeci* ('those

---

[51] Str. *Geog.* 2.5.7. See also Russo, *Forgotten Revolution*, p.69; Heron (*Dioptr.* 35) also takes 1° of latitude being equal to 700 *stadia* as he also accepts the figure of 252,000 *stadia* for the circumference of the Earth.
[52] Russo, *Forgotten Revolution*, p.69.
[53] Plut. *Mor.* 925D.

who dwell opposite'), and the *antipodes* ('the other foot') which generally correlate with Europe/Near East, Asia, Sub-Saharan Africa, and Australasia, respectively (Fig. 80).[54]

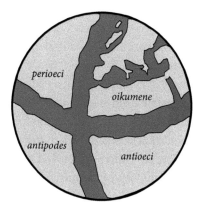

Fig. 80 A representation of the globe according to Crates of Mallus.

However, it was the smaller world of Posidonius/Ptolemy, rather than the larger Earth of Eratosthenes, Hipparchus or Cleomedes, which became the accepted view. This may have been, in part, due to the wide acceptance of Ptolemy's model of the working of the heavens which would remain in place for another 1,300 years. Importantly, this notion of a smaller globe with other inhabited continents would have far-reaching implications for centuries to come.

Throughout the Middle Ages, for example, both Christian and Muslim scholars attempted to estimate the size of the Earth, particularly because they had no idea of the size of the *stade* that had been used in the ancient calculations. This resulted in a variety of sizes being in circulation.[55] For example, while the work of Ptolemy was widely accepted by many, there were still others, such as Hermannus Contractus (1013–1054) who favoured the 252,000 *stadia* figure for the circumference of the Earth that had been attributed to Eratosthenes in the Roman-period texts.[56] In 1406, Jacopo Angelo translated Ptolemy's *Geography*, with its 180,000 *stadia* circumference, into Latin,

[54] Str. *Geog.* 1.4.2–5.
[55] Ptol. *Geog.* 7.5.
[56] Str. *Geog.* 2.5.10; see also Pliny (*HN* 2.161) who suggests that the antipodes is inhabited, and recounts how some people wonder why the inhabitants of such a place do not fall off and they would be upside down from the perspective of someone in Europe.

and brought the original text from Constantinople to Europe.[57] This sparked a European interest in further exploration and a greater understanding of the world. Ptolemy's *Geography* was then first put into print between 1475 and 1477, which allowed it to be more widely distributed.[58] Not much later, Christopher Columbus intentionally used the smaller of the estimates for the size of the world, like those of Posidonius and Ptolemy, to try to convince people to support his expedition to sail directly westward from Europe to Asia—a distance which he said was 3,000 miles.[59] Such a claim bears all the hallmarks of the globe of Crates of Mallus from sixteen centuries earlier, which did not account for the Americas. Additionally, back in the first century BC, Posidonius himself had suggested that if a ship sailed due west from the Strait of Gibraltar it would reach India.[60] Such values and claims must have been a great inspiration for the explorers and adventurers of the late fifteenth century. As Nicastro points out, the works of Posidonius and Ptolemy worked well for Columbus' entreaties to the ruling houses of Europe as they suggested a world that was 30 per cent smaller than that offered by Eratosthenes.[61] Interestingly, Columbus did reach land after a 3,000-mile voyage, it was just not Asia as he had expected. Nicastro offers that our knowledge of the size of the world did not really catch up with Eratosthenes until true ocean-going ships were developed, nearly 2,000 years after Eratosthenes' time, and explorers like Ferdinand Magellan (1480–1521) circumnavigated the globe and proved that the Pacific Ocean was much bigger than had been originally believed.[62] Had Columbus based his estimations and expectations on the larger, and more accurate, size of the world that had been determined by Eratosthenes, rather than the smaller circumference of Posidonius and Ptolemy, the history of the world might be very different.[63]

Eratosthenes died at the age of eighty and was buried in Alexandria.[64] A funerary epigram for Eratosthenes by Dionysius of Cyzicus states:

---

[57] See: Evans, *Ancient Astronomy*, pp.65–66; for outlines of some of the Arabic calculations of the circumference of the Earth, see: Al-Biruni, *The Determination of the Coordinates of Cities*, 218.1–223.15.

[58] Hermannus Contractus, *De Utilitatibus Astrolabii Libri Duo*, 2.3–2.4.

[59] J. Larner, 'The Church and the Quattrocentro Renaissance in Geography' *Renaiss. Stud.* 12.1 (1998), pp.26–39; A. Grafton and A. Shelford, *New World, Ancient Texts* (Harvard, Harvard University Press, 2015), p.50.

[60] H.N. Stevens, *Ptolemy's Geography: A Brief Account of all of the Printed Editions down to 1730* (London, Stevens, Son and Stiles, 1908), pp.1, 39.

[61] See: S.E. Morrison, *The European Discovery of America: The Southern Voyages AD1492–1616* (Oxford, Oxford University Press, 1974), p.30.

[62] Str. *Geog.* 1.4.6, 2.3.6.

[63] Nicastro, *Circumference*, p.180.

[64] Nicastro, *Circumference*, pp.24, 181.

A mild old age, no darkening disease, extinguished your light, Eratosthenes son of Aglaus, and, your high studies over, you sleep the appointed sleep. Cyrene, your mother, did not receive you into the tombs of your fathers, but you are buried on the fringe of Proteus' shore [i.e. Alexandria], beloved, even in a strange land.[65]

The *Suda* relates how, with his eyesight failing, Eratosthenes gave up food and starved himself to death.[66] It seems likely that, following a lifetime of study and observation, the thought of no longer being able to work was more than Eratosthenes could bear. Eratosthenes was succeeded by his pupil, Aristophanes of Byzantium, who in turn had Hipparchus as one of his pupils. Yet despite the criticisms and revisions of Hipparchus, the recalculations of Posidonius, and the incorrect reporting of his results by Cleomedes, Eratosthenes left behind a legacy of highly accurate observation and calculation in his determination of the circumference of the Earth. As Fisher points out: Eratosthenes' use of 'painstaking measurements instead of speculations' in his own time is what he should be given full credit for, and that 'we apply the best available measuring techniques of our time as he applied the best available measuring techniques in his time.'[67] This sentiment is echoed by Nicastro who says that regardless of how accurate Eratosthenes was, we still have reason to admire his achievement.[68] Such comments imply that Eratosthenes' methodology was somewhat crude and his results inaccurate. Yet the review of the available evidence has shown that through the use of a 180 m *stade*, Eratosthenes was the first person to be able to calculate the circumference of our world to within a margin of error of <1 per cent. Thus, Eratosthenes not only 'made the world', as Sagan puts it, but the rejection of his accurate measurements indirectly contributed to the discovery of the 'New World'—which is no small legacy indeed.

[65] In another interesting connection between Columbus and Eratosthenes, Columbus' son, Hernando (aka Ferdinand) used the sizeable wealth that his father had gained from the New World in an attempt to gather all of the known books into one collection—following the ideals set out by the first Ptolemies and, in effect, creating another 'Great Library' like the one that had been established in Alexandria more than seventeen centuries earlier (see: E. Wilson-Lee, *The Catalogue of Shipwrecked Books* (New York, Simon and Schuster, 2018), pp.179–200, 259–278). A 2,000-page codex containing summaries of around 15,000 books in Columbus' collection, the so-called *Libro de los Epitomes* (Book of Summaries), was discovered in a Copenhagen Library in 2013. See: J. Brean, 'Professor Discovers Centuries-Old attempt by Christopher Columbus' Son to Summarise Every Book in Existence' *National Post* 12 April 2019.
[66] *Suda s.v.* Ἐρατοσθένης.
[67] Dion. Cyz. *On E.*: Πρηΰτερον γῆράς σε, καὶ οὐ κατὰ νοῦσος ἀμαυρὴ ἔσβεσεν εὐνήθης δ' ὕπνον ὀφειλόμενον, ἄκρα μεριμνήσας, Ἐρατόσθενες· οὐδὲ Κυρήνημαῖά σε πατρῴων ἐντὸς ἔδεκτο τάφων, Ἀγλαοῦ υἱέ· φίλος δὲ καὶ ἐν ξείνῃ κεκάλυψαι πὰρ τόδε Πρωτῆος κράσπεδον αἰγιαλοῦ.
[68] *Suda s.v.* Ἐρατοσθένης.

References

Ancient And Medieval Sources

Achilles Tatius (trans. S. Gaselee) (London, W. Heinemann, 1917).

Achilles Tatius, *Leucippe and Clitophon* (trans. S. Gaselee) (Cambridge, Loeb Classical Library—Harvard University Press, 1969).

Aelian, *Historical Miscellany* (trans. N.G. Wilson) (Cambridge, Loeb Classical Library—Harvard University Press, 1997).

Aelian (Tacticus), *The Tactics of Aelian* (trans. C.A. Matthew) (Barnsley, Pen & Sword, 2012).

Aeneas Tacticus/Asclepiodotus/Onasander (trans. Illinois Greek Club) (Cambridge, Loeb Classical Library—Harvard University Press, 2001).

Aetius, *Placita* (trans. J. Mansfield and D.T. Runia) (Leiden, Brill, 2020).

Al-Biruni, *The Determination of the Coordinates of Cities* (trans. J. Ali) (Beirut, American University of Beirut Press, 1967).

Ammianus Marcellinus, *History Vol.II* (trans. J.C. Rolfe) (Cambridge, Loeb Classical Library—Harvard University Press, 1940).

Appian, *Roman History Vol.III: Civil Wars* (trans. B. McGing) (Cambridge, Loeb Classical Library—Harvard University Press, 2019).

Appian, *Roman History Vol.IV: Civil Wars* (trans. B. McGing) (Cambridge, Loeb Classical Library—Harvard University Press, 2020).

Apollonius Rhodius, *Argonautica* (trans. W.H. Race) (Cambridge, Loeb Classical Library—Harvard University Press, 2009).

Aphthonius, *Progymnasmata* (ed. I. Petzholdt) (Lipsiae, A.F. Boehme, 1839).

Aratus's *Phenomena*, Cleomedes' *On the Circular Motions of the Celestial Bodies*, and Nichomachus' *Introduction to Arithmetic* (Ukraine, Central Scientific Library of V.N. Karazin Kharkiv National University (manuscript #14757), c.1300).

Archimedes, *The Method* (ed. T.L. Heath) (Cambridge, Cambridge University Press, 1912).

Archimedes, *Sand Reckoner* (ed. T.L. Heath) (Cambridge, Cambridge University Press, 2009).

Aristarchus, *On the Sizes and Distances of the Sun and Moon* (ed. F. Commandino) (Pisa, Camillum Francischinum, 1572).

Aristophanes, *Birds/Lysistrata/Women at the Thesmophoria* (trans. trans. J. Henderson) (Cambridge, Loeb Classical Library—Harvard University Press, 2000).

Aristophanes, *Clouds/Wasps/Peace* (trans. J. Henderson) (Cambridge, Loeb Classical Library—Harvard University Press, 1998).

Aristophanes, *Frogs/Assemblywomen/Wealth* (trans. J. Henderson) (Cambridge, Loeb Classical Library—Harvard University Press, 2002).

Aristotle, *Athenian Constitution/Eudemian Ethics/Virtues and Vices* (trans. H. Rackham) (Cambridge, Loeb Classical Library—Harvard University Press, 1935).

Aristotle, *De Caelo* (Oxford, Oxford University Press, 1936).

Aristotle, *Metaphysics Volume I: Books 1–9* (trans. H. Tredennick) (Cambridge, Loeb Classical Library—Harvard University Press, 1933).

Aristotle, *Metaphysics Volume II: Books 10–14/Oeconomica/Magna Moralia* (trans. H. Tredennick and G.C. Armstrong) (Cambridge, Loeb Classical Library—Harvard University Press, 1935).

Aristotle, *On the Soul/Parva Naturalia/On Breath* (trans. W.S. Hett) (Cambridge, Loeb Classical Library—Harvard University Press, 1957).

Aristotle, *Nicomachean Ethics* (trans. H. Rackham) (Cambridge, Loeb Classical Library—Harvard University Press, 1926).

Aristotle, *Politics* (trans. H. Rackham) (Cambridge, Loeb Classical Library—Harvard University Press, 1932).

Aristotle, *Poetics*/Longinus, *On the Sublime*/Demetrius, *On Style* (trans. S. Halliwell, W. Hamilton Fyfe, and D.C. Innes) (Cambridge, Loeb Classical Library—Harvard University Press, 1995).

Arrian, *Anabasis of Alexander Vol.I* (trans. P.A. Brunt) (Cambridge, Loeb Classical Library—Harvard University Press, 1976).

Arrian, *Anabasis of Alexander Vol.II* (trans. P.A. Brunt) (Cambridge, Loeb Classical Library—Harvard University Press, 1983).

Arrian, *Quae Existant Omnia* (eds. A.G. Roos and G. Wirth) (Leipzig, Teubner, 1968).

Arrian, *Scripta Minora et Fragmenta* (ed. A.G. Roos) (Berlin, De Gruyter, 2002).

Arrian, *Tactical Handbook/Expedition Against the Alans* (trans. J.G. DeVoto) (Chicago, Ares, 1993).

Athenaeus, *The Deipnosophists Vol. VI* (trans. C.B. Gulick) (Cambridge, Loeb Classical Library—Harvard University Press, 1959).

Athenaeus, *The Learned Banqueters Vol.II* (trans. S.D. Olson) (Cambridge, Loeb Classical Library—Harvard University Press, 2007).

Athenaeus, *The Learned Banqueters Vol.VI* (trans. S.D. Olson) (Cambridge, Loeb Classical Library—Harvard University Press, 2010).

Autolycus, *On a Moving Sphere and On Risings and Settings* (ed. J. Mogenet) (Louvain, Bibliothèque de l'Université Bureaux et Recueil, 1950).

Autolycus, *On a Moving Sphere and On Risings and Settings* (trans. F. Bruin and A. Vondjidis) (Beirut, American University of Beirut, 1971).

Caesar, *Civil War* (trans. C. Damon) (Cambridge, Loeb Classical Library—Harvard University Press, 2016).

Cicero, *On Old Age/On Friendship/On Divination* (trans. W.A. Falconer) (Cambridge, Loeb Classical Library—Harvard University Press, 1923).

Cicero, *On the Nature of the Gods/Academics* (trans. H. Rackham) (Cambridge, Loeb Classical Library—Harvard University Press, 1933).

Cicero, *On the Orator Book 3/On Fate/Stoic Paradoxes/Divisions of Oratory.* (trans. H. Rackham) (Cambridge, Loeb Classical Library—Harvard University Press, 1942).

Cicero, *On the Republic/On the Laws* (trans. C.W. Keyes) (Cambridge, Loeb Classical Library—Harvard University Press, 1928).

Clement of Alexandria, *Writings of, Vol.II* (trans. W. Wilson) (Edinbrugh, T&T Clark, 1869).

Cleomedes, *De motu circulari coporum cealestium* (ed. H. Ziegler) (Lipsiae, B.G. Teubneri, 1891).

Cleomedes, *Lectures on Astronomy* (trans. A.C. Bowen and R.B. Todd) (Berkeley, University of California Press, 2004).

Comicorum Graecorum Fragmenta (ed. G. Kaibel) (Berlin, Weidmann, 1899).

Corpus of Ptolemaic Inscriptions (eds. A.K. Bowman, C.V. Crowther, S. Hornblower, R. Mairs, and K. Savvopoulos) (Oxford, Oxford University Press, 2021).

Curtius (Quintus Curtius), *History of Alexander Vol. II* (trans. J.C. Rolfe) (Cambridge, Loeb Classical Library—Harvard University Press, 1962).

Demosthenes, *Orations Vol.I* (trans. J.H. Vince) (Cambridge, Loeb Classical Library—Harvard University Press, 1930).

Didymos, *On Demosthenes* (trans. P. Harding) (Oxford, Oxford University Press, 2006).

Dio Cassius, *Roman History Vol.IV* (trans. E. Cary and H.B. Foster) (Cambridge, Loeb Classical Library—Harvard University Press, 1916).

Dio Chrysostom, *Discourses 37–60* (trans. H.L. Crosby) (Cambridge, Loeb Classical Library—Harvard University Press, 1946).

Diodorus Siculus, *Library of History Vol. I* (trans. C.H. Oldfather) (Cambridge, Loeb Classical Library—Harvard University Press, 1933).

Diodorus Siculus, *Library of History Vol. II* (trans. C.H. Oldfather) (Cambridge, Loeb Classical Library—Harvard University Press, 1935).

Diodorus Siculus, *Library of History Vol. III* (trans. C.H. Oldfather) (Cambridge, Loeb Classical Library—Harvard University Press, 1939).

Diodorus Siculus, *Library of History Vol.IV* (trans. C.H. Oldfather) (Cambridge, Loeb Classical Library—Harvard University Press, 1946).

Diodorus Siculus, *Library of History Vol.VIII* (trans. C. Bradford Welles) (Cambridge, Loeb Classical Library—Harvard University Press, 1963).

Diodorus Siculus, *Library of History Vol. IX* (trans. R.M. Geer) (Cambridge, Loeb Classical Library—Harvard University Press, 1947).

Diodorus Siculus, *Library of History Vol. X* (trans. R.M. Geer) (Cambridge, Loeb Classical Library—Harvard University Press, 1954).

Epiphanius, *Treatise on Weights and Measures* (ed. J.E. Dean) (Chicago, University of Chicago Press, 1935).

Eusebius, *Chronicle* (trans. J. Karst) (Leipzig, J.C. Hinrich, 1911).

Eusebius, *Ecclesiastical History Vol.II* (trans. J.E.L. Oulton) (Cambridge, Loeb Classical Library—Harvard University Press, 1932).

Eusebius, *Preparation for the Gospel Vol.II* (trans. E.H. Gifford) (Eugene, Wipf & Stock, 2002).

Eutocius of Ascalon, *Commentary* (trans. M. Decorps-Foulquier and M. Federspiel) (Berlin, DeGruyter, 2017).

Florus, *Epitome of Roman History* (trans. E.S. Forster) (Cambridge, Loeb Classical Library—Harvard University Press, 1929).

Galen, *Historia Philosophiae* (ed. K.G. Kühn) (Hildesheim, Olms, 1965).

Galen, *Institutio Logica* (trans. J.S. Kieffer) (Baltimore, The Johns Hopkins University Press, 1964).

Gellius, *Attic Nights Vol.II* (trans. J.C. Rolfe) (Cambridge, Loeb Classical Library—Harvard University Press, 1927).

Geminus, *Elementa Astronomiae* (ed. C. Manitius) (Lipsiae, Teubneri, 1897).

Geminus, *Introduction to the Phenomena* (eds. J. Evans and J.L. Berggren) (Princeton, Princeton University Press, 2006).

Greek Anthology (The) Vol.II (trans. W.R. Paton) (Cambridge, Loeb Classical Library—Harvard University Press, 1919).

Greek Anthology (The) Vol.IV (trans. W.R. Paton) (Cambridge, Loeb Classical Library—Harvard University Press, 1918).

Greek Lyric Vol.II: Anacreon, Anacreontea, Choral Lyric from Olympus to Alcman (trans. D.A. Campbell) (Cambridge, Loeb Classical Library—Harvard University Press, 1988).

Greek Mathematics Vol. I: From Thales to Euclid (trans. I. Thomas) (Cambridge, Loeb Classical Library—Harvard University Press, 1939).

Greek Mathematics Vol. II: From Aristarchus to Pappus (trans. I. Thomas) (Cambridge, Loeb Classical Library—Harvard University Press, 1957).

Hellenica Oxyrhynchia (trans. P.R. McKechnie and S.J. Kern) (Wiltshire, Aris and Phillips, 1993).

Herodian, *Technici Reliquia* (ed. A. Lentz) (Leipzig, BG Teubneri, 1870).

Herodotus, *The Persian Wars Vol.I* (trans. A.D. Goodley) (Cambridge, Loeb Classical Library—Harvard University Press, 1971).

Herodotus, *The Persian Wars Vol.III* (trans. A.D. Goodley) (Cambridge, Loeb Classical Library—Harvard University Press, 1922).

Heron of Alexandria, *Vermessungslehre Und Dioptra* (ed. H. Schone) (London, Forgotten Books, 2018).

Hesiod, *Theogony/Works and Days/Testimonia* (trans. G.W. Most) (Cambridge, Loeb Classical Library—Harvard University Press, 2018).

Hesychius Alexandrinus, *Lexicon Vol. I* (Amsterdam, Adolf M. Hekkert, 1965).

The Hibeh Papyri (trans. B.P. Grenfell and A.S. Hunt) (London, Egypt Exploration Fund, 1906).

Hipparchus, *Commentary on Aratus and Eudoxus' 'Phenomena'* (trans. C. Manitius) (Lipsiae, B.G. Teubneri, 1894).

Hippolytus, *Refutation of All Heresies* (trans. M.D. Litwa) (Atlanta, SBL Press, 2016).

Homer, *Iliad Vol.I* (trans. A.T. Murray and W.F Wyatt) (Cambridge, Loeb Classical Library—Harvard University Press, 1924).

Homer, *Iliad Vol.II* (trans. A.T. Murray and W.F Wyatt) (Cambridge, Loeb Classical Library—Harvard University Press, 1925).

Homer, *Odyssey Vol.I* (trans. A.T. Murray and G.E. Dimock) (Cambridge, Loeb Classical Library—Harvard University Press, 1919).

Homer, *Odyssey Vol.II* (trans. A.T. Murray and G.E. Dimock) (Cambridge, Loeb Classical Library—Harvard University Press, 1919).

Horace, *Odes and Epodes* (trans. N. Rudd) (Cambridge, Loeb Classical Library—Harvard University Press, 2004).

Hyginus, *Poeticon Astronomicon* (ed. T. Muncherus) (Amsterdam, Joannis à Someren, 1681).

Hypsicles, *Anaphorikos* (ed. J. Mantellius) (Paris, Ex Officina Cramosiana, 1657).

Jerome, *Chronicon* (trans. M. Donalson) (Lewiston, Edwin Mellen Press, 1996).

Josephus, *Against Apion* (trans. H. St. J. Thackeray) (Cambridge, Loeb Classical Library—Harvard University Press, 1926).

Josephus, *Jewish Antiquities Vol.V* (trans. R. Marcus) (Cambridge, Loeb Classical Library—Harvard University Press, 1943).

Josephus, *The Jewish War Vol.II* (trans. H. St. J. Thackeray) (Cambridge, Loeb Classical Library—Harvard University Press, 1927).

Justin, *Epitome of the Philippic History of Pompeius Trogus Vol.II* (trans. J.C Yardley) (London, Clarendon Press, 2012).

Letter of Aristeas (The) (trans. M. Hadas) (Eugene, Wipf & Stock, 1951).

Livy, *History of Rome Vol.XII* (trans. E.T. Sage and A.C. Schlesinger) (Cambridge, Loeb Classical Library—Harvard University Press, 1938).

Lucan, *The Civil War* (trans. J.D. Duff) (Cambridge, Loeb Classical Library—Harvard University Press, 1928).

Lucian, *Vol. VI* (trans. K. Kilburn) (Cambridge, Loeb Classical Library—Harvard University Press, 2019).

Lucretius, *On the Nature of Things* (trans. W.H.D. Rouse) (Cambridge, Loeb Classical Library—Harvard University Press, 1924).

Lydus, *de Mensibus* (trans. R. Wuensch) (Leipzig, Teubner, 1898).

Macrobius, *Saturnalia Vol.I* (trans. R.A. Kaster) (Cambridge, Loeb Classical Library—Harvard University Press, 2011).

Macrobius, *Somnium Scipionis* (trans. W.H. Stahl) (New York, Columbia University Press, 1990).

Marcian of Heraclea, *Periplous of the External Sea* (trans. W.H. Schoff) (Philadelphia, Commercial Museum, 1927).

Martianus Capella, *The Marriage of philology and Mercury* (ed. J. Willis) (Berlin, De Gruyter, 1983).

Medicorum Graecorum Opera quae Exstant Vol.XV—Galen (ed. C.G. Kühn) (Lipsiae, Cnoblochii, 1828).

Medicorum Graecorum Opera quae Exstant Vol.XVII—Galen (ed. C.G. Kühn) (Lipsiae, Cnoblochii, 1828).

Musici Scriptores Graeci: Aristoteles, Euclides, Nicomachus, Bacchius, Gaudentius, Alypius (ed. C. Janus) (Lipsiae, Tuebneri, 1895).

Nicomachus, *Introduction to Arithmetic* (trans. M.L. D'Ooge) (London, Britannica, 1994).

Palladius, *Opus Agriculturae* (ed. J.C. Schmitt) (Leipzig, Teubner, 1898).

Pappus, *Book 4 of the Collection* (trans. H. Sefrin-Weis) (Berlin, Springer, 2010).

Pappus, *Collection Vol.II* (trans. F. Hultsch) (Berolini, Weidmannos, 1877).

Pappus/Theon of Alexandria, *Commentaires de Pappus at de Théon d'Alexandrie sur l'Almageste Vol.II* (ed. A. Rome) (Rome, Biblioteca Apostolica Vaticano, 1936).

Pausanias, *Description of Greece Vol. I* (trans. W.H.S. Jones) (Cambridge, Loeb Classical Library—Harvard University Press, 1918).

Pausanias, *Description of Greece Vol. III* (trans. W.H.S. Jones) (Cambridge, Loeb Classical Library—Harvard University Press, 1988).

Philo, *On Abraham/On Joseph/On Moses* (trans. F.H. Colson) (Cambridge, Loeb Classical Library—Harvard University Press, 1935).

Philoponus, *On Aristotle Meteorology 1.1–3* (trans. I. Kupreeva) (London, Bloomsbury, 2014).

Plato, *Charmides/Alcibiades I and II/Hipparchus/The Lovers/Theages/Minos/Epinomis* (trans. W.R.M. Lamb) (Cambridge, Loeb Classical Library—Harvard University Press, 1927).

Plato, *Cratylus/Parmenides/Greater Hippias/Lesser Hippias* (trans. H.N. Fowler) (Cambridge, Loeb Classical Library—Harvard University Press, 1926).

Plato, *Euthyphro/Apology/Crito/Phaedo* (trans. W.R.M. Lamb) (Cambridge, Loeb Classical Library—Harvard University Press, 1924).

Plato, *Laches/Protagoras/Meno/Euthydemus* (trans. C. Emlyn-Jones and W. Preddy) (Cambridge, Loeb Classical Library—Harvard University Press, 2017).

Plato, *Laws* (trans. R.G. Bury) (Cambridge, Loeb Classical Library—Harvard University Press, 1926).

Plato, *Lysis/Symposium/Gorgias* (trans. W.R.M. Lamb) (Cambridge, Loeb Classical Library—Harvard University Press, 1925).

Plato, *Republic* (trans. P. Shorey) (Cambridge, Loeb Classical Library—Harvard University Press, 1970).

Plato, *Theaetetus/Sophist* (trans. H.N. Fowler) (Cambridge, Loeb Classical Library—Harvard University Press, 1921).

Plato, *Timaeus/Critias/Cleitophon/Menexenus/Epistles* (trans. R.G. Bury) (Cambridge, Loeb Classical Library—Harvard University Press, 1929).

Pliny (the Elder), *Natural History Vol. I* (trans. H. Rackham) (Cambridge, Loeb Classical Library—Harvard University Press, 1938).

Pliny (the Elder), *Natural History Vol. II* (trans. H. Rackham) (Cambridge, Loeb Classical Library—Harvard University Press, 1942).

Pliny (the Elder), *Natural History Vol. IV* (trans. H. Rackham) (Cambridge, Loeb Classical Library—Harvard University Press, 1945).

Pliny (the Elder), *Natural History Vol. X* (trans. D.E. Eichholz) (Cambridge, Loeb Classical Library—Harvard University Press, 1962).

Plutarch, *Lives Vol. II—Themistocles and Camillus/Aristides and Cato Major/Cimon* and Lucullus (trans. B. Perrin) (Cambridge, Loeb Classical Library—Harvard University Press, 1914).

Plutarch, *Lives Vol. III—Pericles and Fabius Maximus/Nicias and Crassus* (trans. B. Perrin) (Cambridge, Loeb Classical Library—Harvard University Press, 1967).

Plutarch, *Lives Vol. IV—Alcibiades and Coriolanus/Lysander and Sulla* (trans. B. Perrin) (Cambridge, Loeb Classical Library—Harvard University Press, 1916).

Plutarch, *Lives Vol. V—Agesilaus and Pompey/Pelopidas and Marcellus* (trans. B. Perrin) (Cambridge, Loeb Classical Library—Harvard University Press, 1968).

Plutarch, *Lives Vol. VII—Demosthenes and Cicero/Alexander and Caesar* (trans. B. Perrin) (Cambridge, Loeb Classical Library—Harvard University Press, 1967).

Plutarch, *Moralia Vol.II* (trans. F.C. Babbitt) (Cambridge, Loeb Classical Library—Harvard University Press, 1928).

Plutarch, *Moralia Vol.III* (trans. F.C. Babbitt) (Cambridge, Loeb Classical Library—Harvard University Press, 1931).

Plutarch, *Moralia Vol.IV* (trans. F.C. Babbitt) (Cambridge, Loeb Classical Library—Harvard University Press, 1936).

Plutarch, *Moralia Vol.V* (trans. F.C. Babbitt) (Cambridge, Loeb Classical Library—Harvard University Press, 1936).

Plutarch, *Moralia Vol.XII* (trans. H. Cherniss and W.C. Helmbold) (Cambridge, Loeb Classical Library—Harvard University Press, 1957).

Plutarch, *Moralia Vol.XIII* (trans. H. Cherniss) (Cambridge, Loeb Classical Library—Harvard University Press, 1976).

Plutarch, *Moralia Vol.XV* (trans. F.H. Sandbach) (Cambridge, Loeb Classical Library—Harvard University Press, 1969).

Plutarch, *Morals* (trans. G.N. Berardakis) (Leipzig, Teubner, 1893).

Plutarch (Pseudo-Plutarch), *Morals* (trans. W.W. Goodwin) (Boston, Little Brown & Co., 1874).

Polyaenus, *Stratagems of War Vol. I* (trans. P. Krentz and E.L. Wheeler) (Chicago, Ares Publishers, 1994).

Polybius, *The Histories Vol.V* (trans. W.R. Paton, F.W. Walbank and C. Habicht) (Cambridge, Loeb Classical Library—Harvard University Press, 2012).

Polybius, *The Histories Vol.V* (trans. W.R. Paton and F.W. Walbank) (Cambridge, Loeb Classical Library—Harvard University Press, 2012).

Proclus, *Hypotyposis Astronomicarum Positionum* (ed. C. Mantinus) (Berlin, de Gruyter, 2010).

Pseudo-Callisthenes, *The Romance of Alexander the Great* (trans. M. Wolohojian) (New York, Columbia University Press, 1969).

Pseudo-Democritus, *Natural and Secret Questions* (trans. M. Martelli) (London, Routledge, 2019).

Ptolemy, *Almagest* (trans. R.C. Taliaferro) (London, Britannica, 1989).

Ptolemy, *Geography* (trans. J.L. Berggen) (Princeton, Princeton University Press, 2000).

Ptolemy, *Harmonics* (trans. J. Solomon) (Leiden, Brill, 1999).

Ptolemy, *The Analemma* (ed. J.L. Heiberg) (Berlin, Leopold Classic Library, 2017).

Quintilian, *Institutio Oratoria* (trans. H.E. Butler) (Cambridge, Loeb Classical Library—Harvard University Press, 1920).

(The) Rhind Mathematical Papyrus Vol.I (trans. A.B. Chace) (Oberlin, Mathematical Association of America, 1927).

Scholia in Aristophanem Vol.1.1a (ed. W.J.W. Koster) (Groningen, B. Boekhuis, 1975).

Scholia in Aristophaneum Vol.3.1a (ed. M. Chantry) (Groningen, Egbert Forsten, 1999).

Seneca, *Moral Essays Vol.II* (trans. J.W. Basore) (Cambridge, Loeb Classical Library—Harvard University Press, 1932).

Seneca, *Natural Questions Vol.II* (trans. T.H. Corcoran) (Cambridge, Loeb Classical Library—Harvard University Press, 1972).

Servius (Maurus Servius Honoratus), *In Vergilii Carmina Comentarii* (eds. G. Thilo and H. Hagen) (Leipzig, B.G. Teubner, 1881).

Sextus Empiricus, *Opera Vol.III: Adversus Mathematicos* (ed. J. Mau) (Berlin, De Gruyter, 1961).

Simplicius, *On Aristotle, Physics 1–3.4* (trans. P.M. Huby and C.C.W. Taylor) (London, Duckworth, 2011).

Simplicius, *On Aristotle, On the Heavens 1.1–4* (trans. R.J. Hankinson) (London, Duckworth, 2001).

Simplicius, *On Aristotle, On the Heavens 2.10–14* (trans. I. Mueller) (London, Bloomsbury, 2001).

Stephanus of Byzantium, *Ethnicorum* (eds. A. Meinekii and T. Prior) (Berolini, G. Reimeri, 1849).

Stobaeus, *Anthologium* (ed. K. Wachsmuth and O. Hense) (Berlin, Weidmann, 1884).

Strabo, *The Geography of Strabo Vol.I* (trans. H.L. Jones) (Cambridge, Loeb Classical Library—Harvard University Press, 1917).

Strabo, *The Geography of Strabo Vol.IV* (trans. H.L. Jones) (Cambridge, Loeb Classical Library—Harvard University Press, 1927).

Strabo, *The Geography of Strabo Vol.V* (trans. H.L. Jones) (Cambridge, Loeb Classical Library—Harvard University Press, 1928).

Strabo, *The Geography of Strabo Vol.VI* (trans. H.L. Jones) (Cambridge, Loeb Classical Library—Harvard University Press, 1929).

Strabo, *The Geography of Strabo Vol.VII* (trans. H.L. Jones) (Cambridge, Loeb Classical Library—Harvard University Press, 1930).

Strabo, *The Geography of Strabo Vol.VIII* (trans. H.L. Jones) (Cambridge, Loeb Classical Library—Harvard University Press, 1932).

Sylloge Inscriptionum Graecarum, 3rd edn. (ed. W. Dittenberger) (Lipsiae, S. Hirzel, 1915).

Syncellus, *Chronographia* (ed. B.G. Niebuhrii) (Bonn, E.D. Weberi, 1829).

Theocritus, Moschus, Bion (trans. N. Hopkinson) (Cambridge, Loeb Classical Library—Harvard University Press, 2015).

Theodoret, *Graecarum Affectionum Curatio* (ed. I. Raeder) (Lipsiae, B.G. Teubneri, 1904).

Theon of Smyrna, *Exposition* (trans. J. Dupis) (Paris, Librairie Hachette & Co., 1892).

Theophrastus, *Enquiry into Plants Vol. I* (trans. A. Hort) (Cambridge, Loeb Classical Library—Harvard University Press, 1968).

Theophrastus, *Enquiry into Plants Vol. II* (trans. A. Hort) (Cambridge, Loeb Classical Library—Harvard University Press, 1916).

Theophrastus, *Characters*/Herodas, *Mimes*/*Sophron and Other Mime Fragments* (trans. J. Rusten and I.C. Cunningham) (Cambridge, Loeb Classical Library—Harvard University Press, 2003).

Thucydides, *History of the Peloponnesian War Vol. I* (trans. C.F. Smith) (Cambridge, Loeb Classical Library—Harvard University Press, 1969).

Valerius Maximus, *Memorable Doings and Sayings* (trans. D.R. Shackleton Bailey) (Cambridge, Loeb Classical Library—Harvard University Press, 2000).

Vegetius, *Epitome of Military Science* (trans. N.P. Milner) (Liverpool, Liverpool University Press, 1993).

Vitruvius, *On Architecture Vol.I* (trans. F. Granger) (Cambridge, Loeb Classical Library—Harvard University Press, 1931).

Vitruvius, *On Architecture Vol.II* (trans. F. Granger) (Cambridge, Loeb Classical Library—Harvard University Press, 1934).

Xenophon, *Hellenica* (trans. C.L. Brownson) (Cambridge, Loeb Classical Library—Harvard University Press, 1918).

Xenophon, *Memorabilia*/*Oeconomicus*/*Symposium*/*Apology* (trans. E.C. Marchant and O.J. Todd) (Cambridge, Loeb Classical Library—Harvard University Press, 2013).

Modern Sources

Aaboe, A. and de Solla Price, D.J. 'Qualitative Measurements in Antiquity: The Derivation of Accurate Parameters from Crude but Crucial Observations' in A. Koyré (ed.), *L'aventure de la Science, Mélanges* (Paris, Hermann, 1964), pp.6–10.

Anderson, A. 'Alexander at the Caspian Gates' *TAPA* 59 (1928), pp.130–163.

Andronikos, M. *Olympia* (Athens, Ekdotike Athenon, 1999).

Angel, J.L. 'A Racial Analysis of the Ancient Greeks: An Essay on the Use of Morphological Types' *Am J Phys Anthropol* 2 (1944), pp.329–376.

Angel, J.L. 'Skeletal Material from Attica' *Hesperia* 14 (1945), pp.279–363.

Bagnall, R.S. 'Alexandria: Library of Dreams' *Proceedings of the American Philosophical Society* 146 (2002), pp.348–362.

Bagordo, A. (ed.) *Fragmenta Comica: Telekleides* (Heidelberg, Verlag-Antike, 2013).

Balzac, C.-L., Cécile, F-C., and de Charbol de Volvic, G-J-G. (eds.), *Description de l'Égypte, ou Recueil des observations et des recherches qui ont été faites en Égypte pendant l'expédition de l'armée française* (Paris, l'Imprimerie Royale, 1812).

Barnes, R. 'Cloistered Bookworms in the Chicken-Coop of the Muses: The Ancient Library of Alexandria' in R. MacLeod (ed.), *The Library of Alexandria: Centre of Learning in the Ancient World* (London, IB Tauris, 2004), pp.61–77.

Bechler, Z. 'Aristotle Corrects Eudoxus: Met 1073b39–1074a16' *Centaurus* 15 (1970), pp.113–123.

Bell, S. *A Beginner's Guide to Uncertainty in Measurement* (Middlesex, National Physical Laboratory, 1999).

Ben-Menahem H. and Hecht, N.S. 'A Modest Addendum to the Greek Metrological Relief in Oxford' *AntJ* 65 (1985), pp.139–140.

Bennett, G.G. 'The Calculation of Astronomical Refraction in Marine Navigation' *J. Inst. Navig.* 35 (1982), pp.255–259.

Bennett, J., Donahue, M., Schneider, N., and Voit, M. *The Cosmic Perspective* (San Francisco, Pearson, 2014).

Berger, E.H. *Die geographischen Fragmente des Eratosthenes* (Leipzig, Teubner, 1880).

Berger, E.H. *Geschichte der wissenschaftlichen Erdkunde der Griechen* (Leipzig, Von Veit & Co., 1903).

Berrey, M. *Hellenistic Science at Court* (Berlin, De Gruyter, 2019).

Berve, H. *Das Alexanderreich auf Prosopographischer Grundlage Vol.I* (München, Beck, 1926).

Bilić, T. 'Apollo, Helios, and the Solstices in the Athenian, Delphian and Delian Calendars' *Numen* 59 (2012), pp.509–532.

Bilić, T. 'The Island of the Sun: Spatial Aspect of Solstices in Ancient Greek Thought' *GRBS* 56 (2016), pp.195–224.

Borchardt, L. 'Altägyptische Sonnenuhren' *ZÄS* 48 (1910), pp.9–17.

Bosworth, A.B. *A Historical Commentary on Arrian's History of Alexander Vol.I* (Oxford, Clarendon Press, 1980).

Bosworth, A.B. *Conquest and Empire: The Reign of Alexander the Great* (Cambridge, Cambridge University Press, 2008).

Boutsikas, E. and Ruggles, C. 'Temples, Stars and Ritual Landscapes: The Potential for Archaeoastronomy in Ancient Greece' *AJA* 115.1 (2011), pp.55–68.

Bowen, A.C. 'The Exact Sciences in Hellenistic Times: Texts and Issues' in D. Furley (ed.), *Routledge History of Philosophy Vol.II: From Aristotle to Augustine* (London, Routledge, 1999), pp.287–319.

Bowen, A.C. 'Cleomedes and the Measurement of the Earth: A Question of Procedures' *Centaurus* 45 (2003), pp.59–68.

Boyce, M. and Grenet, F. *A History of Zoroastrianism Vol.III: Zoroastrianism under Macedonian and Roman Rule* (Leiden, Brill, 1991).

Britton, J.P. *Models and Precision: The Quality of Ptolemy's Observations and Parameters* (New York, Garland, 1992).

Broneer, O. *Isthmia Vol.I: Temple of Poseidon* (Princeton, American School of Classical Studies in Athens, 1971).

Bruins, E.M. (ed.) *Codex Constantinopolitanus, Palatii veteris Vol.3* (Leiden, Brill, 1964).

Bruins E.M. 'The Egyptian Shadow Clock' *Janus* 52 (1965), pp.127–137.

Burkert, W. *Structure and History in Greek Mythology and Ritual* (Berkeley, University of California Press, 1979).

Burkert, W. *Greek Religion* (Cambridge, Harvard University Press, 1985).

Burns, A. 'Hippodamus and the Planned City' *Historia* 25 (1976), pp.414–428.

Cameron, I. *Mountains of the Gods* (London, Century, 1984).

Carman, C.C. and Evans, J. 'The Two Earths of Eratosthenes' *Isis* 106 (2015), pp.1–16.

Cary, M. 'A Constitutional Inscription from Cyrene' *JHS* 48.2 (1928), pp.222–238.

Cerri, G. 'Le scienze esalte nel mondo antico' *MedAnt* 1.2 (1998), pp.363–380.

Champion, J. *Pyrrhus of Epirus* (Barnsley, Pen & Sword, 2009).

Charitonidou, A. *Delphi* (Aharnes, Hesperos, 1978).

Christian-Meyer, W. 'Der "Pythagoras" in Ägypten am Beginn des Alten Reiches' in *MDAIK* 43 (1987), pp.195–206.

Clarysse, W. 'Greeks in Ptolemaic Thebes' in S.P. Vleeming (ed.), *Hundred-Gated Thebes: Acts of a Colloquium on Thebes and the Theban Area on the Greco-Roman Period* (Leiden, Brill, 1995), pp.1–19.

Cohen, G.M. *The Hellenistic Settlements in the East from Armenia and Mesopotamia to Bactria and India* (Berkeley, University of California Press, 2013).

Connolly, P. *Greece and Rome at War* (London, Greenhill Books, 1998).

Cooper, G.M. 'Astrology: The Science of Signs in the Heavens' in P.T. Keyser and J. Scarborough (eds.), *Oxford Handbook of Science and Medicine in the Classical World* (Oxford, Oxford University Press, 2018), pp.381–407.

Crawford, D.J. *Kerkeosiris: An Egyptian Village in the Ptolemaic Period* (Cambridge, Cambridge University Press, 1971).

Cuomo, S. *Ancient Mathematics* (London, Routledge, 2005).

d'Anville, J.B.B. 'Écclaircissements géographiques sur l'ancinne Gaule' *Precedés d'un traité des mesures itinéaires des romains, et de la lieue gauloise* (Paris, chez la Veuve Estienne, 1741).

d'Anville, J.B.B. 'Mémoire sur la mesure du schène égyptien, et du stade qui servant à le composer' *Mém. Acad. Inscript. et belles-lettres* 26 (1759), pp.82–91.

d'Anville, J.B.B. 'Discussion de la mesure de la terre par Ératosthène' *Mém. Acad. Inscript. et belles-lettres* 26 (1759), pp.92–100.

Daiber, H. (trans.), *Die Vorsokratiker in arabischer Überlieferung* (Wiesbaden, Franz Steiner, 1980).

Dain, A. *Sylloge Tacticorum quae elim 'inedita Leonis tactica' dicebantur* (Paris, Société d'édition 'Les Belles lettres', 1938).

Davidson, J.A. 'Peisistratus and Homer' *TAPA* 86 (1955), pp.1–21.

Davies, T.S. 'An Enquiry into the Geometrical Character of the Hour Lines upon the Antique Sundials' *Trans. R. Soc. Edinburgh* 8 (1818), pp.72–122.

Defense Mapping Agency, *Supplement to Department of Defense World Geodetic System 1984 Technical Report: Part I—Methods, Techniques, and Data Used in WGS 84 Development* (Washington DC, US Naval Observatory, 1987).

Dekoulakou-Sideris, I. 'A Metrological Relief from Salamis' *AJA* 94.3 (1990), pp.445–451.

Depuydt, L. 'The Egyptian and Athenian Dates of Meton's Observation of the Summer Solstice (−431)' *AncSoc* 27 (1996), pp.27–45.

Dicks, D.R. *The Geographical Fragments of Hipparchus* (London, Athlone Press, 1960).

Diller, A. 'The Ancient Measurements of the Earth' *Isis* 40 (1949), pp.6–9.

Diller, A. 'Julian of Ascalon on Strabo and the Stade' *Cl. Phil.* 45 (1950), pp.22–25.

Dimitrijević, M.S. and Bajić, A. 'Mythological Origin of Constellations and their Description: Aratus, Pseudo-Eratosthenes, Hyginus' in L.Č. Popović, V.A. Srećković, M.S. Dimitrijević, and A. Kovačrvić (eds.), *Proceedings of the XII Serbian-Bulgarian Astronomical Conference* (Belgrade, Astronomy Society 'Rudjer Bošković, 2020), pp.129–138.

Dinsmoor, W.B. 'Archaeology and Astronomy' *PAPS* 80 (1939), pp.95–173.

Dinsmoor, W.B. 'The Basis of Greek Temple Design in Asia Minor, Greece and Italy' *Atti VII—Congresso Internazionale di Archologia Classica I* (Rome, L'Erma di Bretschneider, 1961), pp.358–361.

Dinsmoor, W.B. *The Architecture of Ancient Greece: An Account of its Historic Development* (New York, Biblo and Tannen, 1973).

Dolan, W.W. 'Early Sundials and the Discovery of Conic Sections' *Math. Mag.* 45.1 (1972), pp.8–12.

Donlan, W. and Thompson, J. 'The Charge at Marathon: Herodotus 6.112' *CJ* 71 (1976), pp.393–343.

Dörpfeld, W. 'Beiträge zur antiken Metrologie 1: Das solonisch-attische Syetem' *Ath. Mitt.* VII (1882), p.277.

Dreyer, J.L.E. *A History of Astronomy from Thales to Kepler* (London, Constable, 1953).

Droysen, J.G. *Histoire de l'hellénisme* (Paris, E.Leroux, 1883).

Dueck, D. *Geography in Classical Antiquity* (Cambridge, Cambridge University Press, 2012).

Dürrback, F. and Jardé, A. 'Fouilles de Délos' *BCH* 29 (1905), pp.250–252.

Dutka, J. 'Eratosthenes' Measurement of the Earth Reconsidered' *Arch. Hist. Exact Sci.* 46.1 (1993), pp.55–66.

Engels, D.W. *Alexander the Great and the Logistics of the Macedonian Army* (Berkeley, University of California Press, 1978).

Engels, D. 'The Length of Eratosthenes' Stade' *Am. Journ. Phil.* 106.3 (1985), pp.298–311.

English, S. *The Army of Alexander the Great* (Barnsley, Pen & Sword, 2009).

Erskine, A. 'Culture and Power in Ptolemaic Egypt: The Museum and Library of Alexandria' *G&R* 42.1 (1995), pp.38–48.

Evans, J. *The History and Practice of Ancient Astronomy* (Oxford, Oxford University Press, 1998).

Fermor, J. 'Timing the Sun in Egypt and Mesopotamia' *Vist. Astron.* 41 (1997), pp.157–167.

Fernie, E. 'The Greek Metrological Relief in Oxford' *AntJ* 61 (1981), pp.255–261.

Fine, J.V.A. *The Ancient Greeks: A Critical History* (Cambridge, Cambridge University Press, 1983).

Firsov, L.V. 'Eratosthenes's Calculation of the Earth's Circumference and the Length of the Hellenistic Stade' *VDI* 121 (1972), pp.154–175.

Fischer, I. 'Another Look at Eratosthenes' and Posidonius' Determinations of the Earth's Circumference' *Q. Jl R. Astr. Soc.* 16 (1975), pp.152–167.

Fischer, K. 'Zur Lage von Kandahar an Landverbindungen zwischen Iran und Indien' *Bonner Jahrb.* 167 (1967), pp.129–252.

Floyer, E.A. 'Primitive Sundials in Upper Ancient Egypt' *Athenaeum* (1895), p.3545.

Foxhall, L. and Forbes, H.A. 'Σιτομετρεία: The Role of Grain as a Staple Food in Classical Antiquity' *Chiron* 12 (1982), pp.41–90.

Fraser, P.M. *Cities of Alexander the Great* (Oxford, Oxford University Press, 1996).

Freeth, T. 'Decoding an Ancient Computer' *SciAm* 301.6 (2009), pp.76–83.

Fuller, J.F.C. *The Generalship of Alexander the Great* (Hertfordshire, Wordsworth, 1998).

Gabriel, R. *Philip II of Macedonia—Greater than Alexander* (Washington DC, Potomac Press, 2010).

Gabriel, R. and Metz, K. *From Sumer to Rome: The Military Capabilities of Ancient Armies* (Connecticut, Greenwood Press, 1991).

Gatty, A. *The Book of Sundials* (London, George Bell & Sons, 1900).

Geus, K. and Tupikova, I. 'Astronomy and Geography: Some Unexplored Connections in Ptolemy' in F. Pontani (ed.), *Certissima Signa: A Venice Conference on Greek and Latin Astronomical Texts* (Venezia, Edizioni Ca' Fascari, 2017), pp.61–73.

Gibbs, S.L. *Greek and Roman Sundials* (New Haven, Yale University Press, 1976).

Gillings, R.J. *Mathematics in the Time of the Pharaohs* (New York, Dover, 1982).

Girard, P.S. 'Sur la coudée septénnaire des anciens Égyptiens et les différents étalons qui en ont été retrouvés jusqu'à présent' *Mem. Acad. Sci. Inst. Fr.* 9 (1830), pp.591–608.

Goldstein, B.R. 'Eratosthenes on the "Measurement" of the Earth' *HIST MATH* 11 (1984), pp.411–416.

Goldstein, B.R. and Bowen, A.C. 'A New View of Early Greek Astronomy' *Isis* 74.3 (1983), pp.330–340.

Goldstein, B.R. and North, J.D. 'The Introduction of Dated Observations and Precise Measurement in Greek Astronomy' *Arch. Hist. Exact Sci.* 43.2 (1991), pp.93–132.

Gosselin, P.F.J. *Recherches fur la Geographie Systematique et positive des Anciens* (Paris, National Institute of France, 1798).

Gow, M. *Measuring the Earth: Eratosthenes and His Celestial Geometry* (Berkeley Heights, Enslow, 2010).

Grafton, A. and Shelford, A. *New World, Ancient Texts* (Harvard, Harvard University Press, 2015).

Gratwick, A. 'Alexandria, Syene, Meroe: Symmetry in Eratosthenes' Measurement of the World' in L. Ayres (ed.), *The Passionate Intellect: Essays on the Transformation of Classical Traditions Presented to Professor I.G. Kidd* (New Brunswick, Transaction, 1995), pp.177–202.

Guadognoli, M., Fober, G., and Terry, P. 'Accuracy of Pace Count as a Distance Estimation Procedure' *Mil. Psychol.* 2–3 (1990), pp.183–191.

Guest, R., Miguel-Hurtado, O., Stevenage, S., and Black, S. 'Exploring the Relationship Between Stride, Stature and Hand Size for Forensic Assessment' *J Forensic Leg Med* 52 (2017), pp.46–55.

Gulbekian, E. 'The Origin and Value of the Stadion Unit used by Eratosthenes in the Third Century B.C.' *Arch. Hist. Exact Sci.* 37.4 (1987), pp.359–363.

Güterbock, H.G. 'The Hittite Version of the Hurrian Kumarbi Myth: Oriental Forerunners of Hesiod' *AJA* 52.1 (1948), pp.123–134.

Gysembergh, V. 'Aristotle on the "Great Year", Eudoxus, and Mesopotamian "Goal Year" Astronomy' *Aion* 35 (2013), pp.111–123.

Hahn, R. *Anaximander and the Architects: The Contributions of Egyptian and Greek Architectural Technologies to the Origins of Greek Philosophy* (Albany, State University of New York Press, 2001), pp.200–210.

Hammond, N. *The Genius of Alexander the Great* (London, Duckworth, 1997).

Hansman, J. 'The Problems of Qūmis' *J.R. Asiat. Soc.* 100.3–4 (1968), pp.111–139.

Hanson, V.D. 'Hoplite Technology in Phalanx Battle' in V.D. Hanson (ed.), *Hoplites: The Classical Greek Battle Experience* (London, Routledge, 1991), pp.63–84.

Harding, P. (ed.), *From the End of the Peloponnesian War to the Battle of Ipsus* (Cambridge, Cambridge University Press, 2012).

Hayduck, M. (ed.), *Commentaria in Aristotelem Graeca* XIV (Berlin, Reimer, 1901).

Heath, T. *The Thirteen Books of Euclid's Elements* (New York, Dover, 1956).

Heath, T. *A History of Greek Mathematics* (New York, Dover, 1981).

Heath, T. *Aristarchus of Samos: The Ancient Copernicus* (New York, Dover, 2004).

Heckel, W. and Jones, R. *Macedonian Warrior—Alexander's Elite Infantryman* (Oxford, Osprey, 2006).

Heckel, W. and Trittle, L. *Alexander the Great: A New History* (Oxford, Wiley, 2011).

Henneberg, M. and Henneberg, R.J. 'Biological Characteristics of the Population Based on Analysis of Skeletal Remains' in J.C. Carter (ed.), *The Chora at Metaponto: The Necropoleis Vol.II* (Austin, University of Texas Press, 1998), pp.503–537.

Hetherington, N.S. 'Plato's Place in the History of Greek Astronomy: Restoring *both* History and Science to the History of Science' *J. Astron. Hist. Herit.* 2 (1999), pp.87–110.

Hignett, C. *Xerxes' Invasion of Greece* (Oxford, Clarendon Press, 1963).

Hiller, E. *Eratosthenis carminum reliquiae* (Leipzig, Teubner, 1872).

Hon, G. 'Is There a Concept of Experimental Error in Greek Astronomy?' *Br. J. Hist. Sci.* 22.2 (1989), pp.129–150.

Honigman, S. *The Septuagint and Homeric Scholarship: A Study in the Narrative of the Letter of Aristeas* (New York, Routledge, 2003).

Houtum-Schindler, A. 'Notes on Some Antiquities Found in a Mound Near Damghan' *J.R. Asiat. Soc.* 9 (1877), pp.425–427.

How, W.W. and Wells, J. *A Commentary on Herodotus Vol.II* (Oxford, Clarendon Press, 1912).

Hultsch, F.O. *Griechische und römische metrologie* (Berlin, Weidmannshce Buchhandlung, 1882).

Humphrey, J.H. *Roman Circuses: Arenas for Chariot Racing* (Berkeley, University of California Press, 1986).

Hunt, S. *The Oxyrhynchus Papyri Part.X* ((London, Egypt Exploration Society, 1914).

Ioppolo, G. 'La ta vola délie unità di mi su re ad mercato augusteo di Leptis Magna' *QAL* 5 (1967), pp.89–98.

Irby-Massie, G.L. and Keyser, P.T. *Greek Science in the Hellenistic Era* (London, Routledge, 2002).

Isler, M. 'The Gnomon in Egyptian Antiquity' *Journal of the American Research Center in Egypt* 28 (1991), pp.155–185.

Jackson, A.V.W. *From Constantinople to the Home of Omar Khayyam* (New York, Macmillan, 1911).

Jeremiah, E. 'Not Much Missing: Statistical Explorations of the *Placita* of Aetius' *Philos. Antiq.* 148 (*Aëtiana IV*) (2018), pp.279–373.

Johnstone, S. 'A New History of Libraries and Books in the Hellenistic Period' *Class. Antiq.* 33.2 (2014), pp.347–393.

Jones, A. 'Greco-Roman Sundials: Precision and Displacement' in K.J. Miller and S.L. Symons (eds.), *Down to the Hour: Short Time in the Ancient Mediterranean and Near East* (Leiden, Brill, 2020), pp.125–157.

Jones, M.W. 'Doric Measure and Architectural Design 1: The Evidence of the Relief from Salamis' *AJA* 104.1 (2000), pp.73–93.

Kegerreis, C. 'Setting a Royal Pace: Achaemenid Kingship and the Origin of Alexander the Great's *Bematistai*' *AHB* 31.1–2 (2017), pp.39–64.

Kiepert, H. 'Zur Topographie des alten Alexandrien' *Zeitschr. d. Gesell. f. Erdkunde zu Berlin* 7 (1872), pp.337–349.

Kline, M. *Mathematical Thought from Ancient to Modern Times* (Oxford, Oxford University Press, 1990).

Lambert, W.G. and Parker, S.B. *Enûma Eliš. The Babylonian Epic of Creation* (Oxford, Oxford University Press, 1966).

Larner, J. 'The Church and the Quattrocentro Renaissance in Geography' *Renaiss. Stud.* 12.1 (1998), pp.26–39.

Larsen, J.A.O. 'Notes on the Constitutional Inscription from Cyrene' *CP* 24.4 (1929), pp.351–368.

Laskar, J. 'Secular Terms of Classical Planetary Theories using the Results of General Theory' *A&AT* 157 (1986), pp.59–70.

Lasky, K. *The Librarian who Measured the Earth* (New York, Little Brown Books, 1994).

Lazenby, J.F. *The Defence of Greece 490–479BC* (Warminster, Aris and Phillips, 1993).

Lehmann-Haupt, C.F.F. *Das altbsbylonische Maß und Gewichtssystem als Grundlage der antiken Gewichts, Münz und Maßsysteme* (Leiden, Ausz. Aus den Akten des 8 internat. Orientalistenkomgresses, 1896).

Lehmann-Haupt, C.F.F. 'Stadion (Metrologie)' in G. Wissowa and W. Kroll (eds.), *Paulys Real-Enccyclopädie* (Stuttgart, J.B. Metzlersche, 1929), pp.1961–1963.

Letronne, J.A. 'Mémoire sur cette question: Les Anciens ont-ils exécuté une mesure de la terre postérieurement à l'établissement de l'école d'Alexandrie?' *H. M. Inst. R. Fr.* 6 (1822), pp.261–323.

Letronne, J.A. *Recherches critiques: historiques et géographiques sur les fragment d'Heron d'Alexandrie, ou de systéme métriques égyptien* (Paris, l'Imprimeire Nationale, 1851).

Lewis, M. 'Greek and Roman Surveying and Surveying Instruments' in R.J.A. Talbert (ed.), *Ancient Perspectives: Maps and Their Place in Mesopotamia, Egypt, Greece and Rome* (Chicago, University of Chicago Press, 2012), pp.129–162.

Lewis, N. *Greeks in Ptolemaic Egypt: Case Studies in the Social History of the Hellenistic World* (Oxford, Clarendon, 1986).

Lewis, N. 'Greek and Roman Surveying and Surveying Instruments' in R.J.A. Talbert (ed.), *Ancient Perspectives: Maps and Their Place in Mesopotamia, Egypt, Greece and Rome* (Chicago, University of Chicago Press, 2012), pp.129–162.

Liampi, K. *Der makedonische Schild* (Bonn, Rudolf Habelt, 1998).

Lloyd, A.B. *Herodotus Book II: Commentary 1–98* (Leiden, Brill, 1976).

McKenzie, J. *The Art and Architecture of Alexandria and Egypt 300BC–AD700* (New Haven, Yale University Press, 2007).

McKenzie, J.S., Gibson, S., and Reyes, A.T. 'Reconstructing the Serapaeum in Alexandria from Archaeological Evidence' *JRS* (2004), pp.73–121.

McKirahan, R.D. *Xenophanes of Colophon: Philosophy Before Socrates* (Indianapolis, Hackett Publishing Company, 1994).

Macan, R.M. *Herodotus: The 7th, 8th and 9th Books Vol.I* (New York, Arno Press, 1973).

MacLeod, R. 'Introduction: Alexandria in History and Myth' in R. MacLeod (ed.), *The Library of Alexandria: Centre of Learning in the Ancient World* (London, IB Tauris, 2004), pp.1–15.

Magdolen, D. 'The Solar Origin of the "Sacred Triangle" in Ancient Egypt?' *SAK* 28 (2000), pp.207–217.

Mansfield, J. 'Cosmic Distances: "Aetius" 2.31 Diels and Some Related Texts' *Phronesis* 45.3 (2000), pp.175–204.

Mansfield, J. and Runia, D.T. 'Liber 2 Caput 31' *An Edition of the Reconstructed Text of the Placitia with a Commentary and a Collection of Related Texts* (Leiden, Brill, 2020), pp.1103–1112.

Matthew, C.A. *A Storm of Spears: Understanding the Greek Hoplite at War* (Barnsley, Pen & Sword, 2012).

Matthew, C.A. *An Invincible Beast: Understanding the Hellenistic Pike-Phalanx at War* (Barnsley, Pen & Sword, 2015).

Mattingly, H.B. 'The Athenian Coinage Decree' *Historia* 10 (1961), pp.148–188.

Mattingly, H.B. 'Epigraphy and the Athenian Empire' *Historia* 41 (1992), pp.129–138.

Mattingly, H.B. 'New Light on the Athenian Standards Decree' *Klio* 75 (1993), pp.99–102.

Mazza, L. 'Plan and Constitution: Aristotle's Hippodamus: Towards an 'Ostensive' Definition of Spatial Planning' *TPR* 80 (2009), pp.113–141.

Meeus, J. *Astronomical Algorithms* (Richmond, Willmann-Bell, 1998).

Meiggs, R. and Lewis, D. *A Selection of Greek Historical Inscriptions to the End of the Fifth Century BC* (Oxford, Oxford University Press, 2004).

Mendell, H. 'Reflections of Eudoxus, Callippus and their Curves: Hippopedes and Callippopedes' *Centaurus* 40 (1998), pp.177–275.

Meyboom, P.G.P. *The Nile Mosaic of Palestrina: Early Evidence of Egyptian Religion in Italy* (Leiden, Brill, 1995).

Michaelis, A. 'The Metrological Relief at Oxford' *JHS* 4 (1883), pp.335–350.

Minge, J-P. *Patrologiae cursus completes, sive Bibliotheca universalis [Vol. 127–162] (Ed. 1844–1864)* (Paris, Hachette Livre/BNF, 2012).

Moorman, E.M. *Ancient Sculpture in the Allard Pierson Museum Amsterdam* (Amsterdam, Allard Pierson Series, 2000).

Morrison, S.E. *The European Discovery of America: The Southern Voyages AD1492–1616* (Oxford, Oxford University Press, 1974).

Moussas, X. 'The Antikythera Mechanism: The Oldest Mechanical Universe in its Scientific Milieu' in D. Valls-Gabaud and A. Boksenberg (eds.), *The Role of Astronomy in Society and Culture: Proceedings of IAU Symposium No.260, 2009* (Cambridge, IAU Publishers, 2011), pp.135–148.

Munn, M. 'From Science to Sophistry: The Path of the Sun, the Shape of the World, and the Place of Athens in the Cosmos' in A. Pierris (ed.), *Physis and Nomos: Power, Justice, and the Agonistical Ideal of Life in High Classicum—Proceedings of the Symposium Philosophiae Antiquae Quartum Atheniense*, 4–12 July 2004 (Patras, David Brown Book Co., 2007), pp.111–133.

Murray, J. *A Handbook for Travellers in Lower and Upper Egypt; Including Descriptions of the Course of the Nile through Egypt and Nubia, Alexandria, Cairo, The Pyramids, Thebes, The Suez Canal, The Peninsula of Mount Sinai, the Oases, the Fayoom, Part I. Seventh Edition* (London, John Murray, 1888).

Murray, O. 'The Letter of Aristeas' *BSA* 54 (1987), pp.15–29.

Nassau, J.J. *Practical Astronomy* (New York, McGraw-Hill, 1948).

Neugebauer, O. *The Exact Sciences in Antiquity* (New York, Dover Publications, 1969).

Neugebauer, O. *A History of Ancient Mathematical Astronomy* (Berlin, Springer, 1975).

Neugebauer, O. and Parker, R. *Egyptian Astronomical Texts Vol.I* (Providence, Brown University Press, 1960).

Neugebauer, O. and Parker, R. *Egyptian Astronomical Texts Vol.II* (Providence, Brown University Press, 1966).

Newhall, S.H. 'Peisistratus and His Edition of Homer' *PNAS* 43.19 (1908), pp.491–510.

Newton, R.R. 'The Sources of Eratosthenes' Measurement of the Earth' *Q. Jl R. Astr. Soc.* 21 (1980), pp.379–387.

Nicastro, N. *Circumference: Eratosthenes and the Ancient Quest to Measure the Globe* (New York, St. Martins Press, 2008).

North, J. *Cosmos: An Illustrated History of Astronomy and Cosmology* (Chicago, University of Chicago Press, 2008).

O'Neil, W.M. *Early Astronomy from Babylon to Copernicus* (Sydney, Sydney University Press, 1986).

Osborne, C. *Presocratic Philosophy: A Very Short Introduction.* (Oxford, Oxford University Press, 2004).

Panchenko, D. 'Thales and the Origin of Theoretical Reasoning' *Configurations* 1.3 (1993), pp.387–414.

Papathanassiou, M.K. 'Reflections on the Antikythera Mechanism Inscriptions' *Adv. Space Res.* 46 (2010), pp.545–551.

Parker, R.A. 'The Calendars of Ancient Egypt' *SAOC* 26 (1950), pp.1–83.

Parker, R.A. 'Ancient Egyptian Astronomy' *Philos. Trans. R. Soc.* 276 (1974), pp.51–65.

Peltz, U. 'Der Makedonische Schild aus Pergamon der Antikensammlung Berlin' *Jahrb. Berlin Museen* 43 (2001), pp.331–343.

Pfeiffer, R. *History of Classical Scholarship from the Beginnings to the End of the Hellenistic Age* (Oxford, Clarendon Press, 1968).

Pinotsis, A.D. 'A Comparative Study of the Evolution of the Geographical Ideas and Measurements until the Time of Eratosthenes' *A&AT* 24 (2005), pp.127–138.

Pinotsis, A.D. 'The Significance and Errors of Eratosthenes' Method for the Measurement of the Size and Shape of the Earth's Surface' *J. Astron. Hist. Herit.* 9 (2006), pp.57–63.

Pipunyrov, V.N. *Istoriya Chasov s Drevnejshix Vremen do Nashix Dnej [Story Hours from Ancient Times to the Present Day]* (Moscow, Nauka, 1982).

Powell, M.A. 'Maße und Gewichte' in A. Bramanti, E. Ebeling, and M.P. Streck (eds.), *Reallexikon der Assyriologie und Vorderasiatischen Archäologie VII* (Berlin, DeGruyter, 1990), pp. 457–517.

Rawlins, D. 'Eratosthenes' Geodesy Unraveled: Was there High-Accuracy Hellenistic Astronomy?' *Isis* 73 (1982), pp.259–265.

Rawlins, D. 'Eratosthenes' Too-Big Earth and Too-Tiny Universe' *Dio* 14 (2008), pp.3–12.

Richardson, W.F. *Numbering and Measuring in the Classical World* (Bristol, Bristol Phoenix Press, 2004).

Rohr, R.R.J. *Sundials: History, Theory and Practice* (New York, Dover, 1996).

Roller, D.W (ed.), *Eratosthenes' Geography* (Princeton, Princeton University Press, 2010).

Rome, A. 'Le problème de la distance entre deux villes dans la dioptra de Héron' *Ann. Soc. Sci. Bruxelles, Ser. A* 42 (1923), pp.234–258.

Roy, A.E. and Clarke, D. *Astronomy Principles and Practice* (Bristol, Institute of Physics Publishing, 2003).

Russo, L. 'The Astronomy of Hipparchus and His Time: A Study based on Pre-Ptolemaic Sources' *Vist. Astron.* 38.2 (1994), pp.207–248.

Russo, L. *The Forgotten Revolution: How Science was Born in 300BC and Why It Had to Be Reborn* (trans. S. Levy) (Berlin, Springer, 2000).

Sagan, C. *Cosmos* (London, MacDonald & Co., 1980).

Schaefer, B.E. and Liller, W. 'Refraction Near the Horizon' *Publ. Astron. Soc. Pac.* 102 (1990), pp.796–805.

Schaldach, K. 'The Arachne of the Amphiareion and the Origin of Gnomonics in Greece' *J. Hist. Astron.* 25 (2004), pp.435–445.

Schechner, S. 'The Material Culture of Astronomy in Daily Life: Sundials, Science and Social Change' *J. Hist. Astron.* 108 (2001), pp.189–222.

Schwarz, K.P. 'Zur Erdmessung des Eratosthenes' *AVN* 82 (1975), pp.1–12.

Scott, N.E. 'An Egyptian Sundial' *MMAB* 30.4 (1935), pp.88–89.

Shcheglov, D. 'Hipparchus' Table of Climata and Ptolemy's Geography' *Orb. Terr.* 9 (2007), pp.159–192.

Shcheglov, D.A. 'The So-Called 'Itinerary Stade' and the Accuracy of Eratosthenes' Measurement of the Earth' *Klio* 100.1, 2018, pp.153–177.

Sheppard, R. (ed.) *Alexander the Great at War* (Oxford, Osprey, 2008).

Shipley, G. *The Greek World after Alexander 323–30BC* (London, Routledge, 2000).

Sidoli, N. 'Heron's *Dioptra* 35 and Analemma Methods: An Astronomical Determination of the Distance Between Two Cities' *Centaurus* 47 (2005), pp.236–258.

Slapsak, B. 'The 302mm foot measure on Salamis?' *DHA* 19.2 (1993), pp.119–136.

Snodgrass, A.M. *Arms and Armour of the Greeks* (Baltimore, The Johns Hopkins University Press, 1999).

Standish, J.F. 'The Caspian Gates' *G&R* 17 (1970), pp.17–24.

Stevens, H.N. *Ptolemy's Geography: A Brief Account of all of the Printed Editions down to 1730* (London, Stevens, Son and Stiles, 1908).

Stoneman, R. 'Romantic Ethnography: Central Asia and India in the Alexander Romance' *AncW* 25.1 (1994), pp.93–107.

Symons, S. *Ancient Egyptian Astronomy: Timekeeping and Cosmography in the New Kingdom* (Leicester, University of Leicester—unpublished thesis, 1999).

Symons, S. and Khurana, H. 'A Catalogue of Ancient Egyptian Sundials' *J. Hist. Astron.* 47 (2016), pp.375–385.

Szabó, A. and Maula, E. *Enklima: Untersuchungen zur Frühgeschichte der griechischen Astronomie, Geographie und der Sehnentafeln* (Athens, Akademie Athen—Forschungsinstitut für Griechische Philosophie, 1982).

Taisbak, C.M. 'Posidonius Vindicated at All Costs? Modern Scholarship Versus the Stoic Earth Measurer' *Centaurus* 18 (1974), pp.253–269.

Tanner, R.G. 'Aristotle's Works: The Possible Origins of the Alexandria Collection' in R. MacLeod (ed.), *The Library of Alexandria: Centre of Learning in the Ancient World* (London, IB Tauris, 2004), pp.79–91.

Tarn, W.W. *Hellenistic Military and Naval Developments* (Cambridge, Cambridge University Press, 1930).

Tarn, W.W. *Alexander the Great Vol.II* (Chicago, Ares, 1981).

Taub, L. *Science Writing in Greco-Roman Antiquity* (Cambridge, Cambridge University Press, 2017).

Theodossiou, E., Manimanis, V.N., Mantarakis, P., and Dimitrijrvic M.S. 'Astronomy and Constellations in the *Iliad* and *Odyssey*' *J. Astron. Hist. Herit.* 14 (2011), pp.22–30.

Thibodeau, P. 'Anaximander's Spartan Sundial' *CQ* 67.2 (2017), pp.374–379.

Thomas, R. and Masson, A. 'Altars, Sundials, Minor Architectural Objects and Models' in A. Villing, M. Bergeron, G. Bourogiannis, A. Johnston, F. Leclère, A. Masson, and R. Thomas (eds.), *Naukratis: Greeks in Egypt* (London, British Museum, 2015), pp.2–14.

Tomlinson, R.A. 'The Town Plan of Hellenistic Alexandria' in P. Schmidlin, G. Monaco, M. Trojani, and D. Said (eds.), *Alessandria e il Mondo Ellenistico-Romano* (Rome, L'Erma di Bretschneider, 1995), pp.236–240.

Todd, R.B. 'Cleomedes' in V. Brown, P.O. Kristeller, and F.E. Cranz (eds.), *Catalogus Translationum et Commentariorum: Medieval and Renaissance Latin Translations and Commentaries Vol.VII* (Washington DC, Catholic University of America Press, 1992), pp.1–11.

Tscherikower, V. 'Die hellenistischen Städgrundüngenvon Alexander dem Grossem bis auf die Römerzeit' *Philologus Suppl.* 19.1 (1927), pp.1–216.

Tupikova, I. 'Ptolemy's Circumference of the Earth' (Berlin, Max Planck Institute for the History of Science, 2014).

Tupikova, I. 'Eratosthenes' Measurements of the Earth: Astronomical and Geographical Solutions' *Orb. Terr.* 16 (2018), pp.221–254.

Tupikova, I. and Soffel, M. 'Modelling Sundials: Ancient and Modern Errors' in M. Geller and K. Geus (eds.), *Productive Errors: Scientific Concepts in Antiquity* (Berlin, Max Planck Institute for the History of Science, 2012), pp.1–22.

Tzifopoulos, Y. ''Hemerodromoi' and Cretan 'Dromeis': Athletes or Military Personnel? The Case of the Cretan Philonides' *Nikephoros* 11 (1998), pp.137–170.

US Department of the Army, *Army Field Manual 3-25.26: Map reading and Land Navigation Handbook* (Eugene, Doublebit Press, 2019).

Vari, R. 'Die sog. Inedita Tactica Leonis' *BZ* 27 (1927), pp.241–270.

Van Donzel, E.J. and Schmidt, A.B. *Gog and Magog in Early Christian & Islamic Sources: Sallam's Quest for Alexander's Wall* (Leiden, Brill, 2010).

van Hook, L. 'On the Lacedaemonians Buried in the Kerameikos' *AJA* 36 (1932), pp.290–292.

Viedebantt, O. 'Eratosthenes, Hipparchos, Poseidonios' *Klio* 14.14 (1915), pp.207–256.

Vodolazhskaya, L.N. 'Reconstruction of Ancient Egyptian Sundials' *AAATec* 2 (2017), pp.1–18.

Wallace, R.W. 'Private Lives and Public Enemies: Freedom of Thought in Classical Athens' in A.L. Boegehold and A.C. Scafuro (eds.), *Athenian Identity and Civic Ideology* (Baltimore, Johns Hopkins University Press, 1994), pp.127–155.

Warry, J. *Warfare in the Classical World* (Norman, University of Oklahoma Press, 1995).

Waugh, A.E. *Sundials: Their Theory and Construction* (New York, Dover, 1973).

Wilson, H.H. *Ariana Antiqua* (London, East India Co., 1841).

Wilson-Lee, E. *The Catalogue of Shipwrecked Books* (New York, Simon and Schuster, 2018).

Woods, M. *In the Footsteps of Alexander the Great* (London, BBC Worldwide, 2001).

Electronic Sources

Brean, J. 'Professor Discovers Centuries-Old attempt by Christopher Columbus' Son to Summarise Every Book in Existence' *National Post* (12 April 2019), viewed 04 August 2021, https://nationalpost.com/news/professor-discovers-centuries-old-attempt-by-christopher-columbuss-son-to-index-every-book-in-existence.

Google, *Google Earth version 9.3.114.1* (2020), viewed 27 Apr 2020, https://earth.google.com/web/search/Alexandria,+Egypt/@28.40652716,27.04311183,243.68989055a,1789297.35587776d,35y,359.99686928h,0t,0r/data=CigiJgokCRq1yyDuDkdAEWiawIWm4UZAGdwQcq_Rvk9AIUOybWs6i09A.

Google, *Google Maps* (2020), viewed 28 Apr 2020, https://www.google.com/maps/@34.3021534,62.303482,11z.

Hoffmann, T, *Suncalc*, Suncalc.org (2021), viewed, 21 Dec 2021, https://www.suncalc.org/#/-31.1895,152.9707,15/2021.12.22/12:46/1/3.

Sagan, C., Druyan, A., and Soter, S. *Cosmos* (Studio City, Cosmos Studios, 2000).

SMB digital Online Collections Database, *Sundial, tenons modernly added*, SMB digital (2021), viewed 12 Aug 2021, http://www.smb-digital.de/eMuseumPlus?service=direct/1/ResultLightboxView/result.t1.collection_lightbox.$TspTitleImageLink.link&sp=10&sp=Scollection&sp=SelementList&sp=0&sp=0&sp=4&sp=Slightbox_3x4&sp=0&sp=Sdetail&sp=0&sp=F&sp=T&sp=0.

Stoa Consortium, *Suda Online: Byzantine Lexicography*, University of Kentucky (2019), viewed 10 Mar 2020, sc.uky.edu.

The Met, *Shadow Clock 306–30BC*, Metropolitan Museum of Art (2021), viewed 6 August 2021, https://www.metmuseum.org/art/collection/search/576278.

Thompson, R. *Australian Government Bureau of Meteorology: Space Weather Services— The Equation of Time* (2021), viewed 19 June 2021, https://www.sws.bom.gov.au/Educational/2/1/14.

Walkup, N 'Eratosthenes and the Mystery of the Stadia' *Convergence* (2010), viewed 15 April 2020, https://www.maa.org/press/periodicals/convergence/eratosthenes-and-the-mystery-of-the-stadia-introduction.

Western Sydney University, *Intellectual Property Policy* (2020), viewed 19 August 2020, https://policies.westernsydney.edu.au/view.current.php?id=85.

Worthington, I. (ed.) *Die Fragmente der griechischen Historiker (Brill's New Jacoby Online)* (2020), viewed 11 June 2020, https://referenceworks.brillonline.com/browse/brill-s-new-jacoby.

Index